W9-ALJ-185

THE DELPHIC BOAT

THE DELPHIC BOAT

WHAT GENOMES TELL US

ANTOINE DANCHIN

Translated by Alison Quayle

HARVARD UNIVERSITY PRESS
Cambridge, Massachusetts, and London, England

2002

Printed in the United States of America

Library of Congress Cataloging-in-Publication Data
Danchin, Antoine.
[Barque de Delphes English]
The Delphic boat : what genomes tell us /
Antoine Danchin ; translated by Alison Quayle.
p. cm.
Includes index.
ISBN 0-674-00930-4 (alk. paper)
1. Genomes. I. Title.

QH447 .D3613 2002
572.8′6—dc21 2002027273

CONTENTS

THE DELPHIC BOAT

Among the questions asked by the Oracle of Delphi was this enigma: If we consider a boat made of planks, what is it that makes the boat a boat? This question is more than just a mind game: as time passes, some of the planks begin to rot and have to be replaced. There comes a time when not one of the original planks is left. The boat still looks like the original one, but in material terms it has changed. Is it still the *same boat?* The owner would certainly say yes, this is my boat. Yet none of the material it was originally built from is still there. If we take the boat to pieces, it is reduced to a pile of planks—but they are not the same ones as at the beginning! If we were to analyze the components of the boat, we would not learn very much—a boat made from planks of oak is different from a boat made from planks of pine, but this is fairly incidental. What is important about the material of the planks, apart from their relative stability over time, is the fact that it allows them to be shaped, so that they relate to one another in a certain way. The boat is not the material it is made from, but something else, much more interesting, which organizes the material of the planks: the boat *is the relationship* between the planks. Similarly, the study of life should never be restricted to objects, but must look into their relationships. This is why a genome cannot, must not, be regarded as simply a collection of genes. It is much more than that. But is this what the genome programs are exploring? A brief history will tell.

PROLOGUE

First it spread like a rumor, and then it became a cliché: for years now we have been hearing about "sequencing the human genome." This mysterious expression is obviously meant to speak for itself, but what "sequencing" means is not immediately clear. What it involves is determining the order—the sequence—of the chemical motifs that make up the chromosomes. As for "genome," both the mass media and the scientific literature have discussed in detail some of the consequences of knowing the complete human genome, but our knowledge of heredity is so recent that many people have not even had a chance to get to know the basic vocabulary, let alone the concepts involved. If we want to look beyond these references to a great modern technological achievement, or even to a new way to cure a great many diseases (more or less associated with genes), we find almost nothing on the fundamental reasons *why* these whole genome sequencing projects, whose objective is to know the entire chemical makeup of living organisms, are so extraordinarily worthwhile.

Louis Pasteur, the father of microbiology, was honored in 1995, the hundredth anniversary of his death. But in the same year a revolution occurred. The laboratory led by Craig Venter, in the United States, published the complete genome sequences of two pathogenic bacteria. Was this no more than a technical feat, or were we looking at a decisive change of direction, which would change the future of genetics forever? What this book shows is that we are now at a turning point: never again will we understand life, or organize the way we think about it, in quite the same way. Primarily, this book will reveal how our knowledge of whole genomes not only makes life easier to understand but also clearly places life totally beyond the reach of commonsense prediction, showing that it is fundamentally different from everything we are used to thinking about in the mechanistic world of physics and chemistry. Life is *creative,* in the strongest possible sense. And we can begin to understand how an essential trait of the processes at work is the *symbolic* nature of the interactions between the objects living things are made of. The major control functions that operate almost everywhere in life "capture" and keep the structures that implement them. Evolution selects those organisms that have been able to create major functions capable of coping with a future that is always unpredictable. So it is not generally structure that dictates function, but the other way round. Knowing the sequence of genomes enables us to begin to see how this process—capturing architectures for the benefit of the functions essential to life—actually operates.

Why do we sequence genomes? First, to compile a catalogue of the genes of a living being. All the main domains of science have evolved from a starting point where a particular area of the physical world was subjected to some kind of classification, taxonomy, or system. For instance, Dmitri Mendeleev devised the periodic table to catalogue the atoms of the universe, we have compiled a catalogue of the stars we see in the night sky, and we continue to classify plants and animals, as we try to organize our knowledge of the immense diversity of life. In the same way, before we can understand the molecular basis of life, we have first to establish a complete chemical description of the cell. To do this we must identify all the small basic molecules that work together to build the cell and give it life. In fact most of these small molecules have already been identified in model organisms. But they are manipulated by other molecules, and these are enormous. We are still a long way from having identified all these macromolecules, and in many cases we have not characterized their function in the cell. Sequencing an entire genome is a step in this direction. Using the rules of the genetic code—we will see how these work—not only enables us to study the function of macromolecules, but above all it gives us access to the link that exists between heredity itself and the synthesis of these molecules in the various cells. First of all, then, studying the genome provides us with a list of all the objects needed to support the life of the cell.

Yet we are still a long way from understanding life, for life is unquestionably the result of something very different from the simple juxtaposition of objects in a collection. The fact is that a genome is *not* just a series of genes. Sequencing does not just give us access to the collection of macromolecules essential to life. More than anything else, it brings home to us that the genome is the program or the recipe that enables the cell to be constructed and set to work. Looking more closely, we will see that knowing whole genome sequences opens up new a area of research, similar to the science of decoding texts written in unknown languages or cracking secret codes. We are right at the start of a new era in science. But the first elements of this knowledge, which are already in our hands (and the genetic code is not the least of them), allow us to characterize many of the subjects that interest researchers in molecular genetics: the function, structure, history, and regulation of gene expression.

Beyond chemistry, genome sequencing projects will allow us to do much more than compile a catalogue; they will lead us to the heart of what makes life, provided we are in a position to understand the relationships among the different objects identified. This assumes that we begin by representing to ourselves what life is. The image of a program or of an instruction leaflet for building a model probably illustrates it best. But we also have to understand the machine that reads the program, and its relationship with it. Nature is such that living organisms, all living organisms, build themselves by referring to an original molecule made of four types of similar chemical motifs, joined together like the letters in an alphabetic text. This text is interpreted by a complicated piece of chemical machinery to make objects that manipulate all the components of the organism, transforming them, giving them their shape, putting them into position, making sure there are the right number, and ensuring their stability in space and time. It constitutes the organism's construction and survival plan, or rather program. All by itself, evolving over the three and a half billion years that have passed since life began, it determines the specific characteristics of a given organism. And the most remarkable feature of life is that it is both a process that reads and expresses a program and a process that builds the machine which does this.

I shall begin by telling the stories of the first genome programs, the projects of the last fifteen years of the twentieth century—stories of organisms, of techniques, of men and women and politics. I shall briefly illustrate the concepts and social background associated with genome science, because sometimes discoveries need to be set in their social and political context. A practical interlude will give an idea of how scientists go about obtaining and understanding genome sequences, and how the international scientific community is now managing to access a constantly growing number of whole genome sequences. This was one of the most fascinating scientific quests of the twentieth century, and understanding these genome texts is a major challenge for the twenty-first.

We then come to the heart of the book, as we realize—and this is surely so obvious that we can only be amazed not to find it stated more often—that what counts in life is not objects themselves, but the relationships between them. The central concern of this book is to highlight some of these relationships and show their original properties. In particular we shall look closely at the way the system of relationships making up a living cell is transmitted from one generation to the next. This will lead us to investigate in depth the meaning of the information transmitted from generation to generation, and that of the genetic program, and we will not be satisfied with the vagueness of broad descriptions rooted in what I think is a deep misunderstanding of the relationship between physics and biology. We will reach a conclusion, which for me was the greatest and the most wonderful of all surprises as the outcome of the genome program to which I contributed: there is a map of the cell in the chromosome. Superimposed on the genetic program, there is a geometrical program, and this resolves the enigma of the machine that constructs the machine.

Finally, we will use these premises, based on our modern analytical knowledge and illuminated by genome studies, to define what life is. And we will illustrate the creation of biological functions with real examples, setting the stage for a reassessment of a number of philosophical questions about what life is, and forming the basis for an understanding of some of the experimental questions that will shape the future of genome studies.

I

EXPLORING THE FIRST GENOMES

In the mid-1980s a group of scientists, in particular some of the best-known names in biochemistry, began to advocate launching a gigantic molecular biology research program. The plan was to embark on sequencing the entire human genome. This would mean no less than determining the sequence of 46 human chromosomes (22 pairs of almost identical chromosomes plus the 2 sex chromosomes, X and Y) to build up a text with more than 3 billion letters, those now-familiar A's, C's, G's, and T's. A comparison with a page of this book, which contains fewer than 2,000 characters, gives an idea of the scale of the enterprise.

Early discussions on the project were often marked by political battles quite alien to the idea we usually have of science. At first it was portrayed as a matter of pulling off a technical exploit, using equipment rather like that used in high-energy physics or in space research. In the mid-1990s, when these programs were becoming a reality, the idea of sequencing genomes was still sometimes presented as an end in itself; yet clearly that would be of little interest, unless perhaps as a purely technological objective, with the aim of inventing new DNA sequencing methods. Genome sequencing is in fact terribly repetitive and painstaking work, and it seems hard to understand the enterprise unless we bear in mind that the main aim is to study the information the genomes contain. The letters of the text represent the chemical motifs, called *nucleotides* or *bases,* that carry the genetic information of every organism and that pair up—A with T and C with G—between the two strands of the famous double helix. We can identify genes and their products from the sequence of these bases by reading the genetic code that nature uses, in which each set or *codon* of 3 successive bases represents one of 20 amino acids, the building blocks of proteins. With a very few exceptions, the genetic code is universal; the correspondence between codons and amino acids is the same in bacteria as in humans. This is clearly not a trivial statement: it means that when we know the sequence of a gene, we know the sequence of the protein that corresponds to any given part of it. Used together with other biological or biochemical knowledge, this fact often makes it possible to predict the nature, function, and regulation of the corresponding protein and its location in the cell. One of the first objectives of the genome programs is to produce a catalogue of all the proteins coded for by the genome in question, by deciphering the text of the DNA, so as to draw up an inventory of all the functions required for life. With this information we can begin to interpret the meaning of genomes and what they tell us about life, its evolution and origin.

Philosophers and historians of science have made it clear that much progress in knowledge has come from the systematic accumulation and classification of related objects.[1] Astronomy really grew out of the mapping of the heavens, establishing what is fixed and what is not, what is similar and what is not. Biology really began with collections of anatomical specimens, of species of plants in herbariums, of insects and birds. Even mathematics owes a considerable debt to the collection of examples of classes of abstract objects or geometrical figures, which were then related to one another and generalized. No one would dream of laughing at the painstaking work of the monks who, by copying manuscripts over the centuries, preserved the memory and knowledge acquired by those who had gone before them. Genome sequencing is like the preliminary work of archaeologists, bringing to the surface the essential material on which all their future work will be based. It is our extraordinary luck, beyond our dreams just a short time ago, to have the chance to fulfill this essential precondition, which will enable us to get a glimpse of what life is.

But we have to recognize that scientists' motives are often quite remote from the pure quest for knowledge, and that large-scale, high-profile projects are stakes in a game that goes far beyond mere scientific considerations. We should remember—without drawing the negative conclusions that Paul Feyerabend did, for example[2]—that social reality can have a significant impact. Science plays an important role in our society, often going hand in hand with power or glory. Scientific exploits are the main driving force behind the excessive influence of the visual media over society, supported by sections of the press, including some primary sources of scientific information. Easily impressed by technical feats and by numbers with strings of zeros after them, the mass media have seized on the most grandiose sequencing projects without regard for their scientific content. This state of affairs may well account for the lack of enthusiasm sometimes shown by scientists dedicated rather too exclusively to their own research, who fear that the limited funds available to them will be diverted to what they consider fruitless ends. In fact, genome sequencing programs are very good value for money, but more importantly, the overall information discovered has provided completely unexpected knowledge about life, and this in itself entirely justifies whole-genome programs. I shall tell the stories of some of them, with their anecdotes (*how* all this is done will be described later). But we must bear in mind that this is just one brief stage in the story of a profound change in genetics.

Once the publication of the first whole-genome sequences had been announced, the atmosphere changed rapidly, not only among specialists in molecular genetics but also among fieldworkers, such as those involved in the epidemiology of major diseases. At a meeting at the headquarters of the World Health Organization in Geneva in September 1996, the future of genome programs was discussed with reference to their applications in medicine: as an aid in diagnosis, in developing new drugs, and in designing the new generation of vaccines. At the end of the millennium it was announced that the first draft of the human genome sequence had been completed, while the genomes of two model animals, the nematode *Caenorhabditis elegans* and the geneticists' favorite model, the fruit fly *Drosophila melanogaster* (the tiny yellowish flies with red eyes that you see flying around the fruit bowl in summer), were known almost in their entirety. By mid-2002, over 550 genome programs were under way or completed, enabling scientists not only to get an idea of the genetic diversity of microbes but to discover many aspects of the evolution of species and, perhaps, the origins of life.

There are many scientific reasons why genome sequencing is justified, especially genomes besides our own. Indeed, although our inevitable anthropocentrism makes us regard the human genome as a priority, it is particularly ill suited to scientific study, even if it is obviously valuable in helping us understand human origins and certain diseases. It is essential to know the genomes of the organisms that serve as models, and very early on, scientists around the world drew up a list of these. To begin with, there were the major models used in microbiology: first the bacterium *Escherichia coli,* also known as the colibacillus. Most of the established knowledge in molecular genetics was drawn from the study of this organism, and some strains are also important because they cause outbreaks of serious illness. Then there is *Saccharomyces cerevisiae,* or brewer's yeast, also used to make bread, and perhaps the longest-domesticated organism of all. Another bacterium can be added to this list: *Bacillus subtilis,* much used in the food industry, is particularly known as a model for all bacteria that produce spores when faced with a situation in which they are unable to reproduce.

After the microbes, *Drosophila melanogaster* would be the obvious insect to choose. Thomas Hunt Morgan and Alfred Sturtevant's work on this fruit fly in the first quarter of the twentieth century laid the foundations for the whole of genetics, and it is also an invaluable model for the study of cell differentiation. Two microscopic fungi also suggest themselves. Work on *Neurospora crassa,* complementary to Boris Ephrussi's study of *Drosophila* eye pigments at the Institute of Physico-Chemical Biology in Paris (analyzed with insight by Richard Burian and his colleagues), enabled the relationship between a gene and its product, a protein, to be understood.[3] The second, *Aspergillus nidulans,* is a model organism of interest to medicine and industry. And there certainly ought to be a mammal on the list: doubtless it would be the laboratory mouse, *Mus domesticus.* Sometimes called *Mus laboratorius,* since the species used in laboratories is beginning to show distinct differences from the wild species, the mouse is the best model known for mammalian genetics, even compared with the rat *Rattus norvegicus,* because it is small, easy to breed, and reproduces quickly. Finally, there was a model proposed by Sydney Brenner at Cambridge at the end of the 1960s, although it is very little known outside the scientific community. *Caenorhabditis elegans* is a minute nematode worm whose cellular biology is very well known, from the egg to the adult.[4]

Strangely, when work on sequencing the human genome began in the mid-1990s, it was not envisaged that all the organisms on this list should also be sequenced; far from it—and in fact research into some of these genomes had barely started or was not even contemplated. On the contrary, sequencing began on fifty or so other genomes. This was the result of a combination of scientific and strategic choices, the influence of fashion, and the pursuit of the sound-bite effect, which determined the orientation of genome research until the end of the millennium.

Microbes, Genetics, and Genomes

Microbiology held a place of honor for almost a century. Louis Pasteur's discoveries, first on the fermentation of beet sugar, and then on the diseases of beer and wine, as well as the work of Robert Koch, which enabled extremely dangerous pathogens to be identified, demonstrated both the ubiquity and the apparent simplicity of these multicellular organisms. It was impossible to ask "What is life?" without asking "What are microbes?" However, classical, formal genetics was based not on the study of these organisms,[5] but on the study of plants (beginning with Mendel and his peas) and on the *Drosophila*, the universal animal model. Because of this, for a long time bacteria seemed to be outside the usual laws of genetics, especially where sexuality was involved, and they remained excluded from this essential approach to life.

When it became clear, not long after the Second World War, that bacteria were also a suitable subject for genetics and could be studied much more quickly and easily than flies, microbiology became the spearhead of genetics. In the 1960s every university worthy of the name had its department of microbiology. And it was there, as well as in other institutes specializing in genetics research, such as the John Innes Institute at Norwich in England, the Institute of Physico-Chemical Biology and the Pasteur Institute in Paris, or the Cold Spring Harbor Laboratory near New York City, that *molecular biology* was born, to study genes and the way their expression is controlled, or *regulated*. The story of this remarkable period is now well known. It gave us a first revolution, that of molecular biology,[6] and we are now witnessing a second, that of *genomics*, with the study of whole genomes. Yet as the new millennium begins, microbiology seems to have lost this favored status. If only because, in a democracy, financial backers are inevitably swayed by public opinion, the fashion now (and scientists are very sensitive to fashion) is to be interested only in the way cells differentiate into organisms or the way the nervous system develops. All over the world, departments of microbiology, once so flourishing, have disappeared, to be replaced by centers devoted above all to the study of multicellular organisms, especially animals.

However, the beginnings of a silent revolution have gradually revived interest in microbiology. Organisms were classically divided into those whose cells have no nucleus, called *prokaryotes*, from the Greek prefix πρό *(pro)*, indicating an anterior or primitive state, and κάρυον *(karyon)*, a kernel or nucleus, and those with nucleated cells, or *eukaryotes*, from the Greek εὖ *(eu)*, meaning "well," and κάρυον. Two distinct lineages in the way heredity is organized correspond to these architectures: the genome structures of prokaryotes and eukaryotes show marked differences in their general properties. But in the late 1970s the microbes, already very varied, gained a new family, a new domain of living things as distant from the kingdom of the bacteria already known as they were from plants and animals. Carl Woese was the first to identify them at the molecular level. Like the bacteria that had already been described, they had no nucleus, but Woese discovered that they could not be classified with other known bacteria, even taking into account species known to be very distant from one another. He called these bacteria *archaebacteria,* as they seemed to him to be nearer the origin of life than those already known. The fact that his laboratory at the University of Illinois at Urbana is sponsored by the National Aeronautics and Space Agency (NASA), with its interest in life in space, implicitly confirms this belief. He proposed a division into three rather than two major families of living organisms: three great kingdoms, the *eubacteria* or Bacteria, the *eukaryotes* or Eukarya, and the *archaebacteria* or Archaea. The usual distinction between prokaryotes and eukaryotes thus lost at least a part of its evolutionary significance, to become no more than a general characteristic that would not immediately indicate a given cell's place in evolution. This entirely new domain of microbiology is becoming increasingly important because these bacteria normally live in extremely hostile environments, where we can scarcely imagine that life would be able to survive—environments that are scaldingly hot (water under pressure at more than 100°C), very acid (in volcanic springs where the pH is less than 1), or very salty (in the Dead Sea, where the salt level is at saturation point, or on piles of salt in saltworks, for instance). But the main reason for the renewal of interest in microbiology is undoubtedly the resurgence of old diseases and the appearance of new ones. And unfortunately, at the turn of the millennium, the old deadly demons that plague the human mind have revived an evil use of microbes in what is now called bioterrorism.

Microbes make up 50 percent of the protoplasm on Earth, and they alone would survive a major chemical, nuclear, or cosmic catastrophe, such as the impact of a giant meteorite. Microbes are not only our past, but our future as well. They are certainly as diverse as other organisms, a fact that adds a special interest to their study. Their commercial value is obvious; a large part of what we eat depends on them. Very often the best way to preserve food is to ferment it, and this can be done using a wide variety of microbes. Finally, of course, the cell is the unit of life. We cannot understand life without understanding the cell, and microbes are the ideal organism for studying it because they are individual cells themselves. Given all this, as well as the fact that most microbes have small genomes (fewer than 10 million base pairs), it is not surprising that the exploration of the first genomes has led to a major renewal of interest in the study of bacteria and other single-celled organisms. Besides, while the dogged researchers lining up their rows of bases were once considered fair game, now that a large number of genomes have been sequenced or are well under way, few people remember their own original lack of enthusiasm.

Genome sequencing made rapid progress. The first genome sequence known was that of a virus called ΦX174, a parasite of *E. coli*, sequenced by Frederick Sanger and his colleagues in 1977. This genome, a single strand of DNA joined in a circle, has 5,386 bases. Five years later it was the turn of a second virus that infects the same organism, the lambda virus. This is a member of the class of bacterial viruses discovered by Félix d'Hérelle in 1917, which he called *bacteriophages*, from the Greek *φαγεῖν (phagein)*, "to eat," because they are parasitic on bacteria. Studies on this virus by André Lwoff and François Jacob at the Pasteur Institute were the basis of the concepts central to our understanding of gene expression. Sanger and his colleagues identified the order of the nucleotides in its double strand of DNA, made up of 48,502 base pairs. From this date on, progress accelerated, first with viruses such as the cytomegalovirus, with its 229,354 base pairs, sequenced in 1990. This opportunist pathogen frequently infects humans and has become famous because it causes serious illness in AIDS patients.

Then in 1992 the 315,339 base pairs of chromosome III (out of 16) of baker's yeast, *Saccharomyces cerevisiae,* were sequenced by a consortium of European laboratories. Even though it was quickly forgotten, so blasé have we become, and so obsessed with the ephemeral, I would like to begin with this success, as it was the first with a nucleated cell, and because it convinced the international scientific community that this new approach to genetics, by determining the complete sequence and studying the genome of a model organism, in other words what we now call genomics, was well founded.

Sequencing the Genome of the Yeast *Saccharomyces cerevisiae*

However far we look back into the past, we find fermentation among mankind's agricultural methods, used to preserve foodstuffs and to produce alcoholic drinks. Brewer's yeast, a variant of which is also used to make bread, was thus one of the first living things to be domesticated. Numerous genetic studies have been carried out on this useful, inoffensive microbe, which reproduces very quickly. What is more, like differentiated multicellular organisms, yeast has what makes the heart of animal and vegetable cells: a well-formed *nucleus; mitochondria,* the organelles that manage the cells' energy resources; and the complicated apparatus of membranes and tubes that distributes the gene products around the cell, secretes them into the external environment, and separates the daughter cells during cell division. Although yeast can reproduce by simple cell division as bacteria do, it also has a mode of sexual reproduction, as more complex organisms have. Finally, it has recently been discovered that fungi—and yeast is a microscopic fungus—are not close to plants, as was long thought, but to animals instead. Many aspects of the cellular life of yeast are similar to those of animal cells. In this context, it becomes clear just how much there was to gain from knowing the complete genome of this relatively simple organism.

However, "simple" is indeed relative; in fact it was a major challenge. At the beginning of the 1990s, sequencing the cowpox virus or the cytomegalovirus, each of which has around a quarter of a million base pairs, was considered to be a feat that only a very few laboratories in the world could carry out successfully. The yeast *Saccharomyces cerevisiae* has a genome of more than 12 million base pairs, coding for more than 6,000 genes, spread out over 16 chromosomes—and that is without counting the highly repetitive section containing the DNA coding for ribosomal RNA. The credit for realizing that this sequencing was possible, and for organizing it by coordinating the work of 643 scientists in a large number of laboratories (92 in Europe), goes to André Goffeau of the Catholic University of Louvain-la-Neuve in Belgium, with the strong and enthusiastic support of the geneticist Piotr Slonimski, a world-class specialist in yeast mitochondria and former student of Boris Ephrussi; Steve Oliver, a researcher at the University of Manchester in England; and Bernard Dujon, a former student of Slonimski.

The beginning of the story is more political than scientific. In 1986 André Goffeau was asked to devise a biotechnology research program for the entire European Community, which would bring the different countries together, and which no one member state could easily achieve alone. He had just read an article by Renato Dulbecco, who was proposing to sequence the human genome in order to explore the causes of cancer in humans in a different way from the notoriously inefficient methods commonly used in cancer research.[7] As a yeast specialist, and knowing that there was a commercial interest in this organism, Goffeau proposed to sequence its genome. It was thus clearly a politically inspired project right from the beginning, implicitly linked with economic issues, and one that depended on the creation of appropriate technology. It is easy to understand why initial reactions were skeptical or even frankly negative. In this difficult context, Goffeau formed a consortium of laboratories with the aim of sequencing a first yeast chromosome, and after five years of effort and persuasion, the sequence of chromosome III was published in 1992. At 315 kb, it was the longest continuous segment of DNA whose sequence was yet known.

The article in which it was announced, in the famous British journal *Nature,* bore the names of 147 authors, showing just how difficult an enterprise it was, and placing molecular genetics in the same light as experiments in high-energy physics, where articles with dozens or even hundreds of authors are common. But at the same time it showed that it was possible to organize an effective collaboration among a great many geneticists and biochemists, in which there was no competition, only a real common interest in a scientific objective (and this aspect of the success was not the least of Goffeau's achievements). The first chromosome sequence was over 300 kb long, and those that remained to be sequenced were even longer—chromosomes VII and XV, for example, are 1,150 kb long. After this first success, and benefiting from the method that had made it possible, progress was rapid, requiring fewer researchers to obtain longer sequences.

By 1994 it was becoming clear that the entire yeast genome could be sequenced before the year 2000, and perhaps much earlier. Chromosomes II (807 kb) and XI (563 kb) were sequenced very quickly, still by a group of European laboratories, coordinated by the European Commission. Two American laboratories sequenced chromosomes VIII, V, XII, and part of chromosome IV; a Japanese laboratory sequenced chromosome VI; and a Canadian laboratory chromosome I and part of chromosome XVI. At the same time a laboratory in England directed by Bart Barrell, one of Sanger's former students, decided to go it alone and sequence several chromosomes (IX, XIII, IV, XVI) either entirely or in part. Thus by the spring of 1996 the entire sequence of the yeast genome was known, most of this achieved by European scientists. This was a remarkable success, in that the sequence involved was more than five times longer than those published the previous year by Craig Venter's TIGR (which we shall look at shortly). Strangely overlooked by the press, this great achievement was announced to the scientific world in September 1996 in Trieste, Italy, during a conference that was to leave its mark on the history of genetics, as James Watson, the American "father" of the double helix, emphasized in his address.

Two main results emerged immediately from this significant work, although they are rarely emphasized. One discovery was that yeast has a remarkably compact genome, containing very little DNA without apparent meaning, and that most of its sequences are for protein synthesis. The average length of protein-coding genes is about 2,000 base pairs. The other observation—confirmed by almost all the other genomes studied—is that a very high proportion of the genes specify completely unknown functions, while the structure of the corresponding proteins is also unknown. At first estimated at about 60 percent, this proportion is dropping all the time as scientists discover that they relate to an understandable function, either by studying the genes at the molecular level through reverse genetics (the planned modification of genes by *in situ* recombination) or by comparing them with other genomes in which some of these genes have been identified. However, more often than not, although they do indeed turn out to resemble something in the data banks, this something is another gene from a different organism, whose function is just as much a mystery.

We need to be aware of the paradox that lies behind this. Scientists have known for a long time that if any mutant from any organism, from bacteria to man, is isolated, and its corresponding gene can also be isolated (cloned), then once the product of that gene (usually a protein) has been identified from the nucleotide sequence (translated by applying the genetic code), it can be checked against everything already known. Because of this constantly repeated observation, by the end of the 1980s it was thought that all the major functions of living things were already known and that there was little to learn in that quarter from sequencing whole genomes. This argument was actually used against whole-genome sequencing projects, on the basis that they would not have much new knowledge to contribute. But the discovery of "unknown" genes has shown, quite unexpectedly, that we had been completely unaware of an enormous part of genetic knowledge. This in itself totally justifies the resources that have been put into getting whole-genome sequencing programs off the ground.

The entire sequence of the yeast genome was published in a supplement to *Nature* in June 1997, more than a year after the end of the sequencing program had been announced. By this time, several more genome sequences were already known. It is time to take a closer look at this race to be the first to sequence the complete genome of an autonomous living organism.

One Hundred Years after Pasteur, the First Bacterial Genomes

The end of sequencing of the first yeast chromosome, chromosome III, was announced in May 1991, at Elounda in Crete. At this meeting Piotr Slonimski wittily drew attention to the European lead over the Americans (who had talked a lot about sequencing the human genome or the *E. coli* genome, without having much to show for it at that stage) by calling these mysterious genes just discovered in such quantities "Elusive, Esoteric, Conspicuous genes," or "EEC genes" for short, clearly based on the old acronym of the European Community. The representative of the National Institutes of Health (NIH), which was financing a large part of the U.S. genome programs, reacted by totally ignoring the European success and presenting a long list of what was to be done in the United States, as if nothing else were important and all other projects around the world had to fall in line with American plans. In one way she was right, because that was how things turned out: the United States, aware of the risk it ran in lagging behind the leaders, increased funds for genome research by enormous proportions, while research in this field was neglected in Europe at both national and EC levels.

However, at the time there was nothing to boast about. Since 1988 word had been going around that at least two American laboratories would have finished sequencing the *E. coli* genome within two years, and that in any case there was no future for any project but the human genome, except as an afterthought (James Watson, who was responsible for the NIH's part in the Human Genome Initiative at the time, was especially of this opinion.) Once this immense task was possible, then sequencing a bacterial genome, for instance, would be of only minor interest. In the United States, the NIH and the Department of Energy (DOE), famous for organizing grandiose nuclear programs, began to compete over control of the financing and above all the data management of genome programs (and especially the sequence data banks). The result was that all sorts of projects were launched in American laboratories, only to be shot down one after another by various officials with allegiances to either of these two agencies. In such circumstances, genome sequencing was making little progress. Luckily, this was not true of technological developments, which were led by a productive cooperation between universities and business and which slowly but surely enabled longer and longer fragments of DNA to be sequenced, particularly by using fluorescent markers. But the first success did come from the United States, and from a completely unexpected quarter, and the first genome known was not that of a model organism, but that of a bacterium studied by fewer than ten laboratories worldwide.

How had this come about? Among the various objectives of whole-genome sequencing projects, a few clear-sighted scientists hoped to discover the existence of rules expressing within the chromosome sequence the rationale for the *collective* expression of genes. Another question was whether each species would show distinct characteristics allowing its particular identity to be clearly recognized. If such an internal consistency did exist, it could be used to recognize the rules for the structuring of nucleic acids and proteins, their translated products, and the personal style of each organism, in the same way as the personal style of an author can be recognized in a written text. By the style of an organism, I mean the same thing as in architecture, where the columns on Greek temples can easily be recognized as Doric, Ionic, or Corinthian because they have different styles, although all have the same function of supporting the roof of the building. This new knowledge could be discovered, it seemed, only by bringing together the expertise of a large number of scientists, especially geneticists working on model organisms, and a knowledge of the genome text. So it would seem logical that the first bacterial genomes known should be those of model organisms, in particular that of the best-understood living thing, the colibacillus *Escherichia coli*. But this was not how things turned out.

In reality, the most significant turning point in our changing perception of how the complete chemical knowledge of living organisms could be accessed came not at the end of the 1980s but in 1995, and *E. coli* was not involved. In that year Craig Venter and his colleagues at The Institute for Genome Research near Washington, known as TIGR (the aggressiveness of the acronym is explicit enough), published in *Science* the sequences of two very small bacterial genomes: *Haemophilus influenzae,* which has a 1.8 Mb genome; and *Mycoplasma genitalium,* remarkable for having the smallest genome known in an autonomous cell, at only 580 kb. (Its genome text would make a book of about 300 pages.)

How had this outsider suddenly appeared in the genome race? Craig Venter was not particularly interested in bacteria. He had been a researcher at NIH, where he had worked on neurotransmitter receptors. But with his interest in technological progress, he was tempted by the challenge of sequencing the human genome very early on, after having been involved in locating the gene for a neurotransmitter receptor on human chromosome 15 right at the beginning of the 1990s. He immediately realized that the scale on which molecular biologists were used to working would have to change if projects of this kind were to be successful. They would have to "think big," on an industrial scale. We can see the size of the problem by calculating how much effort is involved. In the early 1990s it was unrealistic to suggest that a trained technician could sequence more than 50,000 base pairs a year. For a genome of more than 3 billion base pairs, sequencing would involve 60,000 man-years! Clearly, this demanded a considerable (but not impossible) investment in manpower and equipment, and it would be essential to obtain considerable financial support very quickly, in order to establish the necessary infrastructure.

Venter knew that even in the United States, working with public bodies involved a long and difficult struggle with red tape—as could be seen from the consequences of the rivalry between the NIH and the DOE—and that if he wanted rapid success that route was out of the question. He would have to create a tailor-made organization from scratch. In this respect the situation in the United States is much more favorable than in Europe, at least for biotechnology. There is abundant venture capital and, above all, considerable federal or state support for private research, through widely available contracts. It is also relatively easy and cheap to set up a new company, administratively speaking, especially when it comes to promoting new technology such as informatics and biotechnology. Americans are extremely mobile, both geographically and professionally, so it is easy to find people with the necessary skills. But it was also important not to become an entirely commercial enterprise, so as to be able to benefit fully from collaboration with universities and public funding (especially federal funds from the NIH or the DOE). Here Venter showed just how skillful he was. Instead of just one organization, he created two, together with his colleague William Haseltine. Venter was to manage the nonprofit organization, TIGR, while Haseltine would manage the commercial organization, Human Genome Sciences (HGS), which had first industrial property rights over the whole of TIGR's work. TIGR would thus benefit not only from funds from HGS's capital but also from the contracts it entered into with the NIH and the DOE.

Craig Venter also understood intuitively, and very shrewdly, that, faced with a riot of different genome sequencing projects and all the battles and ego-trips they brought with them, it was essential to establish a presence and a reputation for reliability *very quickly.* The human genome was all the news, so he had to do something in this area. But it is huge, with more than 3 billion base pairs, divided into 46 chromosomes. However, the information relating to just protein coding (or the set of genes that are actually expressed) is considerably smaller. At least 95 percent of the sequence in a human genome either relates to other functions than simply coding for a gene; or it represents "archives," sequences that simply continue to be reproduced from one generation to the next because they are not too much of a burden on replication or the gene expression machinery. At the time the number of human genes was estimated at about 80,000, corresponding to 100 million base pairs, or 3 percent of the total length of the genome. This means that for a given portion of a genome sequence, normally only a very small part of that sequence actually codes for a gene, so the likelihood of finding something spectacular, or even interesting, is very small.

Several other characteristics of the human and other large genomes reflect the effects of evolutionary history and pose significant problems for sequencing. In addition to the general process of transcription and translation (the progress from the text of a gene to the text of the protein it codes for, via the text of the "messenger" RNA, which, as its name suggests, carries the genetic information from one to the other), the RNA transcribed in the cell nucleus also goes through a maturation stage in order to become a functional messenger. This is reflected in the gene structure itself. There are in fact inserts in the genes, which Walter Gilbert, one of the inventors of a DNA sequencing technique and director of a laboratory at Harvard, called *introns.* These separate the coding parts, called *exons,* and have to be removed for the messenger to function. A gene in the nucleus is thus rather like a column of text in a magazine, surrounded by columns or pages of advertising: the reader manages to fit the text together by skipping the advertisements. In the human genome, the exons are often only a very small part of the gene: a messenger RNA 1,000 bases long may be coded by a fragment of DNA tens, hundreds, or even 1,000 kilobases long. So exons may be concealed in the middle of a sea of introns, and sequencing DNA fragments, even long ones, may still not be enough to define completely the gene to which that messenger corresponds. One problem that is still unsolved is how to recognize the junction between introns and exons just from the sequence of a fragment of DNA.

Apart from these difficulties inherent in the modular construction of genes, it is extremely difficult to characterize the human genome on the microscopic scale. Because of the enormous variability of human populations (and the nature of the process by which we inherit one set of chromosomes from our father and one from our mother), the detailed structure of each individual's genome can differ considerably, principally outside the strictly coding part of genes (although it is not known *a priori* which part this is). The implication is that a sequence determined in a sequencing program will necessarily be ambiguous. It is also not easy to decide if the genome of just one individual should be sequenced: how should that individual be chosen? But if this is not done, the result will be a patchwork of fragments from several individuals, and, unknown to the scientist, it may combine sequences that are incompatible, in that they specify different variants or *alleles* of genes that cannot coexist. This *polymorphism* (which often comes down to just one base, and is therefore called "single nucleotide polymorphism" or SNP, pronounced "snip") can be valuable for studying the way humans have evolved, identifying individuals, discovering genes implicated in diseases, or analyzing the frequency of particular alleles. The study of SNPs is one of the important projects to have emerged from our knowledge of the human genome. But in terms of understanding the meaning of the genome and making sense of the corresponding text, polymorphism is more a problem than an advantage. Craig Venter realized that without further information, such as a reliable means of recognizing which sequences are used to translate genes into proteins, it would be extremely difficult to distinguish between natural polymorphism and inevitable sequencing errors. In particular he saw that although evolution had managed to cope with this, it would put an intolerable strain on the computer analysis that was to be used to discover the meaning of the genome text. Assembling the sequence fragments also means that they have to be multiplied, so that sequencing a genome normally involves sequencing fragments adding up to ten or fifteen times its total length, a task that is obviously time-consuming and expensive.

So Venter understood two things: that he should look first at gene products rather than at the genes themselves, and that a very strong informatics infrastructure would be essential to genome sequencing. He put these two principles into practice at TIGR. Immediately before the protein, a gene produces messenger RNA, and it is possible to isolate strands of mRNA in the cells of different tissues without much difficulty, because they normally end in a long tail of adenine, one of the DNA bases, called a poly A. It is easy to "fish" these out of a cell extract because they will bind to magnetic beads covered with a polymer of the complementary sequence, a poly T. Even better, since the mid-1970s it has been possible to use reverse transcriptase, an enzyme found in many viruses, to recreate DNA from RNA using a primer. (This enzyme is well known because of its role in the triple therapies used to treat AIDS, where it is an ideal target for antiviral treatments.) This recreated DNA, the complement of the messenger RNA, is called *complementary DNA* or cDNA. Ever since sequencing the human genome had first been considered, there had been suggestions that it would perhaps be better just to sequence the cDNA of certain tissues to begin with, so as to make an initial estimate of the nature and number of human genes.

Venter saw just how profitable this approach could be. He immediately began making collections (banks) of cDNA of all the possible kinds of human tissue (there are about 250). He then set about sequencing the cDNA of all these tissues, in part if not completely. The advantages of this approach are that it is easily done, and especially that these cDNA sequences, intact or not, can simply be cloned and used as sequencing templates, without any need to study the overlap between one sequence and another (to begin with at least), which is essential when determining the text of a genome. With these techniques a sequence of around 500 bases can be obtained directly from each template without any need to check the sequence. This corresponds to 180 amino acids of the protein coded for and represents a significant part of a gene's product (on average a human gene is expected to be 450 amino acids long, but some are much longer). In fact this length could be doubled if the sequence was determined from both ends of the fragment of cDNA inserted into the cloning vector. It is also possible get an idea of how many genes there are from the level of redundancy in the results, and in 1995 Venter estimated the total at around 85,000, a figure that is now highly controversial.

The next problem was with publishing the results. Venter had to show that the approach was viable, but publishing has significant economic consequences, often making it impossible to take out a patent later. At first Venter, perhaps pushed by Haseltine and HGS, wanted to patent all his sequences, and I will say a few words at the end of the book about the debate that followed. This was an error of judgment that turned public opinion, as well as the opinion of the scientific and political communities, against him for the first time (and this happened again five years later). Obviously the sequences can be used for commercial ends (for example, as a starting point for the study of selective targets for new drugs or for the diagnosis of diseases), but certainly not all of them. Then why alienate so many people by proposing to patent them? There is a very simple way to avoid doing so while preserving one's own interests, which is simply to leave out the sequences of particular interest when results are published! With powerful informatics resources, appropriate knowledge can be used to spot these sequences and keep them back. In any case they will be only a few of those obtained, while the number of sequences published will be very large. Creating and sequencing extensive cDNA banks is tedious work, so no one, except perhaps someone who had discovered an interesting sequence by other means (but had for that very reason kept it secret) would notice or point it out. It would remain invisible. I do not know whether this is what Craig Venter decided to do, but his policy of publishing massive quantities of cDNA sequences is compatible with this hypothesis. There are large numbers of them, but are they all there?

This was just one stage, but it was enough to make Venter famous. In 1995 he published a catalogue of more than 45,000 partial cDNA sequences from various human tissues. But something better was needed. He was not yet ready to tackle the complete human genome, but he realized that TIGR's industrial-scale set-up allowed him to contemplate sequencing a complete bacterial genome, provided it was not too big. Hamilton Smith of Johns Hopkins University in Baltimore, close to TIGR, also realized this. He had shared the Nobel Prize with Werner Arber and Daniel Nathans for their discovery of *restriction enzymes,* the enzymes that had made possible the birth of genetic engineering by enabling scientists to cut DNA at specific points, and thus to juggle the "cut and paste" methods that are the basis of molecular biology. As a bacteriologist and biochemist, Smith was familiar with a pathogenic bacterium, *Haemophilus influenzae,* which produces restriction enzymes. It infects the human upper respiratory tract, causing colds, ear infections, bronchitis, and sometimes meningitis, and is particularly dangerous to children. Its genome restriction map (a physical map indicating the distribution of the sites where chosen restriction enzymes cut the DNA) shows that its genome is 1.8 Mb long, almost three times shorter than that of *E. coli* and eight times shorter than yeast. With his usual flair, Craig Venter realized that if he worked quickly, he could soon be the first to have sequenced a complete genome.

And so in April 1995, at a meeting organized by the Wellcome Trust at Dormy House near Oxford, Venter was able to announce that he and his team of about forty had sequenced the entire genome of *H. influenzae* in collaboration with Hamilton Smith. He also announced that he had practically finished the sequence of the smallest known genome of any living organism, that of *Mycoplasma genitalium,* an opportunist pathogen frequently associated with AIDS. The sequences themselves were published in June and September 1995. Even though these were very small genomes, it was still a remarkable achievement, and Venter's subsequent work was no less impressive.

A Cautionary Tale: The *Escherichia coli* Genome

An outline of the story of whole-genome sequencing programs would be incomplete without giving a special place to the genome of *Escherichia coli.* Alongside the organisms listed at the beginning of the chapter, it has a central place in both scientific politics and the history of science. Not only is it the most important model in bacterial genetics; it is the best known and most studied of all living organisms. It reproduces readily in the laboratory and can grow just as well in the presence of oxygen as without it. The strain *Escherichia coli* K-12 was chosen as a research model by microbiologists, more or less by chance, from among a large number of strains isolated around the world, and became the reference strain for thousands of laboratories. This bacterium also served as a reference in the description of an important phenomenon in evolution, bacterial sexuality, a central process by which genes are transferred from one organism to another, not only between bacteria of the same species but also between different species. It has also been an ideal model for the study of viruses, because in nature it is infected by a virus called bacteriophage lambda, which inserts its own genome into the chromosome of the host cell, where it then lies dormant, in a form known as a *provirus.* And of course the study of *E. coli* and its viruses has been the basis for the main concepts of molecular genetics, especially the regulation of gene expression and the information flow from DNA to proteins.

The *E. coli* genome is compact (over 85 percent of the sequence corresponds to protein-coding genes), and at least half of these genes can be associated with a function that has been experimentally characterized. This means that sequencing any part of the genome will reveal a high proportion of genes, so it was clear in advance that there would be a high cost/benefit ratio, which should be important for financial backers. From the biological point of view, *E. coli* is of great interest not only to medicine but also to industry and in agronomics. It lives in the colon of humans and other animals, where it is normally not pathogenic, but virulent strains do exist, as anyone who has heard of the urinary infection colibacillosis will know. It was also responsible for dangerous outbreaks of food poisoning in Japan and Scotland in 1996, caused by the strain O157:H7 (sequenced independently by teams in the United States and Japan in early 2001). Another factor is that many closely related bacteria are highly pathogenic (such as the plague bacillus, the typhoid agent, and the cholera bacillus). *E. coli* is also a remarkably useful tool in biotechnology, as it can easily express high-quality gene products (proteins) from most living organisms in very large quantities. It is the host of choice for cloning all the DNA fragments in sequencing programs, including of course DNA from the human genome. Finally, some related bacteria are highly pathogenic to plants, for instance bacteria that make roots rot and spoil vegetables, and the fire blight that has devastated orchards.

Given all this, it would have seemed logical for this to be the first genome known. As early as the mid-1970s a group of scientists had suggested sequencing the *E. coli* genome as an inaugural program to justify the construction and development of the European Molecular Biology Laboratory, now at Heidelberg in Germany, in association with its parent organization, the European Molecular Biology Organization (EMBO). Close international collaboration would obviously have speeded up the process considerably. A DNA bank had been set up in Japan in 1987 covering and characterizing almost all of the chromosome, and using this bank should have been a priority. But in fact sequencing *E. coli* began with laboratories all over the world independently sequencing their gene of interest, as soon as the necessary sequencing techniques were invented. The two main promoters of the project in the United States were George Church at Harvard and Frederick Blattner at the University of Wisconsin at Madison. The vast size of the project meant that they ought to have collaborated closely right from the start, but because he was young, Church was probably not taken as seriously as he deserved. Besides, the project was doubtless not given all the funds it needed, at least at the launch stage, because it was regarded simply as an offshoot of the general human genome sequencing project, principally promoted by James Watson at the Cold Spring Harbor Laboratory near New York.

When the *E. coli* genome sequencing project was first launched, the polymerase chain reaction or PCR technique, a method for amplifying DNA in a test tube, was not yet practicable in the form that makes it possible to produce long fragments. Blattner developed original techniques for cloning DNA from the *E. coli* chromosome, and also initially decided to use robots to partly automate Frederick Sanger's dideoxy chain termination method using radioactivity. At the end of the 1980s the lack of coordination between the work of Blattner and Church was accentuated by the use of two rival sequencing techniques, the Sanger method on the European side, and the Multiplex technique, invented and advocated by George Church, on the American side. This new method, based on a chemical sequencing method invented by Allan Maxam and Walter Gilbert, was admittedly remarkably clever, and had it been really practicable—if it had produced sequences of a quality comparable to those obtained by the fluorescent dideoxy chain termination method—the sequencing process would indeed have become much faster. Unfortunately, the quality of sequences obtained remained too low for accurate industrial-scale reading of more than about 250 bases of each template sequenced (as against 500 to 700 for the other method, and this number is constantly increasing). Sadly, this entirely canceled out the advantage of the Multiplex approach, especially since problems connected with the difficulty of assembling the fragments sequenced in this way mean that the efficiency of the assembly process varies at least as the square of the length of the templates that can be read. Consequently four times as many templates *n* bases long need to be read compared to templates 2*n* bases long. Church contributed to only a short fragment of the *E. coli* genome, despite having autoradiographic images of templates totaling well over the length of the whole genome, so that in theory he could soon have put them together to make a sequence which, even if not the whole genome, would have been a very large part of it.

As it was, by 1989, 400,000 base pairs (around one-tenth of the total) of the *E. coli* genome were known thanks to the work of a very large number of laboratories. The best-informed scientific journals at the time, no doubt misinterpreting what they were told by the scientists concerned, stated confidently that the *E. coli* genome would be known well before the mid-1990s. Compared to what was being said by those who wanted to carry out the program, this was actually on the pessimistic side, but reality could not match even this cautious view. The actual production of sequences was much lower than had been predicted. In Japan the same overoptimism about the sequencing stage was compounded by slow-moving administrative procedures, and all that could be achieved was to produce sequences to fill in the numerous nonsequenced regions of the chromosome, between two regions determined by one of the string of laboratories sequencing one *E. coli* gene after another for their own research needs. The sequence quality was low, and a great many corrections were needed. The project to sequence the replication terminus region, by a group led by Katsumi Isono at Kobe, did not win any public funding, perhaps because the estimated time required to produce high-quality sequences seemed too long. The paradoxical result of this situation was that at the end of 1995 only 75 percent of the *E. coli* genome was known, the result of combining a very large number of genes sequenced in numerous laboratories, and only 40 percent of the genome was in one stretch. Only Blattner was still continuing his work in the United States, while in Japan T. Horiuchi and Hirotada Mori got together an informal consortium of laboratories, without any dedicated public funding, to sequence half the chromosome of a different strain, mainly around the replication terminus, a region where relatively few genes were known, and where there were consequently a great many gaps.

Blattner's funding ran out at the end of 1995, and after the first assessment by the NIH it did not seem likely that funding would be renewed. It was no longer certain that the *E. coli* sequence would be known in the immediate future! But the almost unanimous reaction of the scientific community concerned (several thousand laboratories work on the genetics of *E. coli*) led to reassessment of the project. Blattner was given further funding on condition that he finish the complete genome sequence before the end of 1996, but without thorough annotation. To achieve this he developed a new technique that enabled him to shear the genome into fragments of about 200 kb, which he could then isolate and shotgun sequence directly, in a similar way to the strategy successfully used by TIGR. He also abandoned the autoradiographic technique in favor of automatic sequencing machines using detection by fluorescence. By the end of 1996 Blattner had almost completely decoded the genome sequence of the reference strain MG1655. The complementary work led by Horiuchi and Mori had produced the sequence of half the chromosome of another strain, W3110, very common in laboratories worldwide, and so the whole chromosome of *E. coli* was reconstituted from a patchwork of sequences. Although 1,800 genes and 2.5 million base pairs of the genome had been known by mid-1995, and more than 2,000 of its genes (of a total of nearly 5,000) had been characterized in varying detail, it was not until 1997 that the sequence was almost completely determined, and even then it had not been validated by the appropriate checks. The last remaining gap had still to be filled, and most of the newly sequenced regions of the genome were not annotated. This was the equivalent of presenting scientists with a long text in a completely unknown language, with no punctuation and no comments. Annotation took Blattner and his colleagues almost another nine months, and the final sequence was not published until late summer of 1997. The Japanese consortium attempted its own annotation, but the lack of coordination between the Japanese and American teams has left our knowledge of this model genome imperfect. It is to be hoped that someone somewhere will take over where they left off, so as to provide the genetics community with the necessary knowledge and tools.

What went wrong? The background to this sequencing project is very confused, and unfortunately most of the material that throws light on the story is found in reports to government institutions or in applications for funding, so it can rarely be quoted explicitly, as it remains confidential. Besides, as these are not scientific publications they are often not archived. This is why so little has been heard about the reasons behind this paradoxical situation whereby, although the text of the *Escherichia coli* genome should have been the first one known, it took more than 10 years to be finished. One reason behind this strange state of affairs was of course the lack of any explicit coordination between the work of Church at Harvard and Blattner at the University of Wisconsin. The implicit competition between the United States and Japan only added to this difficulty. But the main reason is probably that the scientists who embarked on this adventure seriously underestimated the difficulties (and overestimated their ability to solve them). This is understandable in view of the system by which research is funded, which obliges scientists to appear as positive as possible in order to convince the financial backers. But not the least important aspect of this relative failure was a genuine inability to see the technical limits of the enterprise. In particular, they thought that technical improvements in sequencing would develop quickly, in parallel with their own work, and that progress in informatics, especially in software, would come about naturally, without the need for massive investment by the universities. But in both cases researchers, technicians, and engineers have to be especially trained, and this cannot be done overnight. It necessarily takes several years, and the critical mass of human and technical resources needed to carry out these projects was simply not available. What is more, the role of the press certainly had a very negative effect, as journalists systematically presented the scientists' working hypotheses as established facts, thus destroying any hope of applying for adequate financial support.

The success of sequencing programs depends to a large extent both on correct evaluation of the difficulties and on exploring a variety of technical processes in order to achieve the final aim, the complete sequence of a genome, quickly and efficiently. To begin with, at the end of the 1980s, sequencing was still very slow and painstaking, and not surprisingly the project's promoters had concentrated on the DNA *sequencing stage* itself, without realizing that in order to determine a sequence it was obviously necessary to have long DNA fragments to work on, that those fragments had to cover the whole of the genome, and that acquiring those long fragments might actually be difficult. In all the applications for funding for projects of this kind lodged at the time, only the sequencing stage itself is taken into account. It is considered to be the bottleneck, but is only superficially evaluated. For instance, the quantity of individual sequences that can be obtained in one day is calculated without allowing for the fact that it is only "raw" sequence. Normally the fragments to be sequenced are obtained by random fragmentation ("shotgun" sequencing), but the time it takes to obtain enough sequence fragments to cover the *whole* genome needs to be estimated seriously. A quick statistical calculation shows that 90 percent of the sequence required can be obtained very rapidly and easily, but there are two problems. First, there is a very high level of redundancy in the results. Many of the sequenced fragments overlap, so that the total sequencing work done is between five and ten times as much as the final sequence obtained, making the process five to ten times less efficient (although of course it does improve the quality of the result). Second, the remaining 10 percent cannot be obtained in the same way; doing so would imply an exponentially increasing level of redundancy, and above all it would take for granted that cloning the fragments to be sequenced does not introduce any bias. In 1988, for example, there were serious applications for funding that proposed to sequence 40,000 base pairs per week. By this, the applicants clearly meant "finished" sequence; either they were unaware of the overlap problem (when assembling the sequence into long stretches of DNA) or they did not bring the question up in public for fear of spoiling their chances of funding. At this rate, the declared target of sequencing the *E. coli* genome would have been achievable in two to three years. There were even enthusiastic applications that made a point of mentioning partially automated systems, capable of sequencing 10,000 to 15,000 base pairs per day, and possibly even more with the aid of fluorescent detection equipment, or other apparently very promising physical techniques. But at the time it was not possible to improve on a rate of about 40,000 finished bases, without gaps, per person per year, or fifty times more slowly than was hoped (although it is now easy to achieve this at the beginning of the new millennium).

Sequencing the *Bacillus subtilis* Genome

The project to sequence another model bacterium, *Bacillus subtilis,* suffered a different fate, although in some ways it was just as unfortunate. Let us go back to the story of what was still, at the time of its conception in 1987, an adventure, and one in which we at the Pasteur Institute in Paris came to be very closely involved.

The story of sequencing *B. subtilis* actually begins in 1985, when the idea of sequencing the human genome was beginning to be discussed in the United States. At the time, the focus of research in our laboratory was on bringing to light collective rules for the regulation of gene expression, and understanding their hierarchical organization. With the early successes in viral genome sequencing, it seemed reasonable and even necessary to try to understand this aspect of gene functioning—how they function together with one another—by analyzing complete genome texts. This objective assumed that it would be possible to discover them, and implied two technical requirements: experimentation at the workbench (the sequence of several million base pairs would have to be determined *experimentally*) and at the computer (the sequence would have to be analyzed, and obviously it would be impossible to do this manually). To fulfill the first condition, in the summer of 1986 I had met Pierre Prentki, a brilliant young scientist who worked in the United States with David Galas, soon to be responsible for genome programs at the DOE. We had arranged that he should come to the Regulation of Gene Expression Unit at the Pasteur Institute and set up a laboratory for sequencing and analyzing the genetic functions of a model genome. I could not know that he was soon to die tragically.

I was familiar with the second requirement because for several years I had had regular meetings with a group of information scientists and biochemists, who had long been involved in the computational analysis of genetic sequences and biological knowledge in general. In France there were at least two schools of statistical and computational analysis of the problems arising out of molecular genetics. One of these was based on the work of Emil Zuckerkandl at Montpellier, followed by that of Richard Grantham at Lyon. The other was derived from the school of *Drosophila* geneticists with Philippe L'Héritier, Philippe Vigier, and Madeleine Gans, and their colleagues specializing in yeast affiliated with Piotr Slonimski, at the Center for Molecular Genetics at Gif-sur-Yvette. At the time there was also a great deal of talk about "artificial intelligence," but it was hard to tell whether it was just a matter of trying to impress, as is often the case with these fashionable expressions, or if there was really something in it. However, this question could easily be answered by taking the laboratory approach—by trying it out. I discussed this with Olivier Gascuel, then involved in constructing expert systems for medicine, and together we defined a project that would allow us to evaluate in great depth how well a computational approach normally classified under artificial intelligence would deal with the kind of general problems genome programs would be bound to raise.

It was known at the time that it was possible to make bacteria express and secrete human proteins. Because this capacity seemed universal, published models of secretion tended to describe only one general scenario for protein secretion by the cell, whether it took place in bacteria or in animal cells. In these models, a *signal peptide* (sometimes called the "zip code"), at the beginning of the protein, both indicates where secretion is to take place and enables it to happen. But scientists involved in perfecting commercial procedures well knew that although the model was adequate in theory, it was a long way from telling the whole truth. Making *E. coli* secrete a human protein using its own signal peptide is normally very inefficient. Could an *E. coli* signal peptide be described in a way that distinguished it from a human signal peptide? Olivier Gascuel and I set to work and produced a model of the bacterial signal peptide that included 17 specific descriptors, whereas the experts at the time proposed only 3. Each descriptor gives a characteristic of the peptide, such as the position of an amino acid at a certain distance from the cutting site, the distribution of charged amino acids, or the alternation of hydrophilic and hydrophobic residues, or large or small ones. The model indicated that a large number of constraints govern protein secretion in *E. coli,* and that these are different from those operating in human cells.

Did reality reflect what we had predicted? The validity of the study can be judged by the fact that it is regularly invoked to reject applications for patents relating to the secretion of heterologous proteins in *E. coli.* For me, the most convincing aspect of the study was this: what the computer showed us was so striking that we could only wonder why we had not seen it directly. This made it clear that informatics is a real aid to discovery when analyzing biological functions on the basis of collections of gene products. It seemed that the time was right, and I was convinced of the potential of the computational approach, which I called *in silico,* to underline its importance as a complement to *in vivo* and *in vitro* experimentation. I was also convinced that it would be worthwhile to develop informatics substantially, as an aid in making discoveries from genome texts.

These were the reasons that prompted me, when I spoke at the French Society of Microbiology in early 1987, to suggest that we in France should embark on the sequencing of a whole bacterial genome. Because the Americans claimed to be already well advanced in sequencing the *E. coli* genome, we would have to choose a different project. I proposed *B. subtilis* as a model of a very different category of bacteria: their cell envelope has only one membrane, and they produce spores when faced with a difficult situation. The microbiologists at the conference were far from enthusiastic about this proposal. But Simon Wain-Hobson, who had been the first to undertake and to complete the sequencing of the AIDS virus at the Pasteur Institute, was convinced that the idea was viable, and suggested that we try to sequence the short genome of *Chlamydia trachomatis,* the bacterium responsible for a very common sexually transmitted disease that often results in female sterility. During the next few months I took these two proposals to all the organizations that financed research, and especially to the Ministry of Research, but at the time I met with no success.

By coincidence, the biennial reunion of *B. subtilis* specialists took place at San Diego at the end of June 1987. James Hoch, one of the American organizers of the conference, floated the idea that a European-American consortium should sequence its genome. Raymond Dedonder, then director of the Pasteur Institute and a specialist in the sugar metabolism of *B. subtilis,* was convinced that this collective approach would be successful. Returning to France, he remembered my suggestion from the beginning of the year and agreed to the American proposal. This was how, although I was in fact an *E. coli* specialist, I found myself deeply involved in the sequencing program of another model bacterium. Dedonder quickly arranged an international meeting during which it was decided that a consortium of five European laboratories would embark on sequencing this genome as soon as appropriate financing could be obtained.

It all began very well. The Americans were sure they would soon find funding, and declared that they would continue the project even if the Europeans let them down. All that remained was to find the money in Europe. By luck, in November of that year, at a meeting of the scientific board of the Center for Molecular Genetics at Gif-sur-Yvette, I met my fellow board member André Goffeau, who was then setting up the yeast genome sequencing project. After some initial hesitation—European funds were limited, so the two projects would effectively be in competition—we became convinced that they would be mutually beneficial, if both could be funded by the European Community. Raymond Dedonder secured funds for the first exploratory stage of the *B. subtilis* genome sequencing program, under the European Community's Science Program. Dedonder (and later Frank Kunst) was responsible for the administrative side of the project, and I established a sequencing laboratory within the research unit I directed, to be led by Philippe Glaser. In the United States, as a result of squabbles over the priority to be given to the different sequencing projects, funding did not materialize. Under these conditions, a few doubts were enough to cause the European program to be dropped from the first phase of the Biotechnology Action Program (BAP), which followed on from the Science Program.

We ought not to have had any difficulties, however. In 1988 I had been nominated to write the scientific text justifying the inclusion of genome projects in the BAP prior to the program's submission to the European Parliament. In the BAP the *B. subtilis* program was regarded as a useful or even necessary complement to the yeast program, and although this was not enough to secure continued funding, it was possible to extend the Science Program contract by a year.

Then something happened to breathe new life into the project. At the meeting of *B. subtilis* specialists held at the Pasteur Institute in July of that year, our Japanese colleague Hiroshi Yoshikawa made an impassioned speech, saying that he did not understand why Japan had been excluded from the genome sequencing project from the beginning. This reaction was decisive for the future of the project: instead of the United States, where scientists were quarreling among themselves, why not team up with Japan? Yoshikawa was confident of getting the necessary funding. A new proposal submitted to the BAP stated that a consortium of Japanese laboratories would sequence one-third of the genome, while a group of European laboratories (and one Swiss laboratory, financed by the Swiss Federation) would sequence the rest.

Meanwhile the genome program continued in France, thanks to limited but regular support from the Pasteur Institute and a public interest group established in 1990, the Genome Research and Study Group, known by its French acronym of GREG and directed by Piotr Slonimski. The French Institute of Health and Medical Research organized a meeting on genome programs in November 1990. Here, perhaps for the first time in public, the importance of developing a strong conceptual infrastructure in informatics specializing in genomics, clearly distinguished from the informatics useful for structuralists or medical informatics and in parallel with the organization of services aimed at the genetics community, was made clear. Alain Hénaut, then at the Center for Molecular Genetics, organized a Spring School to introduce the basic concepts of molecular genetics to a group of experts in informatics, mathematics, and statistics who were interested in genomes. This move prompted the GREG to form a subcommittee specializing in genomic informatics. It was the only organization taking this original approach at the time, and it was not until the mid-1990s that the rest of the world began to appreciate the need for it—just as the French government, as a result of one of its endless changes of ministers and scientific policy, decided it could do without it!

Early developments in the *B. subtilis* genome sequencing program were presented at Elounda in Crete in May 1991, at the meeting organized by the European Community to present the sequence of the first yeast chromosome, mentioned earlier. Over the next five years the European consortium sequenced two-thirds of the chromosome; the rest was sequenced by the Japanese consortium, using the same reference strain. To put the task into context, in April 1996 it was announced that the genome sequence of baker's yeast was almost finished. Sixteen chromosomes, representing more than 12 megabases, had been sequenced by a consortium of more than 100 laboratories around the world, involving 643 scientists. This exploit was all the more remarkable in that it had been achieved two years ahead of schedule and that the genome sequenced was considerably longer than those of the two bacteria sequenced by TIGR. But the longest contiguous DNA sequences in any of these three were still under 2 Mb. Obtaining long contiguous sequences is a very difficult exercise, as the probability of encountering unclonable DNA or highly repetitive regions increases with the length. Since the genomes of both *E. coli* and *B. subtilis* are longer than 4 Mb, sequencing them completely was bound to be much more difficult, especially at the end of the sequencing process, when the last gaps remained to be filled.

The *B. subtilis* sequence determined by the consortium coordinated by Naotake Ogasawara, of the Institute of Science and Technology at Nara, was finished at the end of 1996. The complete genome sequence was known in April 1997, made available to consortium members and publicly announced in mid-July, and published on November 20. Before the contract ended there was time to validate the quality of the data (obviously not totally consistent, as it reflected the work of 32 laboratories) while an exhaustive annotation of the sequences was carried out. The *B. subtilis* genome was the tenth to be sequenced, but it was the third longest sequence known at that time, and the first from this type of organism, whose genome is particularly difficult to sequence because the clones carrying its DNA are toxic in *E. coli*. And since gene correction and annotation constitute a continuous process, the whole sequence was published in an entirely revised form, 180 base pairs shorter than the original sequence, in mid-April 2001.

The Genome of a Cyanobacterium, *Synechocystis* PCC6803

Formerly known as blue-green algae, cyanobacteria remove oxygen from water and release it into the air, using the sun's energy to fix carbon dioxide gas, present in the atmosphere, in the form of molecules with a carbon skeleton. In order to use light, they synthesize pigments, which often produce a blue-green color; hence their name, from the Greek κυανός *(kyanos)*, meaning "blue." Their ancestors, which appeared around 3 billion years ago, produced our planet's atmospheric oxygen—the first large-scale pollution of Earth's atmosphere by living beings—and they still colonize a very wide range of aquatic and terrestrial ecosystems. With its extremely wide variety of morphological and physiological properties, this group of microorganisms is of great interest in several areas of science. Cyanobacteria are generally considered to be the ancestors of chloroplasts (the organelles that specialize in photosynthesis in algae and higher plants, and give them their color). As they are bacteria, the genetics of their light-using function can be analyzed much more easily and quickly than in plants. Certain species also produce toxins that attack the nervous system or the liver and can cause poisoning in humans and animals. Apart from the fact that they are a main food source for plankton, and therefore affect a significant part of the food chain, they are causing increasingly serious problems for the management of the Earth's water resources. Finally, their unrivaled ability to adapt makes them an excellent model for studying cell functions that integrate morphological, structural, and functional changes in response to quantitative and qualitative changes in light and the nutrients in their environment. Quite a large number of cyanobacteria are studied in laboratories, each because of a specific characteristic.

At the Kazusa DNA Research Institute (KDRI) in Japan, Satoshi Tabata and his colleagues decided to sequence the whole genome of the strain *Synechocystis* PCC6803 (from the Pasteur Institute collection) in October 1994. This strain was chosen from among the hundreds of cyanobacteria known, partly because the study of its genetics was already relatively advanced, and also because this bacterium can be used in *reverse genetics* techniques. This involves using vectors (mini chromosomes or viruses) to import DNA fragments into another organism. *Synechocystis* PCC6803 can thus be transformed by exogenous (foreign) DNA, by inserting genes modified by recombination, so as to study their function. With the use of appropriate vectors it can also be made to express plant genes, making the initial study of these genes much quicker and easier. The KDRI is an institute of the Kazusa DNA Research Foundation, a not-for-profit foundation with a very large budget, created by the Chiba Prefecture near Tokyo. Its objective is to develop research programs of interest to medicine and agronomics from the accumulated knowledge on DNA and genomes in particular. The complete sequence of this genome was determined by May 1996. This was a remarkable performance (especially when compared with what TIGR had achieved by then) that was not given the attention it deserved. At 3.75 Mb, it was the world's longest continuous DNA fragment sequence at the end of 1996. Nearly 3,200 genes have been identified from the genome text.

Genomes by the Dozen

Honor where honor is due: bacterial genome programs owe much to that outsider, Craig Venter. While all this research was being developed, and after the yeast sequence had been placed (relatively unnoticed) in the public domain, Venter and TIGR achieved another first in mid-1996, by publishing the sequence of a bacterium that grows at very high temperatures, the archaebacterium *Methanococcus jannaschii*. This bacterium, fundamentally different from the bacteria found in normal environments, was isolated in 1983 from one of the "black smokers" (which are sometimes white!) found off Baja California at a depth of 2,300 meters. These are jets of scalding hot water found on the ocean bed at points where hot spots in the ocean ridges bear witness to the fact that the solid plates forming the Earth's crust are separating. To understand just why the sequence of this genome, a little over 1.66 Mb, is of such great interest, we have to go back to 1977, when Carl Woese constructed a phylogenetic tree showing the relationships of the bacteria known at the time. He was then studying certain methane-producing species, such as those found in the digestive system of ruminants, and he used the RNA contained in a small subunit of their ribosomes as a means of comparison. This seemed to be just as distinct from ribosomal RNA found in normal bacteria as it was from the very different ribosomal RNA of eukaryotes. For nearly 20 years successive discoveries provided more evidence of the originality and phylogenetic isolation of archaebacteria. But it was the knowledge of the *M. jannaschii* genome that supported the most remarkable observation, which had been suspected since the early 1990s. A large number of genes in archaebacteria do not resemble anything else known, either in bacteria or in eukaryotes. In particular, several essential stages in archaebacterial gene expression are more like those of eukaryotes than those of bacteria. They also have their own original way of carrying out certain steps in the metabolism of small molecules, such as the synthesis of one of the DNA bases, thymine (T). In some cases it was still impossible at the end of the millennium to identify all the major steps in their gene expression and in their core metabolism. But this should not be the case for long, and new discoveries ought to be made one after the other, as a very large number of other bacterial genome sequencing programs are under way.

In June 1997 Craig Venter canceled his agreement with Human Genome Sciences (for $38 million), which meant he did not have to keep the genome sequences produced at TIGR secret for too long. He could not have done this earlier without penalties, because the U.S. law on new companies makes it impossible for a company's director to sell his or her own shares in the first five years. This is a fairly effective way of preventing fraud. But at the end of five years, if the company's capital has appreciated, it is possible for the director to sell the shares and make a fortune. In 1997 TIGR and HGS had been in partnership for five years, and HGS was worth $700 million. Clearly, it was in Craig Venter's interest to place his sequences in the public domain while at the same time realizing his investment in HGS. What is more, publishing the sequences enables much more information to be drawn from them, if only by analyzing the questions that others are asking about them.

Bacteria may be unfamiliar to many people, despite their universal importance, but no doubt this will change in the next few years. So I shall introduce you to some of those that were being sequenced as Venter reached this turning point. On June 24, 1997, TIGR released over 40 million base pairs of various sequences, including those of two complete new genomes, *Helicobacter pylori* (1.7 Mb), the bacterium that causes stomach ulcers (about which I will say more later), and that of another archaebacterium, *Archaeoglobus fulgidus* (2.2 Mb). These were made available to the public on a Web site that also listed sequences for several other important genomes, which, though not quite completed at that time, were already at a very advanced stage. *Deinococcus radiodurans* (3.3 Mb, finished in 1998) is a bacterium that can survive both desiccation and massive irradiation. *Treponema pallidum,* which causes syphilis, was sequenced in collaboration with George Weinstock at the University of Texas at Houston (1.1 Mb, finished at the end of the summer of 1997). *Enterococcus faecalis* (3.2 Mb, completed in 2002) is an opportunistic pathogen, particularly dangerous after surgery. *Mycobacterium tuberculosis* (4.6 Mb, completed in 2001) is the agent of tuberculosis, and a different isolate was sequenced at the Sanger Centre in England and completed in 1998. *Neisseria meningitidis* (2.2 Mb, completed in 2000, at the same time as another isolate sequenced at the Sanger Centre) causes bacterial meningitis. *Thermotoga maritima* (1.86 Mb, finished in 1999) is another bacterium that grows at very high temperatures, and since it is not an archaebacterium, a great deal can be learned from comparing it with archaebacteria that live in the same temperature conditions. *Vibrio cholerae* (3.1 Mb, finished in 2000 after it was discovered that it contains two chromosomes, not just one) causes cholera. In addition to these bacteria, TIGR participated in sequencing eukaryotic organisms: part of chromosome 9 of the model for nonherbaceous plants, *Arabidopsis thaliana* (completed in 2000), as a member of an international consortium similar to the one that enabled the yeast genome to be sequenced; and chromosome 2 of the malarial pathogen *Plasmodium falciparum.* Other *P. falciparum* chromosomes were sequenced by the Sanger Centre and the University of Oxford as part of a consortium financed in part by the World Health Organization.

All these Latin names must seem rather exotic, but most of the organisms whose genetic program has been revealed have been or will be famous (or infamous). Cholera, tuberculosis, meningitis, syphilis, and malaria have all hit human societies hard, and are still rife. But all sorts of other projects emerged from the mid-1990s onward, often more specifically commercial or of interest to the academic world. Clearly, the number will only grow, and genetics will be permanently transformed. In fact there already exist two distinct types of project: those in the public domain, in which genome science is part of a general conceptual progress in our knowledge of living things; and those in the private sector (about which not much is known), which enable commercial groups to develop new products and procedures for pharmaceuticals or the food industry. (There are many other domains where genomics may be involved, such as the cosmetics industry, the environmental industry, and even in certain cases heavy industry.) Many other genome sequences have been published since 1997, but I must just mention *Rickettsia prowasekii*, sequenced at Uppsala in Sweden in 1999, which is of interest not only as a pathogen itself but because it throws light on the origin of mitochondria, the organelles that manage the cell's energy, and to which it is related.

The list goes on and on, to the point where it is impossible to be aware of them all. By mid-2002 more than 500 microbial genome sequencing programs had been undertaken, most of them in university laboratories or not-for-profit research institutes. This figure alone is enough to show that we have witnessed a real revolution in genetics, bringing a complete and irreversible change in the kinds of questions we can ask of biology, from the origins of life to the causes of disease (or even the formation of the Earth's crust).

Pathogenic Bacteria and Disease

Perhaps of greatest interest are pathogenic bacteria—ordinary everyday germs. Their genomes have been decoded more or less all over the world, and there have been many rumors (often correct) about them. For example, several companies are thought to have independently sequenced the genome of *Staphylococcus aureus* in early 1996, but the sequence was not made public until the spring of 2001. We can well understand why it should have been chosen, as this bacterium is one of the main causes of the serious antibiotic-resistant infections that are increasingly common in hospitals, and very difficult to cure. There was also the surprising news that the genome of the bacterium that causes stomach ulcers, *Helicobacter pylori,* had been bought for $22 million by Astra from Genome Therapeutics Corporation in 1995.

The tale of this bacterium is worth telling. In 1982 the Australians Barry Marshall and Robin Warren found that these corkscrew-shaped bacteria were present in the stomach of patients with chronic gastritis. Initially named *Campylobacter pylori* because of their superficial resemblance to another intestinal pathogen, *Campylobacter jejuni* (whose genome was sequenced at the Sanger Centre), they were renamed *Helicobacter pylori* because of their helical shape. From studying the pathological phenomena this bacterium induced in the stomach, Marshall and Warren formed the hypothesis that *H. pylori* is not just an opportunist, present because of preexisting damage to the patient's stomach, but that it is in fact the *cause* of chronic gastritis and ulcers. They reached this conclusion because the bacterium was found in all patients with duodenal ulcers and in 80 percent of patients with stomach ulcers. The remaining 20 percent of patients suffering from stomach ulcers had used large quantities of anti-inflammatory drugs such as aspirin or ibuprofen, which are known to carry a risk of ulceration of the stomach wall, and which would naturally explain their disease.

However, although Marshall and Warren's conclusions were quite clear (especially after they deliberately infected themselves and found that they developed chronic gastritis), their demonstration that *H. pylori* caused ulcers took more than 10 years to be accepted, particularly because of the implicit resistance of pharmaceutical companies that produced antiulcer drugs. Gastritis and stomach ulcers are in fact "ideal" diseases in that they are only rarely fatal but they need long-term treatment. Studies involving human double-blind experiments (one group treated with the drug being tested, and a control group treated with a placebo, with neither doctor nor patients knowing who is getting which treatment) are extremely expensive, so they have to be commercially funded, and lack of funding is enough to prevent the research from taking place. It was thus not generally accepted that *H. pylori* was the cause of the disease until 1994, when a study financed by the NIH concluded that there is indeed a causal link between this bacterium and the ulcers, by showing that it is found in almost all patients, and that treatment with appropriate antibiotics leads to a cure.

Helicobacter pylori is often found in the stomach of mammals. Experiments have shown that it induces chronic inflammation and ulceration (together with the acid secretions of the stomach). *H. pylori* survives in this hostile, extremely acid environment because it produces an enzyme, urease, which breaks down the urea produced during bacterial metabolism to make ammonia. This neutralizes the gastric acid and so allows the bacterium to survive. Because of its corkscrew shape and the way it moves, the bacterium can penetrate the mucous coating that protects the stomach wall. It also degrades this mucus and attacks the cells of the stomach wall, which then become sensitive to the proteolytic enzymes, such as pepsin, secreted in the stomach in the normal way to digest the proteins in food. The bacterium can then attach itself to the surface of the cells of the stomach wall, resulting in inflammation. Finally, for reasons that are not yet fully understood—and it is hoped that analyzing the genome sequence will make this clear—*H. pylori* also increases acid secretion in the stomach, causing inflammation of the first duodenum. In some patients this results in the production of abnormal cells (dysplasia). *H. pylori* then attacks these cells, causing an ulcer and sometimes leading to cancer.

Thus a few weeks after being infected with *H. pylori,* most people develop gastritis. Most of them never have serious symptoms and are not aware that they are infected. It is not yet known why some individuals react and others do not. Some studies show that these bacteria even have a protective effect against some diseases. It is very likely that genetic factors (especially relating to the surface of the cells) and environmental factors (such as stress) favor the development of *H. pylori* and the diseases associated with it. Genome programs have revealed that *H. pylori* exists in a wide variety of strains, some more pathogenic than others. Finally, it is not well known how people become contaminated. There is a strong correlation with social class, ethnic group, and age (it becomes more frequent with age and also with lack of hygiene). However, contamination often takes place in childhood, and the bacterium is then present for life. If we understood what its genome text says, we could perhaps develop a vaccine (by studying the toxins or the molecules on the surface of the bacterium) or find targets for new specific antibiotics. As can well be imagined, the financial stakes are high.

There are many other bacteria of interest to pharmaceutical companies, and one can only guess what is being sequenced by one firm or another. Clearly, only *published* sequences really have any meaning, because only they can be verified and linked up with the knowledge held by the world's geneticists. But the *H. pylori* genome has been sequenced and published by TIGR, and so has that of the syphilis agent. The genomes of other sexually transmissible pathogens are also known, including the gonorrhea bacterium *Neisseria gonorrhoeae* (sequenced at the University of Oklahoma) and *Chlamydia trachomatis,* which apart from its role in female sterility can also cause blindness through trachoma (sequenced at Stanford in California and Genset in France). So are some other pathogens carried by insects, such as the typhus agent, *Rickettsia prowasekii,* spread by lice, and *B. burgdorferi,* spread by tick bites, causing the dangerous, slow-developing Lyme disease. There are of course also pathogenic species of worldwide importance, such as *M. tuberculosis.* TB is on the increase again and now kills more than three million people around the world every year. No less than three research programs—one at the Sanger Centre, financed by the Wellcome Trust, another by GTC, and a third by TIGR—were undertaken in 1996, but unfortunately all these projects used pathogenic strains. Despite extensive discussion at the World Health Organization, no one decided to sequence the genome of the BCG, the reference strain used as a vaccine, in parallel with the virulent strain. This omission meant that progress toward improvements in treatment and vaccination against this worrying disease was not as swift as might be hoped. Apart from the aim of increasing knowledge or simply humanitarian aims, there are political circumstances that result in competition rather than collaboration between the United States and Europe. This explains—but does not excuse—the counterproductive duplication of effort. Fortunately the BCG strain has now been sequenced thanks to the efforts of responsible people, in particular Stewart Cole of the Pasteur Institute in Paris.

As for the leprosy bacillus, *Mycobacterium leprae,* WHO's hesitations should stand as a warning. After a joint sequencing program had been launched, bringing together GTC and Stewart Cole, a pioneer in this bacterium's genomics, in the hope of eradicating this disease, it was discovered that in 10 years the number of people infected had suddenly dropped from 11 million to 1.5 million. The program, already well under way, was halted. But in the mid-1990s it was suddenly found that the numbers involved had to be revised upward because the first and second surveys had not used the same method to estimate the number of victims. The sequencing program thus had to be restarted, still under Stewart Cole, and the genome of *M. leprae* was finally published at the beginning of 2001.

Technical Progress and Funding

In retrospect, automation appears to have played a critical role in the first successes. Research scientists have a different background and approach from scientists in industry, and unlike them perhaps did not pay enough attention at first to considerations such as making sure experimental conditions could be easily reproduced, especially when developing conditions for bacterial cultures and very pure, standardized culture media, but also to all sorts of laboratory facilities such as dust-free rooms with humidity and temperature control. These were the reasons why Frederick Blattner initially used robots for his work on the *E. coli* chromosome. At one time it seemed that Church's Multiplex technique might dethrone the enzyme-based sequencing method. But all the laboratories that persisted in using Multiplex up until the mid-1990s were unsuccessful. An intelligent method is not enough to guarantee success; a whole battery of strategic considerations is necessary, as well as an ability to recognize a dead end in good time. A case in point is the sequencing of *Mycoplasma capricolum* by Patrick Gillevet and Walter Gilbert, who wanted to make it the first genome to be sequenced. Launched at the beginning of the 1990s, it was never finished. It is also behind the repeated early failures of the Genome Therapeutics Corporation (which entered the genome race in parallel with Craig Venter and TIGR) until it changed to the more common method. In fact, as well as the quality problem, there was an inherent weakness: it is not enough just to have a large number of DNA sequence fragments from a genome; they need to be put together, and, even more important, the inevitable gaps remaining after assembly have to be filled in quickly. Yet in 1988 journalists were asserting that the promoters of this technique were so confident that they expected to have finished 90 percent of the *E. coli* genome sequence within a year!

Sequencing large numbers of DNA templates is clearly not the real bottleneck in genome sequencing programs. Although most of the total sequence being studied can be determined quite easily, the last few percent of a sequence are considered to take as long as the first 95 percent. The best approach is to decide *right at the start* of the program how these inevitable gaps will be filled in. Is it necessary to build up a fragment bank covering the entire genome? Which vectors should be used? There is a wide choice of options, and they are not all equivalent. The answer to this question is very important, as it will determine how quickly the sequence will be finished. There are a number of other pitfalls as well: not all the fragments can be cloned easily, and some give rise to incorrect or illegible sequences (the ladder of bands on the gel—illustrated in Chapter 2—may be unclear, or there may be "compressions" where several bands are superimposed and more or less indistinguishable). Different problems arise depending on whether the Multiplex or the

Sanger technique is used. Finally, reconstructing and assembling certain zones may be particularly difficult or even impossible when two different chromosome regions have long identical sequences (imagine a jigsaw puzzle in which two parts of the picture are the same). Two factors have made it possible to speed up the procedure: the availability of software capable of automatically assembling large numbers of fragments, and above all the PCR technique. This has enabled long DNA fragments to be copied accurately since the mid-1990s, but it does take time. Primers have to be synthesized (and ways of checking the quality of the PCR still need to be found, as it sometimes copies the DNA inaccurately), and finally the time it takes to carry out the PCR and to run the gels in between has to be considered. The use of new polymerases capable of proofreading has increased the efficiency and accuracy of PCR remarkably, considerably reducing the time taken to fill in the gaps between sequenced regions. It has also meant that the difficulty in cloning regions of DNA that cannot be easily isolated in the form of an amplifiable template (in a bacteriophage or a plasmid) is much less important.

By the mid-1990s, with the right equipment it was fairly easy for one person to sequence around 200 kb per year, five times more than five years earlier. Craig Venter's group sequenced the *H. influenzae* genome (1.8 Mb) in less than a year by shotgun sequencing of its *entire* chromosome and direct cloning into sequencing vectors. Since then scientists at TIGR have made increasingly efficient use of this shotgun technique and of direct sequencing, without building up complete banks to cover the whole chromosome, to sequence the numerous genomes this institute decides to study. Most laboratories now use the same method. In 1996, at the eighth conference dedicated to genome sequencing and analysis, held at Hilton Head in South Carolina, the various centers taking part in sequencing the human genome were completing an average of one million base pairs per year (finished sequence). By 1997, at the ninth conference, this figure had increased by a factor of 10, and it was clear that the technical limits were still far from being reached. Five years later, about 23 billion base pairs of gene sequences had been deposited in the data library now called the International Nucleotide Sequence Database (INSD), and it was not uncommon to see genomes of a few megabases being sequenced in much less than a year. Production of genome sequences skyrocketed at the turn of the millennium, with more than 60 percent of the draft human sequence being produced in only six months. During this time the international consortium produced 1,000 bases a second of "raw" sequence (not assembled into long contiguous stretches) seven days a week, 24 hours a day.

It would be interesting to know where the funding for all these projects comes from. We have seen the role of the European Union. In the United States, the DOE and the NIH finance these projects, with the NIH being more specifically oriented to human genome sequencing, especially because of the hypothesis that knowing the genome will help to solve the problem of cancer. The DOE has created a Microbial Genome Program with the aim of developing the mass sequencing of microbial genomes. In this program the emphasis is not on health but on the protection of the environment and on energy management. A direct offshoot of the Human Genome Project, it was created in 1994 after being under consideration for several years, and is coordinated with the work of federal agencies such as the National Science Foundation, NASA, and the NIH. Generally speaking, the DOE tends to federalize all the institutions, laboratories, and universities it funds, so as to avoid unnecessary duplication of effort, especially in methodology. During the first year it contributed $3.3 million to the financing of four projects: *Methanobacterium thermoautotrophicum, M. jannaschii, Pyrococcus furiosus* (another archaebacterium), and the pathogen *M. genitalium. M. thermoautotrophicum* was chosen because it grows in wastewater, where it is responsible for a large part of the transformation of carbon dioxide into methane. *P. furiosus* (2.1 Mb), sequenced by the University of Utah (initially using the Multiplex technique, which explains why the project fell behind), is a marine bacterium whose optimum temperature is extremely high, at about 100°C. In the next few years the DOE took an interest in other bacteria from extreme environments, as well as those capable of transforming metallic salts or metals, and photosynthetic bacteria, with the long-term aim of finding ways of exploiting the enormously rich biological diversity of microbes. But the DOE's role in genome projects has gradually been reduced, a fact that is perhaps regrettable given the important role it has played in providing technological support for these programs.

The medical value of genome programs, both in financial terms and in terms of pure knowledge, is clearly very considerable. Besides the NIH, it is perhaps interesting to look at the role of a more recent player on the genomics stage. This is the British charitable institution the Wellcome Trust, whose involvement in genome programs since the mid-1990s has been very significant. The trust, whose aims are to finance biomedical research and the history of science in connection with medicine, draws part of its capital from the sales of the pharmaceutical company Wellcome (particularly because of the global growth of AIDS) and is extremely rich, with an asset base of £15 billion in 2001. In the mid-1990s it was spending £250 million per year on research, two and a half times the total budget of the Pasteur Institute, making it Europe's biggest source of finance for biomedical research, and in 1999/2000, with expenditure at some £600 million, the Wellcome Trust was the world's largest medical research charity.

Together with the famous Medical Research Council (MRC), the Wellcome Trust established the Sanger Centre in 1992 at Hinxton Hall, near Cambridge, initially in order to sequence the human genome and those of *Caenorhabditis elegans* and yeast. These two organizations also took part in founding the European Bioinformatics Institute (EBI) as an outstation of the European Molecular Biology Laboratory at Heidelberg, better known by its acronym EMBL (The Sanger Centre and the EBI share buildings at Hinxton Hall). Since then the Centre has undertaken a large number of other programs in genomics, particularly, in addition to the genomes of several pathogens, that of the chromosome of another yeast, *Schizosaccharomyces pombe,* and several human parasites. With Bart Barrell (whom we met in connection with brewer's yeast), the Sanger Centre also became involved in sequencing mycobacterial genomes, with the help of Stewart Cole. The Centre's funding is ensured by the Wellcome Trust until 2003, by which date it is expected to have sequenced a third of the human genome (around 1,000 Mb), at a cost of around 10 cents per base.

The cost of sequencing was originally estimated at $5 to $10 per base, which would have made all these projects an extremely expensive undertaking. The yeast sequencing program cost more than 2 Euros per base (about $2), to which various national contributions must be added (particularly the contribution from the GREG in France), roughly doubling the cost. For *B. subtilis* it was only half as much, and in 1996 TIGR estimated the cost at less than 50 cents per base (although this depends on the genome, as some are more difficult to sequence and therefore more expensive, for instance bacteria related to *B. subtilis* such as *Staphylococcus pneumoniae* and *S. aureus*). Today the cost is much lower still, at less than 10 cents per base (although this cost does not take into account the previously created human and material infrastructure). Using the shotgun procedure and covering each base seven or eight times, companies can produce the sequence of a 10-megabase genome for $1 million U.S., but this is still often more than doubled when "finishing" is difficult.

However, there is a fundamental element whose cost has not yet been properly estimated: the *in silico* treatment of the information. By comparing genome sequences with each other as they appear, it is becoming possible to establish families of genes that appear to have the same function in the cell. Quite often this means that weak similarities can be used to reveal a basic similarity with genes whose function is already known in bacteria or eukaryotes, so that scientists can, at least initially, work on the hypothesis that they do have that function. This is just how the Rosetta Stone was used to decipher hieroglyphics, and that would have been an ideal job for a computer! Most sequencing centers have now set up their own very powerful informatics laboratories. Provided it is intelligently planned, the fact that the Sanger Centre and the EBI are neighbors could be a shrewd way of ensuring that developments in Europe quickly take the right path.

Sequencing the Human Genome

On February 12, 2001, amid great fanfare, Celera Genomics, based in Rockville, Maryland, and the International Human Genome Sequencing Consortium announced that they had jointly published their long-awaited articles describing the sequencing of the human genome. Although both groups had already stated, at a conference at the White House in June 2000, that they had sequenced 90 percent of the genome, the long-drawn-out negotiations that followed delayed publication by six months. What were the issues behind this?

The project to sequence the human genome was an American initiative. Launched in 1990, and originally planned to last fifteen years, the *Human Genome Project* (HGP) began as a joint project involving the DOE's Human Genome Program and the NIH's National Human Genome Research Institute. Very soon this American project became the driving force behind a more general international effort, involving both government agencies and private organizations, initially not-for-profit foundations but later those with commercial and industrial aims as well. The HGP, whose main principles were established in 1995, set itself the objective of sequencing the human genome chromosome by chromosome, which would allow it to coordinate the work of both small and large laboratories effectively. The story of this program, initially American on the technological side, soon became an international one, but to a very large extent it was dominated by the United States right to the end.

Why Sequence the Human Genome?

The underlying political reasons for the Human Genome Project would not be enough in themselves to motivate the scientific community, and in order to convince the funding agencies, and above all the public, good reasons had to be put forward for undertaking it. The subject was the fascinating one of human heredity. Where wealth is concerned, the words *heritage* and *heredity* generally have positive connotations. But when they are associated with the word *genetic,* the uncomfortable thought of hereditary defects comes to mind, together with the fear of passing them on. Genetics does not have a good press; at least that used to be the case. Nowadays, when genomes are mentioned there is an implicit reference to a positive aspect of heredity, or at least to the hope of curing genetically linked diseases. These are two aspects of the same reality, though they are treated very differently in coffeeshop philosophy. For more than half a century, the science of heredity has been taking giant steps forward, and now it is ready to help us understand how living beings are made and reproduced.

Enthusiastic journalists use words that most of us barely understand to describe the collaborative technical projects undertaken in the name of science over the last fifty years, ranging from the construction of vast facilities for physics (which few understand either) to mankind's first steps on the moon. We need to look beyond this attitude, which by implication reduces human beings to mere mechanical objects and values technical exploits more highly, and try to see the real scientific aim. There are two ways of embarking on a project of exploration. The first is to discover everything that is easily accessible as quickly as possible. Newly discovered continents were first explored along the coast. In the same way, the core of genetics was established by studying large numbers of mutants. Later the new continents are exhaustively explored, even at the cost of the most appalling difficulties, because they may hide unknown treasure. The objective of genome sequencing is an example of this second kind of exploration: it is to find the complete chemical definition of a living organism; to make a full inventory of all the parts it is made from, as well as its construction plan. Of course if this were the only aim, we would surely be disappointed. Just think how difficult it is to put together any fairly complex object, bought as a kit, and then imagine trying to recreate a living thing from its genetic "kit"! But in fact genome sequencing is only the first step in an exploration that has barely begun, and that will lead us to the birth of life, and to the prediction of certain aspects of its future.

The genome sequencing projects dedicated to various organisms thus have two basic elements. First the text has to be established by finding out the sequence of the DNA that makes up the genome, just as an archaeologist discovers manuscripts written in an unknown language. The much more difficult and important task that follows is to understand what the corresponding text means. It goes without saying that the text has to be discovered before we can try to understand it, so the next step was to justify to the public, and especially to the media, the reasons for decoding the text of the human genome.

For most of us, our day-to-day health is one of our most important concerns. The idea that knowing the genome would bring cures for diseases, principally cancer, was therefore used as a justification. It was not necessary to say how the genome text could be used for this; hope was enough to keep the program going. Then, promoted by associations of victims of genetic diseases, the idea appeared that "therapeutic DNA" would enable doctors to cure these diseases. Often dramatic and incurable, they are caused by an anomaly in one or more genes, and the hope is that it will become possible to cure them by replacing the defective genes with imported ones (this is now called gene therapy). It is remarkable how widely this aspect of genetic engineering has become accepted, when the same public is up in arms against the idea of creating transgenic plants.

It soon became clear that many more years' work would be needed before such hopes could be fulfilled. Diagnostic applications of the knowledge of the genome would have to do for the time being. But "diagnostic applications" means, of course, avoiding genetic disease simply by preventing the birth of people with the disease, which comes to the same thing as telling those who are affected that they do not have the right to life. This is a repugnant but increasingly widespread attitude. Where is the dividing line between what is tolerable and what is not? Another objective associated with this one (but based on polymorphism rather than on the sequence itself) is more scientific but still has sociopolitical consequences. This is the study of human polymorphism to understand the origin and migrations of mankind. The genome program has already come up with a dramatic result in this area (although it is not universally accepted): mankind came out of Africa. Not everyone is happy with this . . .

Finally, because a plausible medical justification was needed that was not limited to genetic diseases, which fortunately affect only a small part of humanity, another set of objective (and lucrative) reasons for justifying the program was found. Our great genetic variability means that drugs vary considerably in effectiveness (and above all in toxicity) from one individual to another, because different targets can be affected by the molecules used (molecules that did not exist on Earth before they were created in laboratories), and the nature of these targets can differ from one person to another. If it were possible to find out quickly whether a given molecule would be toxic in a particular individual, the medicine prescribed could be perfectly tailored to that particular patient. This is the aim of pharmacogenomics, which will no doubt be one of the main justifications of the Human Genome Initiative in the years following the publication of the sequence.

In order to study the function of a given gene or the product it specifies, it is normally essential to know its sequence. The fact that the genetic code is universal allows us to work out the sequence of the protein specified by any gene and check this against the data banks. It will often resemble a protein whose function (or part of it) is known elsewhere. Knowing the sequence, we can also often determine which of the protein's amino acids are important for its function, for its stability, or for the way it interacts with the other cell components. Using reverse genetics—replacing a gene with an appropriate variant—we can create mutants with modified properties or express the protein in different kinds of organisms, and thus study its structure and validate the hypotheses made about its function. We can also study how molecules we hope to use as medicines interact with this potential target.

Until very recently, isolating individual genes to study their function was very costly in terms of the work involved, and often took several years of perseverance. There were all sorts of reasons for this: lack of recognized mutants, for instance, or lack of an identifiable *phenotype.* Whereas genotype refers to the genetic information (the potential) carried by the individual's genes as a whole, phenotype normally refers to the result of that information's being expressed in a specific environment, interacting to produce a particular individual with all its physical traits and attributes, both visible and hidden. Focusing more closely, as here, it can also refer to the result of the expression of a single gene in given circumstances. If the scientist is to work on it, the phenotype must be clearly identifiable, and this is not always the case. Other problems might include the toxicity of the gene when it was cloned and expressed in isolation, the masking of its effects because other genes had the same function, and so on. But by definition, sequencing a whole genome implies that *all* the genes of an organism are known. Obviously this can help to resolve the problems faced when each gene is considered individually. Even considering just this simple task of isolating a few genes chosen for their theoretical or practical value, whole-genome programs are very good value for money, and certainly (relatively) less expensive than the work involved in isolating only a tenth of the genes they contain.

Besides all this, as the metaphor of the Delphic boat tells us, a genome is much more than a simple collection of genes. Knowing the genome therefore gives access to unexpected functions, which stem from the character of the genome text as a whole. From the point of view of the replication machinery, the chromosome is first and foremost an organized structure, certainly over a short distance, but probably over long distances as well. And because the genes are expressed collectively, genes with functions that are strongly correlated probably have characteristic sequences or signatures. As the replication machinery looks at it, a genome contains promoters and control regions, which the machinery must recognize (and which ought to be visible with appropriate methods of analysis, if we were intelligent enough to guess what actually happens within the cell). A genome is a collection of coding sequences that can be automatically translated using the universal rules of the genetic code (as a first approximation at least; it can and often must be refined later). The resulting set of gene products has to be distributed around the cell according to an organization system many of whose rules are still unknown. There is a semantic aspect to the sequence, which corresponds not only to its catalytic function, if it is an enzyme, but also to its place in the cell; and this opens up new possibilities for targeting drugs.

However, gene products are also the result of evolution, and studying the way they are related to one another provides information on their function and structure as well as on their origin. Each genome has its own way of using the four DNA bases (for example, it may use G + C pairs more frequently than its neighbor does, and this fact will be reflected in the use of certain words). This style enables the cell to distinguish between what belongs to it and what is foreign matter. This is the first hint of what develops into the immune system in higher organisms, where what is foreign is first identified and then usually eliminated, if it cannot be integrated into the normal functioning of the host. In biotechnology it is essential to recognize this style, in order to make an organism express a gene from another organism in large quantities. It is also essential to distinguish between the host and the pathogenic agent, to create drugs that will have an effect on the microbe without harming the host. The list of properties and functions is endless, and can only grow as we know more genomes, and more about them.

Where to Start? Genomes as Models of the Human Genome

If only for anthropocentric reasons, the human genome is understandably of great interest. But is it the best choice? To judge by the shifts of emphasis in the statements made to justify the human genome sequencing program since it was first proposed, this is not an unreasonable question. We have looked at the role of bacterial models and baker's yeast, but there are many other technologically or medically valuable models that should have been chosen in preference. It seemed vital to begin genome sequencing by starting with models that were as compact as possible, and whose genomes would be amenable to genetic analysis, allowing independent evaluation of the fundamental relationships between gene structure and function, especially in genomes whose internal consistency (the result of evolution's long history) is accessible to computational analysis.

Mammalian genomes are very similar to one another. In order to minimize difficulties caused by the presence of introns in the genes, archives, and polymorphism, it seems sensible to consider an organism whose genotype can be monitored as closely as possible. Inbred strains of laboratory mice provide an ideal model in this respect. The mouse genome, divided into 40 chromosomes, is quite similar to the human genome (even though rodents are a very distinct class of mammal, which in some ways makes them a fairly bad model for humans). The main justification for studying the mouse is the possibility of carrying out real reverse genetics experiments very quickly, using *embryo stem cells,* or *ES cells,* and replacing each gene within them by a modified gene. It is quite obviously out of the question to use this technique on humans for ethical reasons, but it is also impracticable, because a human generation of 25 years is much too long. Although reverse genetics is laborious in practice, it has already been done for many genes, and given the very large number of laboratories involved in this kind of research, theoretically it would be possible to inactivate all the 30,000 mouse genes one after another.

This would of course be very expensive, considering that it takes about six months and $40,000 to inactivate one gene, but on a world scale this is comparable to work envisaged (and actually carried out) in space research and high-energy physics.

There are limits to what this kind of study can achieve, not only because the mouse is not human, but because the effect of inbreeding is to show up the consequences of the presence or absence of a gene only in the fixed genetic context of that particular inbred strain. But because their products are part of complexes, genes rarely function in isolation. Besides, reverse genetics allows genes to be analyzed easily only one at a time, usually by inactivating one without studying the way its expression is regulated. The most important lessons are learned from mutations, whether spontaneous or artificially induced, but in this case it can be slow and difficult to find the corresponding mutated gene. So other complementary models need to be studied, as well as the mouse.

There are other differentiated multicellular organisms that are much simpler, and have already been used as models for embryonic determination and for cell differentiation. The *Drosophila* genome is 20 times shorter than a mammalian genome (though 30 times longer than that of a bacterium). Colonies of mutants can be kept in culture bottles, and more than a century of experimentation has shown that with around 20,000 of these colonies, it is possible to obtain mutants for all the genes controlling the first stages of cell differentiation, when the fate of all the cells of the embryo is decided. This gives an idea of the size of the task in studying a mammal. It is obviously impossible with the mouse, but if the aim is merely to study a vertebrate, the tiny zebra fish (*Danio rerio,* two centimeters long at maturity, and whose young are transparent) could be a suitable subject for study by systematic mutagenesis, as has been done with *Drosophila*. Christiane Nüsslein-Volhard at the Max Planck institute in Tübingen and Nancy Hopkins at MIT took up this daring but fascinating challenge, and the first important results of their experiments were published at the turn of the century.

Although the construction plans of different mammals are similar to those of insects in their broad outlines, there is considerable variation in the detail. Furthermore, there may be major differences from the plan of other animals. This was why Sydney Brenner suggested a complementary model, *Caenorhabditis elegans.* It would make an excellent model for cell differentiation, as its construction plan is firmly fixed, cell by cell, in its distribution in both space and time (including the plan for programmed cell death, known as *apoptosis*). Its genome was sequenced by two laboratories, that of Sir John Sulston at the Sanger Centre, and that of Robert Waterston at Saint Louis, and most of it was known (but far from completely analyzed) before the year 2000.

In parallel with the human genome, there is one other essential choice for the first years of the new millennium. Given the rate at which the human population is growing, the food supply will become a crucial issue. Plants and animals are so different that it seems only natural to consider that they should be studied separately. This is especially obvious with photosynthesis, an essential activity for living things, but one that only plants and a few microbes (mostly cyanobacteria) are capable of. In plants, the interactions between the cell's nucleus and the chloroplasts (thought to be degenerate cyanobacteria living in symbiosis in the nucleus) are particularly interesting. Finally, as they cannot move, plants must have very different defense mechanisms from those of animals. The genome of the thale cress *Arabidopsis thaliana,* at a little over 100 Mb, is more accessible to analysis than a mammalian genome. Through a collaboration program that brought together European laboratories, the Kazusa institute in Japan, and TIGR, the "complete" genome sequence of its five chromosomes was published in 2000.

From DNA Sequencing to the Human Genome Project

As we have briefly seen, the Human Genome Initiative was the result not of scientific deliberation but of a conjunction of technical results, personal ambitions, and political constraints. It involved not only the two principal American federal agencies, the DOE and the NIH, but also many other countries. Scientists around the world participated, independently at first, but then pressure groups formed, leading to better support and improved funding, and to successes and failures here and there. The story of the power struggles and dramatic exploits that, one after the other, left their mark on the way the program was organized, would fill a book. A look at the comments in almost every issue of *Science* and *Nature* over the last five years, as well as the information given on various Web sites, will show the struggle not only between the federal agencies but also between personalities within those agencies, and between countries. Looking back, it is not that easy to reconstruct its history.

On March 14, 2000, Tony Blair and Bill Clinton published a short joint declaration in which they "applaud[ed] the decision by scientists working on the

Human Genome Project to release raw fundamental information about the human DNA sequence and its variants rapidly into the public domain." The declaration concluded with an enigmatic statement in which Blair and Clinton "commend[ed] other scientists around the world to adopt this policy" of rapid publication. It goes without saying that it is unusual for heads of state to intervene in scientists' decisions to publish, but incongruous as this is, the declaration is a salutary if stern reminder that the Human Genome Project is based on a political initiative, not a scientific one.

Immediately after Hiroshima and Nagasaki were crushed by the atomic bombs, the United States initiated a policy of intensive cooperation with Japan in order to hold off the growing threat of Communism. Among other aims, the policy allowed the Americans to salve their conscience by showing an interest in the future of the residents of these martyred cities. Genetics had a central place in the scientific part of this collaboration policy, which explains how the DOE, the federal agency responsible for the United States' nuclear programs, and successor to the Atomic Energy Commission (AEC), came to be involved in research that at first sight would appear to be well outside its natural jurisdiction. The main areas of research were the mechanisms of mutagenesis and the effects of radiation on the genes. In 1947 this effort led to the creation of the Atomic Bomb Casualty Commission (ABCC), financed by the AEC. Genetics made up an important part of its research, to try to evaluate the mutagenic effects of radiation on the descendants of those who survived the atomic bomb.

The mutagenic effects of radiation had been discovered by Hermann Joseph Muller in 1927. In 1946 he was awarded the Nobel Prize for this work and subsequently made appallingly alarmist predictions, with eugenic connotations that were much discussed when the sequencing of the human genome was announced. In 1954 the ABCC published a report by James Neel and William Schull on the first genetic findings on more than 75,000 births in Hiroshima and Nagasaki. The results were reassuring, but they dealt with only the first generation of children born since the bomb, and were based on an analysis that was still rudimentary. The report was published only a year after the structure and mode of replication of the DNA molecule had been discovered. A generation later, more sophisticated studies that analyzed protein mobility in an electric field did not contradict this early work. But to be really sure of what kind of mutations radiation might have caused, it was necessary to find out what happens in the DNA sequence, right down to the level of the nitrogen bases. The expected effect of exposure to radiation is to produce a particular polymorphism in the DNA sequence. Measuring such single nucleotide polymorphisms (SNPs) all along the genome is the basis of one of the most important tasks associated with the Human Genome Initiative. This was not possible at the time of the original studies, but it was clearly a theoretical reason for attempting to discover the genome sequence.

However, in the meantime the political context had changed beyond recognition. In the mid-1980s the Cold War rhetoric gave way to concern about a new adversary. Japan's economic power threatened America's leadership in technology. The federal agencies were mobilized to encourage the setting up of new companies and to protect American intellectual and industrial property.

It is impossible to give a precise date for the beginning of the Human Genome Project. Some writers date it from the Alta summit in Utah in December 1984, organized by the DOE. The aim of this summit, in which James Neel took part, was to discuss what strategies should be used to detect mutations in the generations after Hiroshima and Nagasaki, in the context of the DOE's mission for life sciences. The original motives were soon forgotten. Discussion focused on the state-of-the-art technologies the DOE would be able to deploy, and all sorts of potential models for identifying mutations were reviewed. Direct sequencing of the DNA involved was already considered to be one of the most obvious methods, and in fact the Human Genome Project could not have been imagined without an efficient DNA sequencing method and the constant progress that has been made in this technique. Credit goes to Allan Maxam and Walter Gilbert at Harvard, but above all to Frederick Sanger at the MRC in Cambridge, who discovered how to identify a gene's sequence by reproducing DNA replication in a test tube. "Fluorescent" sequencing, introduced by Leroy Hood's team at Caltech in 1986, was a remarkable improvement. Laboratories in Europe, the United States, and Japan immediately tried their hand at imitating and automating these methods. The technique continued to be improved and developed both by its promoters and by its competitors, and these efforts led to a hundredfold improvement in laboratory performance between 1995 and the end of the century. Hood had set up Applied Biosystems, which specialized in laboratory equipment for molecular biology, in 1981. With a DNA sequencing machine based on the fluorescent technique already on the market by 1987, Applied Biosystems developed at remarkable speed and was bought by PE Biosystems, part of the Perkin-Elmer Corporation. Its model 3700 capillary sequencer came onto the market at the end of 1998, resulting in a considerable acceleration in sequencing speed in laboratories around the world.

The DOE's researchers contributed to another improvement in the techniques of molecular biology, which played an important role in the initial stages. This was the use of "cell sorter" methods, whereby in a mixture of cells, those marked by the presence of a fluorescent molecule can be separated from unmarked cells. It was extended to chromosome sorting, making it possible to purify human chromosomes and to establish specific DNA banks for each chromosome. As there are 22, plus the 2 sex chromosomes, this meant a considerable reduction in the size of sequencing projects. The mainly Japanese consortium that sequenced a draft of chromosome 22, and the French national sequencing center at Evry, near Paris, which sequenced chromosome 14 in collaboration with Leroy Hood, used this method.

Nor would progress have been possible without the systematic development of informatics in terms of both hardware and software. In this area, too, the DOE's contribution was significant. Developments in sequencing were paralleled by developments in computer calculating speed and the exponential growth of their memory capacity. However, the parallel threatened to break down, probably from the turn of the year 1997–98, when the production of sequences began to grow much faster than the development of software used for assembling the sequences and for managing the knowledge associated with them. It had already been clear by 1978 that the sequences would have to be kept elsewhere than just in scientific publications, and that an informatics infrastructure would rapidly become necessary, to allow the scientific community to build the sequences into a continuous text that they could then interpret. A study undertaken by Rockefeller University and the EMBL at Heidelberg led to the idea of the creation of a data bank for gene sequences. It became clear very early on that the possession of this information was of vital importance, with considerable political implications. Frequent discussions, often heated, took place between Europe and the United States about where these data banks would be and how they would be structured. Who would be responsible for sequence quality, its producer or the data bank? Who would produce the annotations? This issue is clearly no small matter: a bad annotation is tantamount to disinformation. It is unfortunately now clear that major annotation errors have spread via data banks through the entire scientific community, resulting in serious errors in interpreting observed biological phenomena. Two banks were established in the late 1970s, in competition but also in touch with each other—one at Heidelberg, the other, the first GenBank, at one of the DOE's laboratories, the Los Alamos National Laboratory (LANL). It was just at this time that the office revolution began (an event whose significance was not understood by many at the time), making microcomputers available to everyone and enabling gene sequences to be analyzed at the desktop.

After the Alta summit at the end of 1984, Robert Sinsheimer, then chancellor of the University of California at Santa Cruz, put forward the human genome sequencing project as a grant application. In May 1985 he brought together a group of well-known scientists to discuss the idea but was unable to raise the funds needed. Independently in 1986 Renato Dulbecco, of the famous Salk Institute, proposed using the human genome sequence to discover the causes of cancer.[8] The same idea was being developed at the same time at France's Center for the Study of Human Polymorphism (Centre d'étude du polymorphisme humain, or CEPH), established by Jean Dausset to collect the entire genetic blueprint of families whose genealogy was well known. Daniel Cohen, a very active researcher at Dausset's laboratory, who had realized the value of the genetic heritage that this unique collection represented, developed an industrial-scale approach that would result in the sequencing of large segments of the genome. Finally, Charles DeLisi proposed independently that this project be carried out at the DOE, which was concerned to see its large-scale operations in nuclear physics coming to an end. DeLisi, who had worked on computational models of biology at the National Cancer Institute, one of the National Institutes of Health, had taken on the task of understanding the meaning of the sequences and had worked on this with researchers from LANL and with Minoru Kanehisa, a Japanese scientist who was to play an important role in the genome annotation programs that followed.

DeLisi was at the time one of the project leaders in biological research at the DOE, which enabled him to cost the project and make the first practical proposals. In 1987 he persuaded the DOE to redirect $5.5 million intended for other projects to the genome program. So it was the DOE that first got explicit recognition for the need for a human genome program in the United States. In 1988, through the influence of New Mexico's senator Pete Domenici, the program was considered by the U.S. Senate and brought into the White House discussions on large-scale scientific projects. David Galas, a pioneer in molecular genetics, soon became a keen supporter. In 1988 the NIH obtained $17.2 million for such projects, while the DOE had to be content with $10.7 million.

In France, Daniel Cohen and Jean Dausset obtained a preliminary budget heading under which to explore the feasibility of the project, using the CEPH's human DNA libraries. More importantly, Cohen managed to persuade the minister of research that the CEPH, with its private structure, could begin a sequencing program more easily than public bodies could, if it had direct help from the ministry. From 1989 onward the CEPH recruited scientists and engineers and purchased robots and industrial equipment in order to begin to map and sequence the human genome on a large scale. At the same time the CEPH and Bertin, a private company (in association with two British partners), obtained Eureka funds for a project to create an industrial supplier for the necessary equipment. This project, called Labimap, was to supply oligonucleotide synthesizers, robots and reactors for automatic plasmid preparation, sets for large-scale molecular hybridization, and miniature electrophoresis gels for sequencing. As Craig Venter would do later, Daniel Cohen already saw quite clearly that genome projects would have to develop molecular biology techniques on a large scale. It would be interesting to analyze Labimap's total failure, as it could have given Europe the equivalent of what Applied Biosystems and PE Biosystems gave the United States.

From this time on, genome projects became a political issue more than a scientific one, and it was not until the mid-1990s, with the first major successes of TIGR, the European Union, and Japan, that science managed to take charge again. Meanwhile, progress was too slow to suit Daniel Cohen. By a happy coincidence, Bernard Barataud, the energetic president of the French Muscular Dystrophy Association, had organized an unexpectedly successful telethon in France in 1987. He planned to use the money collected each year to finance an ambitious program in human genetics. Cohen realized just how far he could turn this to his advantage, and he convinced Barataud that sequencing the human genome would speed up the identification of genetic diseases considerably. Barataud chose Evry, on the outskirts of Paris, as the site for the substantial laboratories that would be needed. The first Genethon was established at the end of 1990, with the first prototypes built by Bertin for Labimap. It very soon became clear that it was too early to sequence the human genome, given the size of the task (a huge number of large chromosome segments have to be cloned, which is very difficult). So to begin with, both in France and elsewhere, the projects were reoriented toward gene mapping (locating markers spread out along the chromosomes). Genethon had three major programs: under Daniel Cohen, yeast artificial chromosome banks (YACs) carrying random fragments of human chromosomes; under Jean Weissenbach, then at the Pasteur Institute, the construction of a detailed physical map; and under Charles Auffray, the creation of a complete set of human complementary DNA. To international astonishment, in the spring of 1992 Daniel Cohen presented the first complete map of chromosome 21 at the annual meeting of the Cold Spring Harbor Laboratory in the United States, and in the fall of the same year he published the first contiguous sequence map of YACs, containing up to 1 megabase of human DNA. This map, made using the computer facilities of INRIA, France's National Institute for Research in Computer Science and Control (with Guy Vaysseix and Jean-Jacques Codani), placed France at the forefront of genomics. This is not the place to discuss the reasons for the rapid collapse of the French lead, except to say that it was largely the result of a serious error of scientific judgment on the part of certain decisionmakers, acting behind the scenes, and of a skilful manipulation of the ministerial structure at the time.

At the same time a battle was going on for the ownership, administration, and scientific management of GenBank between the DOE and the National Center for Biotechnology Information (a division of the National Library of Medicine), financed and now managed by the NIH. After it lost control over GenBank, when genome programs began to be productive the DOE financed a rival bank, Genome Sequence DataBase (GSDB), with the declared intention of collecting and managing data from genome programs. This data bank was managed by the National Center for Genome Resources, a not-for-profit foundation created at the end of 1992 on the initiative of Senator Domenici. However, data entry into the different data banks was not synchronized, and the data they held were inconsistently labeled, putting scientists all over the world in an almost impossible position. Fortunately, a change of political orientation resulted in increased support for GenBank after it had been moved to the National Center for Biotechnology Information (NCBI) at Bethesda. The informal association between GenBank and its European and Japanese counterparts, which had existed since 1990 and which later became official, also brought stability. On the European side were the EMBL, first at Heidelberg and then at the EBI in England, and on the Japanese side the DNA Data Bank of Japan at the National Institute of Genetics (NIG) of the Ministry of Education, Science, and Culture (Monbusho), at Mishima. They effectively became a single data library (the forerunner of the INSD), with entry points at the NCBI, the EBI, and the NIG. But a further rivalry about control over data libraries, between the public genome effort and Celera, was to increase tension again sharply at the beginning of the new millennium.

Clearly, it is not possible to look into the details of these power struggles here, and I mention them merely to show the importance of the impact of politics on scientific development. As often happens, the conflict appeared as the dominant players began to lose influence. This was the case with the DOE, which was witnessing a slowdown in research programs based on nuclear energy, and ran the risk of soon finding itself bled dry financially if it could not put forward to the federal government a long-term program that would be expensive in terms of manpower and funds. But despite appearances, the DOE had very good reasons to be involved in the Human Genome Initiative. Its participation in the technological infrastructure played a key role in the success of genome programs. For all these reasons, individual projects were evaluated in a highly charged atmosphere, hardly likely to encourage either balanced judgments or that national and international collaboration which would certainly have led to the success of the project in a much shorter time.

Meanwhile, the Human Genome Project was getting organized. An informal international association, the Human Genome Organization, shared out as best it could the task of sequencing the human genome, chromosome by chromosome, among laboratories around the world, with a target date of late 2005. But it was "small" genomes that restored the importance of sequencing and made it clear how valuable a knowledge of genome texts could be. The turning point was 1995, when the outsider Craig Venter burst onto the scene. In that year he and his colleagues at TIGR published the sequences of two very small bacterial genomes one after the other, having used a method similar to that used by Daniel Cohen, but more successful.

Then in 1998, in one of those dramatic coups he is so good at, Venter again changed the face of genomics. He is the reason behind the unexpected Blair-Clinton declaration. In mid-1997 he canceled the agreement between TIGR and HGS, freeing himself to scale up his approach to genomics. In early 1998 he announced that together with the Perkin-Elmer Corporation he had created a new company, Celera ("fast" in Latin), with the aim of sequencing the human genome within three years. The plan was to use the shotgun approach, without preliminary separation of the chromosomes, using supercomputers to reassemble the fragments, thanks to a high-speed algorithm invented by Eugene Myers. The sequencing was to be carried out by several hundred of PE Biosystems' capillary sequencers, and the planned coverage of the genome was tenfold, that is, 30 billion bases. In addition, Venter proposed to demonstrate the feasibility of his approach by sequencing the entire *Drosophila* genome (nearly 150 megabases), in collaboration with Gerald Rubin's group at Berkeley, by the end of 1999. They pulled it off—that is, if the sequence can be accepted as "complete" when in reality it is not.

It is worth pointing out again Craig Venter's remarkably clear thinking. It has long been obvious that the *Drosophila* genome should have been chosen in the first place as the model organism. Not only are the genetics of this insect by far the best known in the world, but also its development is, strange as it might seem, remarkably similar to that of vertebrates such as the mouse or man. Venter could thus rely on data obtained from *Drosophila* to help him identify many of the most important human genes, at least as a first approximation. At the same time as he perfected the scaling up of his shotgun technique, he could be preparing to annotate the human genome.

Celera is a private company, and its aim is obviously to make a profit. Venter therefore announced that he would not immediately release his sequences into the public domain, and that in any case any use of his sequences for profit would attract royalties. In the circumstances, the organizations engaged in the Human Genome Project—the Sanger Centre, which planned to produce a third of the sequences, and the groups involved in Europe and Japan—reacted vigorously. They began by speeding up sequence production considerably, aiming to produce a "working draft," a "coverage" of the unassembled genome, by the summer of 2000, and the complete sequence by 2003, two years before the date originally proposed. Very soon the consortium published the sequence of chromosome 22. It also launched a high-profile public debate about the fact that in assembling its sequences, Celera made extensive use of public domain sequences, and that for the company to want to make a profit from this constituted an abuse. In March 2000 letters between Francis Collins and Craig Venter were passed to the national newspapers in an attempt to force Venter to cooperate with the public project and to make his sequences available to researchers throughout the world without charge. It is to this exchange of letters that the Blair-Clinton declaration alludes.

On February 12, 2001, with a joint declaration, the international public consortium and Celera finally published the results of their complementary efforts. The first draft of the human genome sequence was at last available. But whereas the consortium's sequence was deposited in the public data library (now the INSD), Celera's data bank was accessible (for a substantial fee) only via its own computers and could not be distributed worldwide. Ironically, the public-sector data were published in *Nature,* a commercial, profit-making journal, which wrote an editorial on its policy of free access to scientific knowledge, while Celera's data were published in *Science,* the journal of a not-for-profit scientific association, together with a convoluted justification of the fact that the sequence had not been released into the public domain. Despite the fixed smiles, the announcement was almost immediately followed by barbed comments in the media. Why did such hostility greet what should have been a great success?

Apart from a desire for ephemeral glory, the most widely shared value today is the profit motive. The quest for the human genome is a lucrative enterprise, and one area in which PE Biosystems is quietly piling up the profits is through the sale of its sequencers and other laboratory equipment. If only for this reason, the stir created around Celera was an immense commercial success. But what made the public-private quarrel so intense was no doubt the question of the glory. Who should be counted as the real discoverer of the genome sequence? The sequence exists in its own right, no matter who led the consortium that decoded it, so it ought not to be possible to patent a bare sequence. Should the person who discovered it also be counted as the inventor of its potential for use? What the members of the public consortium clearly held against Craig Venter was that he was trying to have it both ways: he wanted both public recognition and private profit. In order to make the profit, he was obliged to limit access to knowledge, which the public considered all the more immoral because Celera's discoveries relied on knowledge produced by the public consortium.

All the arguments, which have sometimes been very heated, center on these questions. But there is also the question of what exactly is a "complete" sequence. With bacterial genomes it was simple: a sequence was announced as complete when it was thought that, apart from a few inevitable errors, the entire sequence was known. But this is not the case with plant and animal genome sequences. If proof is needed, in December 1999 the sequence of the 56 Mb of human chromosome 22 was published, yet in reality the sequence obtained was only 33.5 Mb long. In the spring of 2000 it was the turn of the *Drosophila* genome sequence, but here again "complete" meant only around 120 Mb (with a lot of spurious removal of duplicated sequences normally present in the genome), whereas the entire genome is more than 150 Mb; and it was the same with *C. elegans*. Immediately after the February 12 announcement, many people were concerned about this state of affairs. Several group leaders from the public consortium were heard to say that the genome sequence was a long way from being completed, but that they had to keep up with Celera, because otherwise the publicity would make the public effort look ridiculous, and this would not please taxpayers. From a technical point of view, the anomaly is understandable. Certainly, there are regions that are difficult or impossible to clone, and the highly repetitive regions are virtually impossible to sequence. But we cannot avoid feeling frustrated on learning that "complete" does not really mean complete!

If we look back over a story that continued to twist and turn even as the sequence was announced, we find an explosive mixture of the values that make up science—not only the love of knowledge, of course, but also political rivalry, the search for glory, and the particularly forceful intrusion of the commercial world. In the beginning it was an entirely American game, inspired by the struggle against Communism, then against the technological supremacy, real or imagined, of Japan. It led to 20 years of very strong public support for innovative private business, in a policy that is nowadays imitated in other countries. This makes the Blair-Clinton declaration, which seems to take the opposite view, all the more surprising, as if suddenly the free market and its corollary, the protection of intellectual property, were considered a threat to free access to knowledge.

After Sequencing, What Can We Expect from the Human Genome?

The most surprising result of the sequence as published was surely the reevaluation of the total number of genes it contained. This had been generally thought to be around 80,000 (although some estimates were as high as 150,000). The article in *Science* estimated the number at between 26,000 and 39,000 genes, while the public consortium spoke of 30,000 to 40,000. This compared with 13,000 for *Drosophila*. Of course, the number of genes is not a good measure of an organism's creative capacity, any more than the length of a computer program gives an idea of its inventive power. But it seems that, as always, superficial thinkers will confuse the qualitative with the quantitative, and be astonished.

Just as the map of the heavens is not cosmology, genome texts are not genomics. The aim of this new scientific discipline (the term *genomics* appeared in 1986) is to use what the text tells us to help us understand, as best we can, what life is. From our point of view as genome explorers, it is not the gene sequences themselves that are valuable, but their annotation, the discovery of their meaning, and the inventions that may result from all this. This can be used as a justification for giving an intellectual property value to the discovery of a genome sequence. But does this make sense? It seems to me that patenting genes themselves does not make sense, not for moral reasons—after all, we patent arms, which does not necessarily mean that we agree with their use—but because they are not something that has been "invented." On the other hand, understanding a biological function can lead to the discovery of a therapeutic target and thus to a treatment. Equally, awareness of a function can lead to the discovery of a basis for diagnosis, and, yes, the use of this could be patented.

We have seen some of the hopes raised by the decoding of the human genome sequence. What is the position now? The mapping approach has enabled genes involved in genetic disease to be identified much more quickly, so that diagnostic tests could be developed (with all the moral problems these bring with them). It has brought new approaches to the tracking down and identification of criminals and victims, which had barely advanced since the introduction of fingerprinting. With genetic "fingerprints," produced by amplifying minute quantities of DNA (the root of a hair provides enough), identification is just as precise, but much more sensitive, and less liable to be damaged, either deliberately or accidentally. Hopefully, it will also be possible to develop suitable genetic treatments, either by using DNA itself (DNA medicine or gene therapy) modified in such a way as to suppress, alter, or restore the activity of a gene or a derivative, or alternatively by using gene products—the proteins they specify. But in reality, as with any major exploration, we still do not know what this text really has to offer.

Finally, we should note that the monetary value of the sequencing program depends entirely on the assumption that the systematic analysis of the DNA text will enable us to understand its function, or at least allow it to be used in some way. A large number of companies (particularly in the United States) have been set up on the basis of this hypothesis. According to them, there are huge profits to be made by exploiting the knowledge brought to us by the human genome (and by genomics in general since the beginning of the 1990s). The confidence shown in this hypothesis is demonstrated by the value of shares in the biotechnology industry, which is founded on genomics. This confidence is strongly supported by governments, in the hope that the anticipated success of these industries will generate a great deal of employment (as can be seen, for instance, in the information published by the European Union, which supports some of these projects). Even though it started out as a purely technical project, because of its size the sequencing program has gradually become subject to government appraisal, and has therefore passed into the hands of bureaucrats, with all the attendant advantages and disadvantages. It is now a hybrid project, both academic (on the theoretical side) and commercial (on the industrial side). It is hard to tell which way the balance will tip. But one thing is certain: it will be a great revolution which will leave its mark on the twenty-first century.

What really needs to be monitored is how knowledge of the genomes will be used in the future. That is where the real moral problem lies; but who is paying any attention?

2

THE ALPHABETIC METAPHOR OF HEREDITY

Everything we have said so far assumes that the genome text can be experimentally determined. How can we discuss DNA and the sequence of its bases, unless this can actually be achieved in practice? In order to understand sequencing, we need to look into the mysteries of the technique. This chapter explores how the hereditary memory is transmitted and the main technical principles of genome sequencing, and it throws some light on the factors limiting the conclusions I will offer later. There is no such thing as an infallible technique, and what is more, every technique is associated with a particular conceptual model of the field in which it is used. There are always gray areas of uncertainty, even if the majority of the results are error-free. This chapter will also show that, in principle, the methods of molecular biology are very easy to use—so much so, in fact, that we may wonder how it will be possible to keep a check on certain kinds of experiments, which can be carried out in a garage, or even a hotel room, as has been done by scientists wanting to prove that Japan was disregarding the moratorium on whaling.

The hereditary memory contained in the heart of living cells can be represented as a text written in a four-letter alphabet. We realize of course that a computer model of an airplane *is not the real airplane,* despite the fact that experiments can be carried out on it to discover the way the plane will behave. It is just the same with the genome text: even though we can actually experiment on it (and the preliminary steps in genetic engineering used in biotechnology do exactly this), this alphabetical image of heredity is not the hereditary material itself; it is just a text written in the alphabet A, T, G, and C. Real DNA is a giant molecule, a particular chemical object, which needs water in order to function (it has completely different properties in other solvents), and it has original properties that can be explored using an arsenal of techniques involving physics, chemistry, and the biochemical properties of certain proteins. How do we get from the actual molecule to its representation as a text? Without going into the biochemical detail—which would need a very technical treatise all to itself, and would have little to offer in the way of new concepts—we need to look at some of the main principles of the mechanisms for transmitting heredity. Then we will be able to understand how the genome text, the *sequence* of the corresponding DNA, is discovered.

DNA Replication, or How the Hereditary Memory Is Transmitted

For an organism to be autonomous, to reproduce, develop, and survive, something has to be preserved from one generation to the next—a *memory*. This memory is often called the "genetic heritage" (with the curious connotation of the inheritance of wealth in the word "heritage") or the genome, and it passes on the information needed for the genesis and development of the organism from one generation to the next. The process responsible for transmitting this genetic information is called replication.

The physical material of the genome belongs to a branch of biochemistry, that of nucleic acids, or more specifically of deoxyribonucleic acids (DNA). These are huge molecules in the form of a strand, which may be linear, or may join its ends and form a circle. Linked together in single file are four relatively complicated "motifs," four related members of a chemical repertoire called nucleotides. (They are also known as bases. The distinction between these two terms is explained in the glossary, but we do not need any detailed knowledge of their chemistry here. Just remember that they are represented by four letters of the alphabet, A, T, G, and C, which are the initial letters of their chemical names.) In most cases DNA is a *double* molecule, formed of two strands, whose nucleotides pair up across the center, but not in the way one might expect. A nucleotide on one strand does not correspond to the same nucleotide on the other, but to its *complementary* nucleotide, according to this rule: A is always opposite T, and C opposite G. The nucleotides are joined *along* the strand by strong chemical bonds that are stable in water in normal terrestrial conditions (at least for a relatively long time, compared to the lifespan necessary for the DNA's function to be expressed in the cell). But the chemical bonds that hold the complementary nucleotides together *across* the double helix are weak, and they break and reform easily in water. This difference is essential to DNA replication. The strands are directional—the sequence AAGGCCT is different from the inverse sequence TCCGGAA—and antiparallel—the two strands are "read" in the opposite direction to each other. If we visualize them as represented below, written one underneath the other, the cellular machinery will read the top one from left to right and the bottom one from right to left, as shown by the arrows > and <.

>AAGCTGCGGCCATTTTAGCTTAAGCTGGGCAT>
<TTCGACGCCGGTAAAATCGAATTCGACCCCGTA<

If the sequence of the bases on one of the two strands is known, the sequence and direction of the other strand can therefore be worked out.

The bonds between the bases make each strand twist around the other to form a helical structure similar to a double spiral staircase, like the famous staircase at the Chateau de Chambord in France's Loire valley. The discovery of this *double helix* structure by Maurice Wilkins, James Watson, Francis Crick, and their colleagues in 1953 was a real conceptual revolution.[1] It is the basis of genetics as we know it today, because it enables us to understand how the hereditary memory can be copied exactly from generation to generation. All that is needed is for the strands to separate, and for an appropriate manipulating system to "copy" each strand by linking together the nucleotides present in the immediate environment according to the A–T, C–G rule, so that the correct base is put in opposite each base of the strand being copied. *Copy* is the word normally used, but it is essential to be aware of the distinction between making an identical copy and making a complementary strand. When the DNA molecule as a whole is "copied" in replication, this is done by making a new complementary strand for each original strand. This process produces two identical DNA molecules, each of which has one old strand and one new strand. This is how the genetic information is transmitted from one generation to the next. But how is it concretely put to work by the individual that carries it?

The Genetic Code, Transcription, and Translation

The information carried in the DNA controls the synthesis of *proteins,* macromolecules of a different chemical class. There is a strict correspondence between the memory embodied in the nucleic acids, and the proteins they produce. The chain of DNA bases, which in our alphabetic analogy is seen as a linear text, is *expressed* using a different alphabet, according to a set of "rewriting rules," which can be summed up in the correspondence known as the *genetic code.* This choice of terminology has unfortunately misled some people, who have understood it as implying that the "code" of the genetic code is similar to a computer program or algorithm. In fact it means a simple code in which one symbol or letter stands for another, like the secret code of Mary Queen of Scots, or the dots and dashes of Morse code. Strictly speaking this kind of code should be called a cipher, but *genetic code* has become the standard term.

This central law of biology governs the way the information content of the genome is expressed, and it functions on two levels. On the concrete, biological level, one kind of molecule is succeeded by a different kind of molecule: segments of the chain of DNA bases are ultimately expressed as a linear chain of *amino acids,* to make up proteins. Like DNA, then, proteins are made of chains of basic blocks, but this time there are not 4 but 20 related chemical motifs. On the symbolic level, one linear text is succeeded by a different text: the DNA, represented by a text in the 4-letter alphabet, is expressed using a different 20-letter alphabet, using all the letters except B, J, O, U, X, and Z. The rewriting rules common to both levels are complex, but they break down into several stages. The first stage is *transcription,* in which the fragments of DNA (made up of the nucleotides A, T, G, and C) are copied to make macromolecules of another nucleic acid, ribonucleic acid, or RNA, which is chemically very similar. Like DNA, RNA has an alphabet of 4 letters representing four nucleotides, in this case A, U, G, and C. As in replication, a letter in DNA does not correspond to the same letter in RNA but to a *complementary* letter. A in DNA is transcribed as U in RNA, T in DNA is transcribed as A in RNA, G is transcribed as C, and C as G. Note, too, that the transcription process reads the text in an *oriented* way (just as we read an alphabet in a fixed direction; for instance, the Roman alphabet is read from left to right), but that the transcribed text (the RNA) is usually written in the opposite direction from the original text (the DNA), like a furrow in a plowed field, or a type of ancient Greek inscription called a boustrophedon. So, following the direction of the arrows, the text fragment

>AATGCCTTGAAAGCTAAGC>

is transcribed as

<UUACGGAACUUUCGAUUCG<,

which comes out as

>GCUUAGCUUUCAAGGCAUU>

if we choose to write it in the same orientation as the text it was copied from.

These transcripts, which may be subject to appropriate chemical modification, are mostly used as messages, and for this reason they are called *messenger RNA,* or mRNA. They are then *translated* into proteins by a molecular machine called a *ribosome,* in a second stage that uses a coding mechanism by which each set of three nucleotides (a codon) corresponds to a given amino acid. This correspondence is ensured by an RNA family called *transfer RNA* (tRNA). These molecules have an "anticodon," which recognizes the codon that specifies a particular amino acid, and they carry a molecule of that amino acid, which they add to the growing protein chain. With 4 nucleotides, there are 64 (4×4×4) possible codons, but as there are only 20 amino acids, there is a certain amount of redundancy. On average 3 different codons code for a given amino acid, but this varies between 1 (AUG for methionine, M, or UGG for tryptophane, W) and 6 (for leucine, L, serine, S, or arginine, R). Three codons, UAA, UAG, and UGA, play the role of punctuation marks indicating the *end of translation* of the message. The starting point of a protein coding sequence is normally the codon AUG, and in fact all proteins begin with methionine (which is normally cut off as the first step in the maturation of the protein to its active—*native*—form). Finally, over and above this translation process there is a set of "building regulations," rules that are still not very well understood, and their conceptual form is not yet clear. Depending on the nature of the amino acid chain, these rules cause the corresponding strand, a *polypeptide,* to fold itself up in a much more complex fashion than DNA, forming a specific three-dimensional architecture or *conformation* for each sequence of amino acids, which determines not only the function but also the location of each protein in the cell.

As well as building the cell's tissues, proteins play an *active* role in the processes of life. They act as *catalysts,* accelerating the reactions that take place in metabolism, and making them specific. They thus bring together various objects and transform, transport, or combine them, so it is essential to understand their function. The connection between DNA, protein, and function is a central question for all biologists. It must be stressed, however, that this does not imply that biological functions can be identified on the basis of the DNA sequence alone. Indeed, it cannot be enough, essentially because the relationship between the sequence of a text and the function it refers to is an abstract, *symbolic* one, just as symbolic as the relationship between the letters of the word *life* and its meaning.

Sequencing a DNA Fragment

It is proteins that do the work of DNA replication; in fact the cellular machinery involves quite a large number of different proteins, combined into a complex structure. They break the weak bonds to separate the strands, choose the right nucleotides, manage the chemical energy needed for building the chain, and finally they correct the inevitable copying errors (anyone who has ever had to proofread a text will know how inevitable typographic errors are). The heart of the machinery (before the correction stage) is sometimes no more than one of the enzymes called *polymerases,* which work as catalysts: they enable a chemical reaction to take place, without being used up themselves in the reaction. It is generally very accurate, with about one mistake every 10,000 bases. Often the scientist can do this copying in a test tube, without the backup of nature's proofreading machinery, using just an enzyme called DNA polymerase and the four nucleotides needed for the synthesis or *polymerization* of the new strand from the initial strand. Finding out the sequence of a genome is done by hijacking this DNA replication mechanism and making it work in the test tube.

The fact that a process which normally happens *in vivo* could be carried out *in vitro* gave Frederick Sanger the idea that it would be possible to determine the sequence of the bases on a strand of DNA by modifying this mechanism. Sanger's idea, brilliant because of its conceptual simplicity, was to begin the polymerization reaction at the same place every time, making it stop after a random time, but always after a specific base. This would produce, in the test tube, a mixture containing large numbers of fragments of the sequence, of random length, all starting at the beginning of the strand, and ending after each of the four base types. For instance, with a reaction that would normally produce the following sequence:

GGCATTGTTAGGCAAGCCCTTTGAGGCT>

if it were possible to stop the reaction after any random A, the test tube would contain the following fragments:

GGCA>
GGCATTGTTA>
GGCATTGTTAGGCA>

GGCATTGTTAGGCAA>
GGCATTGTTAGGCAAGCCCTTTGA>

Stopping after any random C would produce these fragments:

GGC>
GGCATTGTTAGGC>
GGCATTGTTAGGCAAGC>
GGCATTGTTAGGCAAGCC>
GGCATTGTTAGGCAAGCCC>
GGCATTGTTAGGCAAGCCCTTTGAGGC>

If these fragments could then be sorted just on the basis of their size (their length) and then placed one above the other in their order of length, this would produce a ladder, whose rungs would correspond to the places where each base appears in the sequence. If the four sets of fragments, one for each kind of base, were sorted out in parallel, and the ladders produced were laid side by side, the sequence of the bases could be read off directly (the third letter is C, the fourth is A, and so on) without the sorting mechanism needing to know anything about the bases, just the *length* of the fragments.

An excellent chemist, Sanger also came up with a simple way of producing such a mixture in practice. The first step is chemically to synthesize a short strand, complementary to the beginning of the strand to be copied. This is the *primer,* and all the newly synthesized strands will begin with it, to make sure they all start at the same point. Next, to make the reaction stop at a given base type, he used chemical analogs of the four nucleotides, which are recognized by the DNA polymerase in the normal, specific way and are paired with the correct base of the strand being copied. The particular chemical structure of the analogs chosen is such that, although they are themselves incorporated into the strand being synthesized, it is *impossible for the polymerization to continue after they have been inserted.* Finally, to produce a statistical (random) mixture of terminated strands, including fragments terminating after every occurrence of each type of base, he included normal nucleotides, which do not stop the reaction when they are incorporated in the growing polymer, together with the chain termination nucleotides, in proportions carefully calculated to produce a mixture containing all the short fragments as well as the long ones. This method is known as the Sanger method, the chain termination method, or the DNA polymerase dideoxy chain terminator method, after the chemical analogs, called dideoxynucleotides.

Suppose the stretch of DNA we want to sequence is this:

5′>AGCCTCAAAGGGCTTGCCTAACAATGCC>3′
3′<TCGGAGTTTCCCGAACGGATTGTTACGG<5′

(We have previously seen the direction of each strand indicated by the arrows < and >. The notation 5′—say "five prime"—and 3′ is a more technical way of indicating the direction, based on the characteristics of one of the two types of molecule that make up the backbone of the strand.)

We will work from the top strand. In fact the only part of it we actually know is a short stretch at the 3′ end of the strand above; the rest is still a mystery:

5′>XXXXXXXXXXXXXXXXXXXXXXAATGCC>3′

However, this is enough to enable us to artificially synthesize a primer, 5′>GGCATT>3′. This will bind to its complementary nucleotides at the 3′ end. If we try to represent what physically happens, it is something like what is shown in Figure 2.1, but as it is inconvenient to write letters upside down, the reaction is conventionally shown with the original strand written back-to-front, from 3′ to 5′, like this:

3′<CCGTAACAATCCGTTCGGGAAACTCCGA<5′
5′>GGCATT>3′

Once the primer has bound to the 3′ end, the polymerase constructs a new complementary strand, by reading the next unpaired nucleotide in the old strand and joining the appropriate complementary nucleotide to the free 3′ end of the new strand. In this case the next free nucleotide in the old strand is a C, so one of the G's floating around in the test tube is attached to the end of the growing strand. Next, the polymerase reads A and attaches T. So far, this has made:

3′<CCGTAACAATCCGTTCGGGAAACTCCGA<5′
5′>GGCATTGT>3′

and so the new chain grows until one of the analog nucleotides (noted A*) is attached instead of a normal A, terminating the reaction, producing the fragment

GGCATTGTTAGGCA*>

Figure 2.1. A schematic representation of the way the primer binds to its complementary nucleotides.

for example, or any other fragment ending in A*. A mixture of fragments will be produced, all ending in A*, like this:

GGCATTGTTA*>
GGCATTGTTAGGCA*>
GGCATTGTTAGGCAA*>
GGCATTGTTAGGCAAGCCCTTTGA*>

This reaction is carried out at the same time as the others, using the analogs G*, C*, and T*.

The next stage is to separate and identify the different elements in the mixture. This is easily achieved by *electrophoresis,* a process that uses a strong electric field to make the different fragments in the mixture migrate through the fine mesh structure of a "fragment separation gel," or "sequencing gel," with small pores of a known average size. A small quantity of the mixture is loaded at the top of the gel, and the electric field is applied. Because the fragments are themselves charged (the DNA molecule has a very strong electric charge, directly proportional to the number of bases it comprises), the fragments migrate through the gel with a snakelike movement called reptation. Clearly, the smaller fragments will pass through the pores more easily than the large fragments. Eventually the fragments are separated according to their exact length, and all that is left to do is identify where they are.

The earliest experiments used a method still often used today, in which either the primer or one of the nucleotides was labeled with a radioactive isotope. The fragments in the gel are therefore radioactive, and it is easy to identify where they are by placing the gel in contact with a photographic emulsion. The chain termination experiment is carried out in four separate test tubes, with an analog of a different nucleotide in each tube. This produces four mixtures, terminated at A, T, G, and C, which are then run side by side through the same gel. The photographic process reveals four ladders or sets of parallel black bands, from which the sequence can be read off directly. It is this image of a series of parallel black bands that has caused many journalists to use the misleading and otherwise incomprehensible metaphor of a "bar code" to describe a fragment of a chromosome. On a 50-centimeter gel, 500 bases can be read with an error rate of around 1 percent. We will not go into detail on this rather complex technical aspect of sequencing, but it does mean that in practice it is difficult, and above all slow and expensive, to determine the sequence of a long DNA fragment with an error rate of less than one in 10,000.

In the mid-1980s, machines were developed that could read the sequence automatically. Instead of radioactive nucleotides, they used fluorescent nucleotides, carrying a chemical substance that emits light at a precise wavelength, when it is illuminated at another, higher-energy wavelength. It is even possible to have different-colored fluorescent nucleotides for each of the four bases. This enables the sequence to be read by loading the samples together, so that the fragments terminated after each of the bases migrate down the same lane (which means that four times as many fragments can be analyzed on the same gel). With automatic reading, the computer representation of the sequence is in color, but the colors are artificial ones, created by the software that analyzes the light signals.

Obtaining DNA Fragments for Sequencing

Up until now, we have taken for granted how the DNA fragments for sequencing are obtained. But the figures show that there is a huge difference between the length of a genome and the length of fragments that can be read on a gel. The virus ΦX174, whose genome was the first to be sequenced, in 1977, was about 6,000 bases long, but bacterial genomes usually have several million base pairs, and the human genome has more than 3 billion base pairs. Only very short lengths can be read on a gel, so the genome must be fragmented before it can be sequenced (from around 400 bases in the mid-1990s, by 1997 new equipment made it possible to read over 1,000 bases, but it has not improved much since then).

There are several different ways of going about this. Random fragmentation, nicknamed the "shotgun" approach, is easy to do, either by ultrasound, using an appropriate wavelength for the size of the DNA, or by forcing the DNA through a tiny opening, in the same way as an atomizer produces a fine spray of perfume. If the genome is not too large, with not much more than

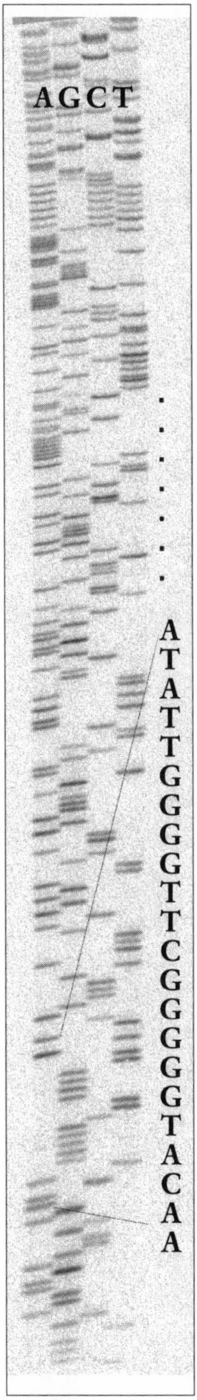

Figure 2.2. Here, a DNA fragment has been sequenced as we have described. The fragments are labeled with a radioactive isotope. When the gel is placed in contact with a photographic plate, the radioactive bands can be seen. The four "lanes" correspond to fragments terminated with a dideoxy A, G, C, and T. As the shortest fragments migrate fastest, they have reached the bottom of the gel. The bases can therefore be read off in order from the bottom upward, as each visible fragment is one nucleotide longer than the one below it.

5 million base pairs, the entire genome can be fragmented and treated directly, without first cloning long segments. This is suitable for small genomes (less than five megabases) and was even used on the human genome by Celera. Alternatively, a *genome bank* can be built up, containing segments of the genome to be sequenced.

A number of methods can be used to obtain these DNA banks. Shotgun fragmentation can be used, or the initial DNA can be fragmented using enzymes that specifically recognize "words" in the text. These enzymes are called restriction nucleases or *restriction enzymes,* and they are relatively simple to use: the genomic DNA is incubated with the enzyme, which cuts it precisely at the place where it recognizes its specific site (for instance, the enzyme *Eco*RI recognizes the site GAATTC and cuts after the G). Fragments can then be inserted into appropriate vectors, units that can be "amplified" by replication within a host cell, normally a bacterium. This selective amplification stage is called *cloning,* because the host cell multiplies to form a colony of millions or billions of individuals, producing a large quantity of each segment inserted (cloned) into the vector. The usual cloning vectors are either viruses or plasmids. Plasmids are autonomous replicating units found in many types of cells. In most cases they are not actually necessary to the cell; sometimes they are parasitic. They are minichromosomes, formed of double-stranded DNA with 1,000 to 500,000 base pairs, and containing all that is needed for their replication, including an origin of replication (a specific site from which replication begins) and various genes that enable them to be recognized and tolerated by the host cell. As they are present within the cell, they are copied when the cell divides, so enabling the given fragment of DNA to be produced in large quantities. One special kind of plasmid is called a cosmid, because these particular plasmids are isolated by inserting them into the coat of a virus, which recognizes a specific region of the plasmid, called "cos," and packages the corresponding DNA. Cosmids produce cloned fragments all of the same length, around 40 kilobases. Other vectors enable much longer segments to be cloned, from 100 kb to nearly 2Mb, and make artificial chromosomes in yeast (yeast artificial chromosomes, or YACs). However, YACs are difficult to handle and often unstable, and minichromosomes from *E. coli* are often used instead. These bacterial artificial chromosomes, or BACs, can contain 100 to 200 kb of DNA and are relatively easy to prepare.

Cloned DNA is thus multiplied inside a living cell. Some fragments may not be clonable, because they cause toxicity in the bacterium used to amplify them for sequencing. For instance, DNA from *Bacillus subtilis,* the bacterium used to make natto (fermented soya) in Japan (and an excellent model of the anthrax agent, *Bacillus anthracis*), is often very toxic when cloned in a vector in the colibacillus *E. coli.* There is a simple reason for this: the host *E. coli* recognizes the signals indicating the start of transcription and translation for *B. subtilis* so strongly and efficiently that its own essential syntheses are overridden in favor of those of the segments cloned in the vector, killing the host. In cases of this kind, purely biochemical methods have to be used instead. Polymerase chain reaction, which we shall look at next, is an excellent illustration of this.

Polymerase Chain Reaction, or PCR

Unlike ordinary physicochemical systems, living organisms can multiply by reproducing, so they can amplify enormously the number of the molecules they are made of. In particular, when a fragment of DNA is cloned into a unit that can replicate, it can easily produce 10^6, 10^9, or even 10^{12} copies of itself. It has been possible to determine the sequence of a large number of whole genomes thanks to this basic property. But although *in vivo* molecular cloning techniques are indispensable to much of molecular genetic research, they are tedious to carry out (and sometimes impossible), and they are time-consuming. The fragments to be cloned have to be isolated and inserted into replicating units; then these units have to be introduced into living cells, which subsequently have to be cultured. Fortunately, there is an extremely effective technique, polymerase chain reaction, or PCR, that enables a given DNA fragment to be amplified *in vitro* instead of *in vivo,* producing a large quantity of the fragment very quickly.[2] This technique is particularly suitable for sequencing purposes.

The fragment to be amplified is incubated together with a special DNA polymerase that is not inactivated at high temperatures, the nucleotides needed for the polymerization reaction, and large numbers of chemically synthesized short fragments (oligonucleotides) complementary to the two ends of the region to be amplified. What is clever about this method is that it involves heating and cooling the mixture in cycles, and that the polymerase used is stable at high temperatures (it is isolated from bacteria that grow in very high temperatures of 100°C and above). When the DNA is heated the strong bonds stay intact, but the weak bonds break, so the two strands separate. Then the mixture is cooled again so that each oligonucleotide binds *(hybridizes)* to its complementary sequence on the ends of the strand to be amplified. In the presence of the polymerase and the nucleotides, the oligonucleotides behave as primers once the mixture has cooled to the appropriate point. Next the mixture is heated so that the newly synthesized strands separate from their template, and then cooled again. The new strands can now become templates themselves, and the cycle begins again.

In principle, at each cycle the amount of the fragment doubles, and at this rate ten cycles would amplify the fragment a thousandfold (2^{10} times). After forty cycles, a very significant quantity will be produced from the fragment (of course it does not double exactly at each cycle, and although the amount produced will be very substantial, and actually visible on the gel, it will be somewhat less than the 2^{40}, or 1 trillion fragments predicted by a doubling at each cycle). This method is extremely powerful, because it can amplify very small amounts of DNA (in theory a *single* molecule is enough). From the time it was invented in 1983 until the beginning of the 1990s, PCR could be used to produce only short fragments, a few hundred base pairs long, and the error rate was high, with about one wrong base in every hundred. But with the use of new DNA polymerases and refinements in experimental conditions, its performance improved considerably, and in the mid-1990s it became possible to amplify segments of about 50 kb by chain polymerization. However, it is still necessary to check carefully that the fragments do not have small deletions caused by the polymerase accidentally "jumping."

PCR is thus a quick, cheap, easily used technique, capable of reproducing a million or more copies of a DNA fragment in a test tube. PCR is important because of its simplicity, partly because practically anyone can do it, taking a very small amount of target DNA from almost any medium, and without any prior purification, amplifying it a billion times in a few hours. The technique has had a major impact in such different domains as pathogen identification, the study of fossil species extinct for millions of years, genetic diagnostics, or crime detection. For our purposes in genome sequencing, PCR makes it possible to amplify and then sequence regions of the genome that cannot be cloned *in vivo*. DNA produced by PCR is thus independent of life.

In Silico *Analysis of Genome Sequences*

At first, experimental techniques were used to look for the meaning of the genome text. Experimental genetics relies on the identification of observable (phenotypic) characters, for which the normal type (also called the *wild type*) is known, and can be compared to the altered types called *mutants*. After isolating mutants—individuals with each modified character—the scientist characterizes both the wild type and the mutant type of the corresponding gene. For a long time this work was done experimentally, using live individuals, and the gene was simply a formal element, whose physical nature was completely unknown beyond the fact that the genes seemed to be linked together in a linear fashion, as if along a strand. Now we have direct access to the DNA, the physical material of the gene, and can determine its sequence. Even better, we now generally know the sequence of a region of DNA that probably corresponds to a gene, *before* we know the gene itself. A branch of computer science, linked to biology and often called *bioinformatics,* has been developed in order to try to meet the increasing (and increasingly specific) needs of biology. Genome analysis is in fact a particular branch of bioinformatics, and must be carefully distinguished from the branch that was developed before genome sequencing programs burst onto the scene. Before then, bioinformatics was used to explore models simulating the physiology of organisms, to analyze images in medicine, or to determine the structure of biological objects by X-ray crystallography (and more recently in nuclear magnetic resonance imaging). For this reason it is preferable to use the term *in silico* genome analysis to refer to the new branch of informatics that has been especially created for the purpose. Alongside *in vivo* experimentation, *in silico* analysis of DNA sequences using computational techniques allows us to explore their meaning via a new kind of experimentation that enables the scientist to answer a number of questions, particularly about hypothetical genes, such as the description of collective signals, the essential traits of a gene or a protein, and a phylogenetic relationship.

Once again, then, it is thanks to the development of a particular technology that genome sequences are accessible. The texts involved are so long that even though the technical means to determine them had long been available, they would have remained shrouded in mystery if informatics had not progressed in parallel with developments in molecular genetics. This progress has been made both in hardware, as computers become more and more powerful, in terms of calculating speed as well as memory size (RAM and ROM), and in software, as the use of algorithms of ever-increasing depth and efficiency keeps pace with the growing capacity of the hardware that runs them. The development of molecular genetics is inseparable from the development of computing, not only conceptually but also in practice.

Informatics is involved on several distinct levels in whole-genome sequencing programs. They form clearly identifiable domains, which are separate but linked. First, informatics contributes to data *acquisition,* by locating the radioactive or fluorescent bands on the sequencing gels, and by overlapping the sequences of fragments appropriately after they have been individually determined. Second, informatics is involved in *exploiting* the data, via the essential exploration of the biological meaning of the sequences. Finally, the data have to be *managed,* and this is no small task, as we will shortly see. Writing out just one genome sequence takes millions or even billions of bytes, a task that has been perfectly possible since the middle of the 1990s but was unrealistic, or at least extremely expensive, in the mid-1980s, and the knowledge associated with the sequence considerably increases the amount of information to be managed, by a factor of 10 at least.

Assembling and Verifying the Genome Text

We have already seen some aspects of the role of informatics in data acquisition. From the mid-1990s it was possible to read 300 to 800 bases on each of 30 to 100 parallel lanes of automatic sequencers. The technique widely employed in systematic sequencing projects involves reading parts of a cosmid fragment (about 40 kilobases) at random, so that each base is covered five or six times. This produces around 1,000 sequences, which then have to be assembled to make up the whole original sequence. Suitable algorithms compare the sequences in pairs, looking for the best possible alignment, then identifying overlapping regions and matching them to reconstruct the original DNA. Early programs, developed at a time when computers were not powerful enough to envisage assembling large numbers of sequences, used algorithms that compared sequences in pairs. However, shotgun sequencing of large genomes involves assembling not just thousands of sequences but tens of thousands, even millions, and with the need to handle these increasing numbers of sequences, it became clear that the methods available were not specialized enough. A point was reached where the computational side was holding sequencing projects back, and the biologist had to make up for the inadequacy of the processing by spending time "helping" the assembly programs. In certain cases reconstituting the whole text from the text of fragments can be not merely difficult but even impossible, particularly when the text includes long repeated regions, a common occurrence with long segments or whole genomes.

All this led several laboratories to develop new algorithms, bringing together the biologist's knowledge, information about the nature of each of the fragments to be sequenced (such as the sequencing strategy used for a given segment of DNA, the source, and the approximate length), and pairwise or multiple-alignment algorithms. As a final stage, so that users can verify the result of the assembly procedure and spot possible errors or inconsistencies, graphic interfaces are used to view and correct the assembly if necessary. This is done either as a whole, by comparing the different segments, or locally, by correcting the raw results of sequencing each fragment. However, the aim of using computational methods is to reduce the user's intervention to a minimum, producing a correct final result directly. An enormous amount of work has been put into this since the mid-1990s, and it was crowned with success when the first draft sequence of the human genome was completed in the year 2000.

The next step in genome sequencing is to check the plausibility of the final sequence, for instance by comparing it with the list of fragments obtained experimentally when the genome is cut by restriction enzymes. If the sequence is known to contain a certain gene, it is possible to check that it is actually included in the final sequence, by using the rule of the genetic code. However some zones in the sequence can be very difficult or even (at least for now) impossible to access, and this can sometimes produce major obstacles, which are not strictly speaking errors. This is why all the large genomes that have been publicly announced as having been "completely" sequenced still in fact contain a not insignificant part that has not yet been sequenced, and is not likely to be in the immediate future. For example, shotgun sequencing can never guarantee that the genome will be assembled without gaps, if only for statistical reasons (this can be calculated quite accurately). *In silico* analysis can also be used to test experimentally determined sequences for validity (in other words, how many mistakes are there in the corresponding text?) by checking whether the text makes sense in biological terms. Just as sequence acquisition exploited the process of DNA replication, sequence validation and interpretation exploit the processes of transcription and translation.

By now we can begin to see what errors might be found in a finished genome sequence. The commonest are insertions or deletions of bases. These can often change the genome text significantly, mainly in the parts that are translated into proteins. As the genetic code reads the DNA text in codons, the same piece of text can be read in three different ways, called *reading frames*. So, depending on where reading begins, the text fragment

TAATGATCCCATCAT . . .

can be read as

. . . T AAT GAT CCC ATC AT . . .

or as

TA ATG ATC CCA TCA T . . .

or as

TAA TGA TCC CAT CAT.

The translation of the text is very sensitive to shifts in the reading frame. The addition or omission of just one letter is enough to change the text of the protein completely. We can see how this works if we imagine a simplified two-dimensional analog of DNA made of only two letters, Y and R, which code for a protein made of three building blocks only: A, B, and C.[3] We then add an extra letter Y to the original text. The "protein" folds in the two dimensions of the page according to the following rule:

A (coded by RR) means straight ahead, B (coded by RY) is turn right, C (coded by YR) is turn left, and YY means the end of translation of the analog gene.

The following sequence:

RRRRRRYRRRRRRRYRRRRRRRRRRRYRRRRRRYRRYRRRRRRRYRRRRRRRRYRRRRRR
YRRRRRRRRRYRRRRRRRRRRYRRRRRRYY

will then be translated as:

A A A C A A A C A A A A A C A A B A C A A A C A A A B A A A C A A A A C A A A
A C A A*

With an insertion, this gives:

RRRRRRYRRRRRRRYRRRRRRRRRRRYRRRRR**Y**RYRRYRRRRRRRYRRRRRRRRYRRRRR
RYRRRRRRRRRYRRRRRRRRRRYRRRRRRYY

which translates as:

A A A C A A A C A A A A A C A A **C C B A A A B A A A A C A A B A A A A B A A A A**
B A A B

The protein formed will fold up as indicated in Figure 2.3, in a very different way from the correct one, where the catalytic center is indicated by the hatched circle. We can see that what appears to be a very small error (perhaps only one base in 1,000) can lead to a seriously mistaken interpretation, which could lead to persisting ignorance about the real function of a protein. Another very common error is inversion of GC into CG, or vice versa, but this normally leads to only local errors in the interpretation of the text. All sorts of other errors can occur, limiting what can be done with the sequenced genome text. But overall the usual error rate is less than one base in 10,000, which is an excellent performance (for genes of an average length of 1,000 base pairs, this still produces one incorrect protein in 10). Luckily, there are many ways of introducing other biological knowledge besides the sequence itself, in order to validate the text.

Finding Genes in the Text

There are different ways of considering a sequence, depending on whether or not we want to take its meaning into account. To begin with, it seems likely that this meaning is ignored from the point of view of replication, which considers only the constraints of the DNA's *physical* structure, when certain strings of nucleotides have particular physical properties. This is probably true of the sequence CTAG, rare in most organisms because it causes a weak link at the TA junction within the DNA molecule, and it is likely to be the case that most sequences have no particular relationship with any special meaning.

But replication is also linked to the biological function of the sequence, because the DNA is not only a text, but also a *substrate* molecule for the action of other molecules, especially during its formation and evolution. The text of a genome is produced, from generation to generation, by synthesizing a replica of a primitive text (which of course we do not know). This replication is carried out by *DNA polymerase,* a complicated enzymatic system made up of several associated proteins, whose own synthesis is specified by a group of genes. We also know that when a new genome is made, this process is often supplemented by the lateral transfer of DNA fragments from one organism to another (often called "horizontal" transfer, to distinguish it from the "vertical" transfer from one generation to the next). It is natural to want to know how these processes have combined to produce the present genome, know-

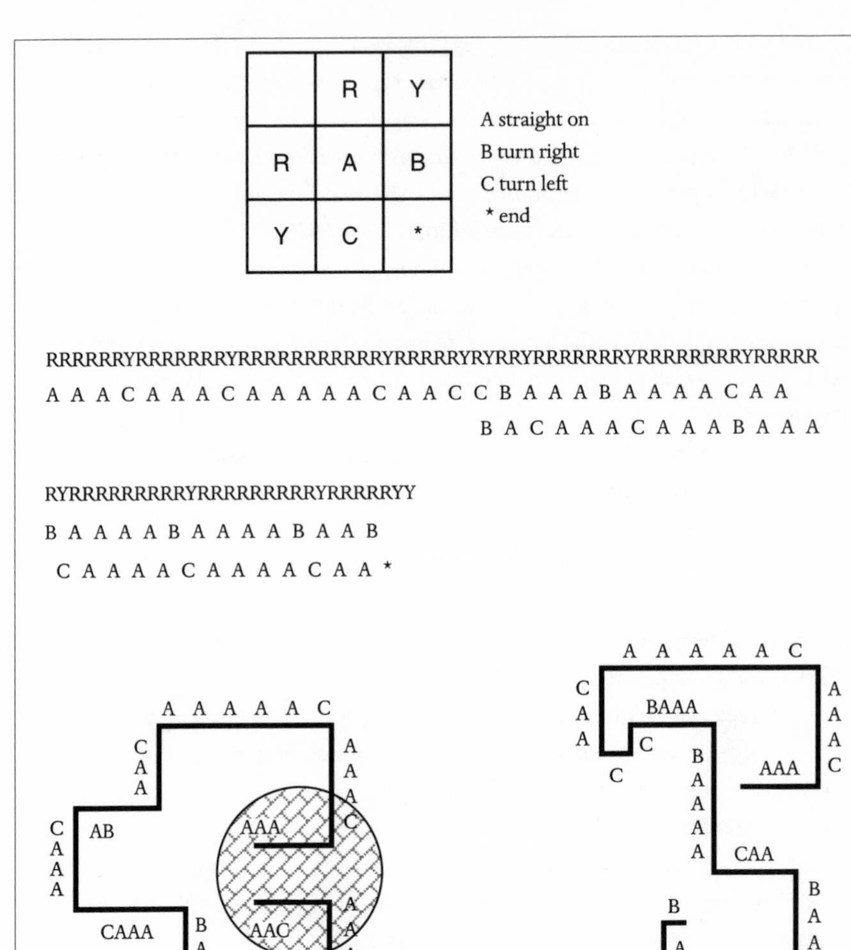

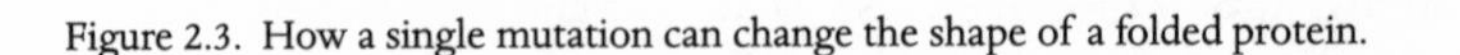
Figure 2.3. How a single mutation can change the shape of a folded protein.

ing as we do that a physicochemical process can never result in a totally faithful replica of the original.

Next, we need to consider what the transcription machinery sees. The whole genome text is not expressed all at the same time. Some genes remain silent as long as their product (usually a protein) is not involved in a process required for the normal development of the organism. The step by which they pass from this silent state to an active one can be seen as the rewriting of a piece of genome text, retaining a 4-letter alphabet, but using basic motifs, ribonucleotides, which are chemically slightly different from the deoxyribonucleotides of DNA. This rewritten, *transcribed* segment is called *messenger RNA* in prokaryotic cells. In cells with a nucleus, *premessenger RNA* is formed first, and then some of the internal segments are cut out, and the cut ends are spliced together like a ship's rope, to form the final messenger. In both cases the messenger is then *translated* into a protein, formally represented as a text not with 4 letters as before, but with 20 letters, the symbols for 20 amino acids.

The observation that only part of the genome is transcribed into (pre)messenger RNA implies that the beginning of the zones to be transcribed must be marked by signals, to make them recognizable. François Jacob and Jacques Monod, who discovered these signals, called them *promoters.* These transcription control zones are hard to recognize, and at the beginning of the millennium we still do not know how to do this accurately, even in bacteria. At best, we know that in most cases promoters contain regions that include the sequence TATA near the site where *RNA polymerase,* the enzyme that transcribes DNA into RNA, begins transcription. Because this is a control process, and because in very complex organisms like mammals, genes must be expressed appropriately in different tissues, the promoter region involved is generally a long one. In each case, the control involves a very sensitive, very specific process for recognizing the corresponding zone in the DNA. Identifying the promoters and the elements they use to exercise their control function is one of the first—and most difficult—aims in decoding the meaning of genome sequences. Equally, it is important to understand where transcription stops. This is more difficult than it first looks, because in prokaryotic organisms (cells without a nucleus, the eubacteria and archaebacteria), several genes can be transcribed together as one mRNA, forming what Monod and Jacob called an *operon.* Of course it is also difficult in the case of eukaryotes, because the exons, the parts of the gene that are to be translated (and thus are found in the mRNA), are separated in the DNA by introns, and neither their position nor their junctions can be deduced a priori.

The mRNA then has to be translated into a protein. Things are a little simpler here, since normally (almost always, in fact, with eukaryotes, and usually in the case of prokaryotes) the codon indicating the beginning of the protein is written AUG in the messenger RNA, and this always codes for the same amino acid, methionine. But how can we recognize the point in the DNA text where the protein text starts? Here, the codon AUG is written ATG, because in DNA nucleotides T takes the place of U. If only for statistical reasons (remember that the alphabet of this text has only 4 letters), the sequence ATG is obviously very common (in a text in which each letter is equally represented it would occur once every 64 letters on average). To try to identify the corresponding regions of the genome, we rely on the fact that proteins are usually made up of a large number of amino acids: on average bacterial proteins have 200 or 300 amino acids, and in eukaryotes, proteins are longer still. The gene has the same number of codons as the protein has amino acids, and the codons are linked together one after another without overlaps. In the mRNA, the end of the gene's coding region is indicated by one of the three "stop" codons, UAA, UAG, and UGA (TAA, TAG, and TGA in the DNA text). So we know in advance that the text of a protein will be coded by a region of nucleotides in multiples of 3, framed at each end by one of the 3 stop codons. To study this kind of region, we can decide on a minimum number of codons, for example 100 codons (300 nucleotides). Of course, shorter sequences may (and indeed do) exist, but they are much more difficult to identify directly from the sequence, because, for purely statistical reasons, the shorter regions of this kind are, the more common they will be.

Because the nucleotides occur in multiples of three, the same text can be read in three different reading frames. For example, the text

TAATGATCCCATCATATGGAGCATGAGCCACACTCATTAGAGTGA

can be read along the strand in three different ways, separating the letters into triplets. The region between two stop codons is called an open reading frame (ORF), and within an ORF the region between an ATG marking the start of translation and the first stop codon downstream is called a coding sequence (CDS).

.T AAT GAT CCC ATG ATA TGG AGC ATG AGC CAC ACT CAT TAG AGT GA..

stop

..TA ATG ATC CCA TGA TAT AGA GCA TGA GCC ACA CTC ATT AGA GTG A..

stop..........ORF............stop

...TAA TGA TCC CAT GAT ATG GAG CAT GAG CCA CAC TCA TTA GAG TGA...

start............................CDS.....................................stop

stop stop.. ORF.....................................stop

If there is a frame that corresponds to a protein, this frame will be noticeably longer than the others, as it will not conform to the statistical distribution of the three different stop codons. If the distribution of the four nucleotides is equal, there will be one stop codon or another about every 20 nucleotides, and if they are randomly distributed across the three reading frames of the same DNA strand, each frame will have an average gap of about 60 nucleotides between two stop codons. So finding stop codons 300 nucleotides apart is obviously rare and thus significant in statistical terms. Of course, when the text is completely unknown, we do not even know which of the two DNA strands is the one that is transcribed and translated. So that gives six reading frames to examine, three for each DNA strand.

Figure 2.4 shows the six possible reading frames in a DNA fragment. The short vertical lines indicate ATG codons, which can mark the start of translation. The tall vertical lines indicate stop codons, and the arrows indicate long regions without any stop codons, where we expect to find a gene. The illustration shows how clearly these can be seen in a DNA fragment. We can also see that genes may be found in several frames, but that they do not overlap.

All this is relatively simple in genomes that have no introns in the genes, because there is a very good chance of finding genes in the long open reading frames. In genomes with introns, the situation is much more complicated, because the exon coding regions may be short, and there is not even any reason why they should be in multiples of three, because the grouping of nucleotides in threes is relevant only at the translation stage. Nor will they necessarily be between two stop codons. In this case, we have to begin by finding a way of spotting the junctions between introns and exons, and find the coding sequences afterward. This problem is still a long way from being solved, although we know a solution must exist, because the cell itself does not (or not often) make mistakes, but cuts out introns and splices together exons with great efficiency and precision.

If we can create effective "descriptors" of a protein-coding region, any base insertion or deletion error (which occurs frequently during the acquisition stage) will immediately be spotted and subjected to a correction process, which ought to be built into the initial sequence data acquisition program. The computational approach needs to be both integrated and dynamic, but this is not yet possible, and in fact conflicts with the way algorithmic models are conventionally used. For this reason, research in molecular genetics is de-

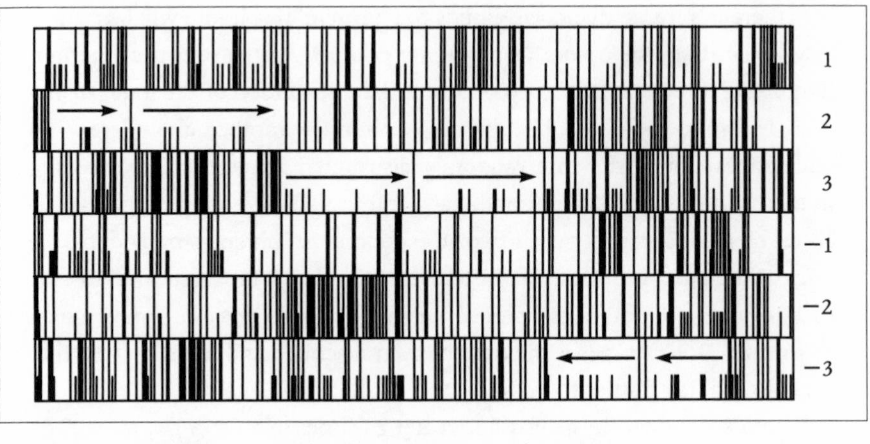

Figure 2.4. Identifying open reading frames in a genome fragment.

veloping through an interaction with state-of-the-art informatics research, which gives priority to exploiting all forms of *knowledge.*

Analyzing the Text

As we are beginning to see, gene identification is difficult—and we are still a long way from predicting gene function! But we can use the knowledge about genes gradually accumulated by geneticists as they analyzed the behavior of mutants of various organisms. There are many genes whose sequence and function are both known. And as all genes have been produced by a process of evolution from ancestral organisms to those we know today, there must often be family relationships between them, which we hope will still be recognizable, at least when the organisms are not too distantly related. In each case, the first steps in exploring a new sequence involve comparing its possible translation products with what is already known. This means first comparing the DNA sequences themselves with all those that are currently known. Next, each of the six possible reading frames is translated, using the genetic code. Of course most of the hypothetical proteins produced by this translation do not actually exist, if only because *in vivo* only one of the six possible reading frames is normally translated. These translated sequences are then compared against those in the DNA and protein data banks (several hundred thousand proteins, of which only a small number have actually been experimentally characterized).

In comparing a sequence with others, we are looking for either a mutation in a gene, or similarities or family relationships between genes, within one species or between several species, or perhaps for particular motifs likely to confer specific properties on the coded proteins. When regions with a strong resemblance are revealed, we can begin to make hypotheses about the location of the gene within the sequence (and to look for its control regions) and about the nature of its transcription and translation products. In all these cases, the methods of analysis have to be applied systematically to large numbers of long sequences, an ideal job for a computer. The first steps in sequence analysis thus depend heavily on the development of comparison algorithms.

Since they are made up of a succession of symbols, the most natural way to compare two sequences would be to try to find the best possible alignment between the two series of symbols. Leaving aside for the moment any constraints related to the physical nature of DNA, it naturally seems that the more identical symbols can be matched up, the better the alignment must be; but this is just an initial approach and is certainly far from complete. This is why we cannot yet easily identify relevant signals in the DNA, where architecture definitely plays a major role by enabling molecules that are "in the know" to recognize particular regions. Furthermore, since two sequences are virtually never completely identical, they must either be aligned overall, introducing insertions, deletions, or one-letter "point" differences where necessary to make them line up, or else the identical parts of the sequences are aligned locally, leaving zones between them that cannot be compared. Which method is used will depend on what information the scientist is looking for. For example, when researching a mutation, a difference between sequences will not be as significant if it still produces the same protein (because the same amino acid can be coded for by different codons) as it would be if it modified one or more of the amino acids, or if it affected the reading frame and resulted in a complete change in the translation from then on, as we saw in the two-letter code analogy. There is a similar distinction to consider when comparing protein sequences, because some amino acids are more interchangeable than others, depending on what type of region is affected within the protein: hydrophilic or hydrophobic zones, for instance (a particularly important criterion in terms of the nature of the protein—whether it is from the cytoplasm or the membrane), or secondary structures, and so on.

Another important aspect of genome text interpretation is comparing the sequences not only with each other but with the prior *knowledge* recorded in the international data banks. The first stages of this process are becoming virtually automatic, which is very important when there are thousands of genes to be identified in a genome. Next, "trees" of relationships are established, making it possible (among other things) to evaluate phylogenetic relationships, which link species through their evolution from common ancestors. At this stage the gene products begin to group themselves into families. It is often possible to formulate descriptors for these families, which can then be used to discover the function of unknown, newly characterized genes. However, it is important not to assume a function too quickly, because in a way all proteins are descended from a small number of common ancestors, even if today they have quite different functions.

We also now know that genomes have an internal consistency, a style, which distinguishes them from one another. The genes have to work together, and the proteins often have to come into contact with one another, and thus have certain characteristics that allow them either to interact or to avoid stable interactions. So the genome text must contain certain fixed sequences of letters (words) that act as collective signals, and other sequences of letters that are very rare, such as CTAG in most organisms. With detailed statistical analysis of the texts of genes and their products, these signals can be identified. It is possible to create descriptors for these, too, via a machine-learning process that involves grouping together examples thought to have a certain property in common, and designing a refinement process that takes account of previous results and improves them as the size of the group increases. When this is possible, the descriptors (or some of them at least) are incorporated into the software used in sequence acquisition experiments. The descriptors are checked against the biochemical, physicochemical, or physiological data already known about the organism being studied, and through analyzing these data it may be possible to identify the biological functions concerned.

The fragments obtained during sequencing must be explored to see if they contain relevant signals. A considerable part of biological informatics research is devoted to creating descriptors for these signals, and exploring the genome texts to do this requires a conceptual infrastructure unique in information science. It involves integrating disparate areas of knowledge in order to predict the structure, function, and regulation of not only the genes but also their products. This would be best done if we were able to build programs that use the same rewriting rules that nature uses, mimicking the recognition processes that allow the cell's machinery to identify the regions to be transcribed and translated, and then folding the protein chain into its proper shape. Of course we are very far from understanding these rules, and computers still use very crude approaches. Evolution has had time to make an immense series of trials and errors before arriving at a certain degree of perfection in the way proteins work, especially the enzymes, whose biochemical activity is particularly precise. This has produced a coherence that is a much richer source of information than can be obtained simply by "*in vitro* mutagenesis," the planned modification of genes in the laboratory. Unfortunately this information—contained in the text of the genes—is difficult to access, in the same way as the meaning of an unknown language is inaccessible without some kind of "intelligent" processing. But it is possible to try, using various computational techniques especially designed for text analysis, or more generally for the analysis of form, or "pattern recognition."

Another aim of genome projects, connected with the aims we have seen, is to gain a better understanding of the rules governing protein folding. Studying the correlation between a protein's sequence and its spatial shape brings advances in this domain, which is still in its infancy, mainly because the sequence of amino acids in a protein is generally not enough to determine its final conformation. *Primary structure* refers to the sequence itself, the result of translating the gene text using the genetic code. *Secondary structure* refers to specific small-scale folding patterns that are common to many different overall structures, and *tertiary structure* is the spatial shape or architecture of the protein as a whole. The tertiary structure can be discovered through a variety of physical techniques, by examining the way the protein's crystals diffract X rays (X-ray crystallography) or, for the smallest proteins, with a technique such as nuclear magnetic resonance. Thanks to the very large number of protein sequences now available, it should become possible to find rules, if not for the way folding takes place, at least for the final architecture of the proteins in their functioning state. But we must not forget that the number of spatial structures discovered experimentally is a very small proportion (a few percent) of the number of known sequences, so the answers will still be very hard to find. It is also not uncommon for a protein to have several different stable tertiary structures—and I will illustrate this later with the "unconventional" infectious agents, prions, which cause bovine spongiform encephalopathy (BSE), or "mad cow" disease. It is what takes place during the transformation from one structure to another, often in larger-scale structures in which several different amino acid chains are combined (this is called a *quaternary structure*), that makes it possible for an enzyme's activity to change. This gives rise to an enormous wealth of possibilities for regulating a biological function.

Answering the questions raised by the existence of genome sequences as very long texts draws on many different domains belonging to informatics, mathematics (number theory, for instance), and statistics. For example, the extraordinary growth in computing power has made it possible to develop statistical techniques that can classify entities *without* having any prior knowledge about them. In particular, France has a school of taxonomic analysis that produces excellent results and has led to the identification of numerous characteristics of genes with common functional properties. It developed as a result of the work done by Vicq d'Azyr at the end of the eighteenth century, and especially pursued by Jean-Paul Benzécri and his colleagues and successors.[4] The techniques used are a special case of the approaches used to recognize form or pattern, a fundamental question in human thought and in biology in general. Using these methods, it is possible to create classes, and then establish correlations between classes of activity already known, and the classes predicted *a priori* by analyzing a large number of objects, such as all the protein sequences from one organism, predicted by applying the genetic code rule to the genome sequence.

In the living cell, the genome text must be recognized by a machine that knows how to read it and express it deliberately. Knowing this, scientists exploring genomes are eager to make use of any method that might lead them to discover what forms need to be recognized, and how they are in fact recognized *in vivo*. Whenever possible, it is very important to consider this inductively, by visualizing as clearly as possible the way biosynthetic processes take place within the cell. This means getting as far away as possible from the usual biochemical image, where the objects in the cell are seen as uniformly distributed, and attempting to build a mental image of the way the cell is organized. This mental reconstruction process is essential because despite the extraordinary progress in microscopes and in techniques that allow objects as small as proteins to be stained while leaving the cell alive, the cells' tiny size means that we cannot always see what is going on inside and how it is organized.

This is why it is particularly important to use statistical methods in which the procedure can produce a model of classification *from data,* in all their complexity, and does not *impose a model on the data a priori.* It must identify structures without introducing any other information than that on which the statistical approach envisaged is based (such as symmetry hypotheses, which suggest that all objects labeled as equivalents have *a priori* the same properties). This "data-driven" approach, common in empirical Anglo-American thinking, does not enable us to understand the meaning of phenomena, but it does allow them to be detected. However, of course, the models must be produced afterward, as we shall discuss later.

Only after pertinent phenomena have been recognized can we work in a generative fashion, and discover what is relevant from a biological point of view. On the one hand there will be a prior model of biological organization (the "hypothesis-driven" approach), and on the other, completely independently, a set of forms or classes derived from the statistical analysis without *a priori* (the "data-driven" approach). The objects in the two models, one involving an explanatory working hypothesis and the other based on the formation of statistically significant classes, can be compared with each other to see if they are compatible. If they are, the corresponding part of the biological model can be considered as validated—at least for the time being—so we can turn to statistical analysis once again and extract the information obtained in the preceding step.

Together with the information storage problems we will look at shortly, data analysis thus represents a huge field in informatics, in terms of both research and application. It can be applied not only to the sequences themselves, but to the search for relationships between all kinds of biological objects such as genes or the various genetic markers—in fact everything found on the physical or genetic map of the organism (the counterpart of the geneticist's early discoveries, such as Morgan's work on *Drosophila*). On the scale of whole organisms, data analysis makes it possible to search for similarities by comparing results from different species, or at the molecular level, to predict protein architecture or to search for genes.

Studying nucleic acid sequences is particularly well suited to data analysis methods, and a great deal of progress has been made since the early 1980s. The simple alphabet of these molecules responsible for the transmission of genetic information makes them very suitable for computer processing. However, although the alphabet has only four letters, not much is yet known about its grammar. Despite this, recent studies have shown the extent to which genes and even in certain cases gene function have been preserved in the course of evolution. This is why model organisms and the sequencing of their genomes are more important than ever, because the basic rules of this grammar can be discovered by comparing them with one another. However, as already mentioned, the major sequencing programs have shown that many genes are unknown, and do not produce easily identifiable phenotypes. One of the major contributions of computational analysis has been to build statistical models capable of suggesting hypotheses that enable the physiological functions of these unknown genes to be explored.

As a first step, multifactorial analysis of the genes collected in well-structured databases shows whether or not they group themselves into several well-identified classes. For example, genes can be classified very distinctly according to the bias in the way each gene uses the genetic code to code for each protein. The same protein (which is of course the same sequence of amino acids) can be coded for by a very large number of genes with different DNA sequences, depending on the way each one uses the genetic code. So some codons may be used in preference to others in the cells of a given organism, and it may be that the proteins in the cell can be classified according to this bias. Experiments have shown that, at least in the genomes of organisms that can reproduce quickly, there are indeed different strategies for using the genetic code, and that gene classes can be drawn up according to this bias. The genes in each class can often be identified, so the scientist can try to find out whether they have some property in common. Once the appropriate classes have been created, the scientist looks for interesting phylogenetic relationships, which point to new ways of looking at the relationships between species, and thus at evolution. Finally, by trying to discover what are the essential elements a cell needs in order to live, we may hope to understand what is at the heart of a living organism, and to ask new questions about the origin of species, and even about the origin of life.

But there remains an essential question, and a very difficult one, at the heart of sequence exploration. Since we are still a long way from having explored all the biological functions of genes, we often find ourselves faced with genes whose structure and function are both unknown. Informatics has an essential contribution to make to the search for their meaning, in the shape of heuristics—incomplete approaches, which work on the basis of an intelligent hunch, much as a chess player does, rather than an entirely and thoroughly formalized approach. In this area, any light that computer science can shed is welcome as an aid to discovery. In a number of cases, the "brute force" methods are better than heuristics because they enable a particular representation of the genome to be exhaustively explored; but they can rarely be actually implemented.

For both biologists and computer scientists, there are specific questions relating to each level of gene expression, from DNA to protein assembly. The immense amount of data, and the complexity and variety of knowledge about each of these levels, represent a series of challenges for researchers in each of these disciplines. Dealing with the vast quantity and variety of the data collected in sequencing projects requires considerations of *scale* that can be crucial, even in the case of relatively conventional methods. All genome sequence analysis programs are truly interdisciplinary, in the sense that neither of the scientific communities can ever be regarded as working *for* the other, with the biologists providing the data or questions or the computer scientists providing the programs. However, uninitiated scientists still often confuse bioanalysts, who simply use these tools, with bioinformatics specialists, who create them. The crux of the interaction doubtless lies in the fact that a return to experimentation, by constructing artificial objects (genes and proteins), is a particularly powerful way to test the creations of both informatics and biology. There are very few similar situations elsewhere in science in which the same degree of generality is possible as in the manipulation of genes *in vitro,* where the results of a computational analysis can be mirrored in reality by producing tangible objects. The use of informatics in genome programs is thus an ideal testing ground for a number of approaches in both information science and genetics, which would otherwise be no more than theoretical approaches.

Annotating Genome Sequences

Let us sum up what we have seen so far. Many different methods of analysis can be used to study genome texts. But our experience with the study of ancient languages shows how difficult it is to decode a text in an unknown language. Champollion's stroke of genius was to guess that the Rosetta Stone told the same story in three languages, and that it would be possible to understand the Egyptian hieroglyphics by connecting the known with the unknown. Luckily, the genome text refers to a core set of biological functions that are similar from one organism to another. This suggests that, as with the

Rosetta Stone, comparing genome texts with each other ought to be a productive approach. If the sections of the text that tell more or less the same story are compared, we ought to be able to make a good guess at their meaning. Here, we have the advantage of knowing that living organisms are descended from common ancestors, so many sequences must themselves have a common ancestor. As they evolve by mutations, usually changing one letter at a time, and sometimes introducing deletions or insertions in the text, aligning the sequences in parallel with each other is an obvious first step.

There are several reasons why annotation is essential. First, it is obvious that when we set out on the adventure of sequencing a genome, it is not with the intention of obtaining a "naked" sequence. A sequence of letters is of no interest in itself. We want to understand what it means. Furthermore, a bare sequence is much more likely to contain errors than an annotated sequence, because if the local meaning has been discovered this means that the sequence must respect a certain number of rules that errors will not necessarily respect. This is borne out by the facts: on inspection, regions that are vaguely annotated or not annotated at all turn out to have a much higher error rate than properly annotated sections. This might be as high as an error per 100 bases, although the high rate of redundancy possible with the new high-speed sequencing machines enables a large number of errors to be corrected. Despite this, even a low error rate has a considerable impact, because a single insertion or deletion of a base changes the reading frame, completely changing the nature of the gene product.

The annotation that makes errors easiest to spot is the identification of protein coding regions. The most common errors are single-base insertions or deletions, which change the reading frame from the point of the error onward, so that all the following codons are incorrect, producing a nonsense translation. We have seen the effect of this on protein structure. To illustrate the effect of small errors on function and meaning, just modifying a few letters in a sentence may change the meaning entirely. A very few insertions, transpositions, and deletions are enough to change the following sentence into something completely different:

The charged particle has shown that induction often works

The changed article now has that introduction of ten words

Apart from the first word, there is absolutely no common meaning, although long strings of letters are exactly the same. The words "that" and "has" are structurally the same in both sentences, but they perform quite different functions.

Generally speaking, geneticists do not have any very effective way of identifying the regions they are interested in, in the bare genome sequence. Most of the methods used rely first and foremost on annotations produced by others, on genes identified individually from analyzing the behavior of mutants; but this can be done only with model organisms, studied by a large scientific community. By a combination of comparison with known sequence banks and with bibliographic databases (*in libro* annotation, to complement *in silico* annotation), these annotations enable at least some of the sequences of interest to be identified. This is why the annotations of model genome sequencing programs are particularly important. In these cases, hundreds or even thousands of researchers have worked on a huge variety of particular features of these models, and have linked them to the functioning or the alteration of a large number of genes. The most interesting (and the best-checked) annotations on the *E. coli* genome, for example, were produced by Kenn Rudd, who was a researcher at the NCBI at the time, and Amos Bairoch, a scientist at the University of Geneva (and director of the world's best protein sequence bank, SwissProt), as well as Monica Riley, of the Woods Hole Marine Biology Laboratory in the United States, a geneticist who worked on the genetics of this organism in the early days. Unfortunately, as has been seen many times over, the considerable effort required to produce a reliable and really useful annotation is almost always underestimated when sequencing projects begin.

How do we go about annotating sequences? In the mid-1990s, annotation was still mainly done manually, almost entirely without automation, by fragmenting the genomic DNA into 5- to 10-kb pieces and comparing them with what was known in the data banks, with the help of programs that searched for similar motifs held in the banks. The coding sequences have to be identified, using specific rules that make it possible to spot correlations specific to protein coding sequences. Advanced statistical techniques are used to do this, taking into account the fact that coding phases have a periodic character of order 3. The search program has to be trained via an appropriate learning stage using coding sequences that have been firmly identified. Similarly, using algorithms that are sometimes very advanced, repeated regions of the genome can be identified, enabling us to approach the question of how it was formed during evolution. (Contrary to what we might perhaps expect, it is very difficult to spot repetitions in a longish sequence *a priori,* unless they are contiguous, and for genomes of a few million bases it becomes virtually impossible, if approximate repetitions are allowed. Here, genome analysis is a source of interesting new problems for algorithmics.)

The next step is to identify and correct any reading frame errors. It is important not to make corrections *automatically;* sometimes the control of gene expression affects the way the translation mechanism sets the reading frame of the mRNA on the ribosome, producing a shift at a specific point in the text to be translated (this is called *programmed frameshifting*). In other cases, an observed change of reading frame corresponds to an evolutionary mechanism that has suppressed the function of a gene, without the gene's being completely destroyed. Genes like this are called pseudogenes, and highly pathogenic bacteria such as the agent of leprosy and the plague agent have evolved toward virulence by losing functional genes in this way.

The only thing that automatic annotation methods can do easily is to point to similarities between sequences and deduce what their function might be. However, this is a dangerous approach (although, with the exponential growth in the number of sequences decoded, we will increasingly be obliged to rely on such similarities). If the most relevant link between structure and function is a symbolic one, the annotation will often be wrong. More seriously, if this incorrect annotation is recorded in the data banks, it will be used in the subsequent annotation of other genomes, spreading the initial mistake. This is in fact the most serious danger facing data banks since the mid-1990s: with the exponential growth in the number of sequences and use of automatic methods, a great deal of research effort has set off down the wrong track because of annotations that have been either misunderstood or misused. Increasingly, automatic annotation procedures will have to include facilities that will enable them to learn, and to validate the commentaries they produce. When the first draft of the human genome text was decoded, these requirements still belonged to the realm of informatics research, and were a long way from being satisfactorily implemented in practice, but significant progress toward improving the procedures can already be seen.

This first stage is very laborious, and yet the annotation is still very incomplete. We are still a long way from gene function, and at most, in some cases, we know what chemical reactions are catalyzed by the predicted enzymes. But we want to know much more than this. It would be interesting to predict the compartmentalization of the proteins in the cell and to predict their function in more detail, for example by studying more closely the annotations of similar proteins held in data banks such as SwissProt. Similarly, it is important to describe not only the transcription promoter regions, but to identify the factors that make this control possible, as well as the conditions in which it is exercised. But control takes place at other sites as well as in the DNA. The spatial shape of the messenger itself and its life span in the cell decide how it is translated and the site where the proteins are synthesized. Not much is yet known about how structures such as introns, which can be cut out, are recognized and removed. The fact that very short exons can sometimes be joined by very long introns, tens or even hundreds of kilobases long, makes this all the more mysterious. How are the ends recognized? How do the exons find each other and link themselves together? Sooner or later this kind of information will need to be included in genome annotations. Annotation clearly has no defined limit, and it can only be enriched as a result of the successive cycles in which more and more information is associated with the genome text. The informatics systems of the future will be able to provide users with both the knowledge associated with biological objects (genes, control regions, cell architecture, and so on) and the methods for studying them.

Managing Vast Quantities of Genome Information

In the last few decades, the development of the techniques used to study living organisms has generated a volume of data that is growing faster and faster. Each genome sequenced represents millions of base pairs and a sum of biological knowledge that is growing exponentially. This information needs to be collected in a standardized, structured way, and accessing the associated knowledge should be as simple as possible. It all has to be stored, available, and easy to use, and the computer is the only tool currently available that is capable of dealing with this triple task. If it were merely a question of archiving data with a well-defined structure, interrelated in a simple fashion, as subscribers are listed in a telephone directory, the problems associated with genome data would be simple to resolve, and there would be no need for a specialized discipline. However, these biological data are connected in complex ways, which are constantly changing as knowledge grows. These links involve many different fields in biology, sometimes quite distant from each other. Thus some problems may be approached via genetics just as well as via molecular biology, or even via structural biology, and in each case the question will be formulated in a very different way. So it is not always possible to know whether some fact already exists elsewhere, or in another form, or what connections it has with other facts that are already known. The computer tools must be designed to interpret the scientist's questions and search for the answers in the right place, as precisely and as fast as possible. Above all, in the case of a long search, the initial question needs to be framed in terms the computer can understand, so that the user is not kept waiting in front of the machine, replying to its questions as and when it needs further details. All this has to be achieved despite the complexity of the systems and domains involved in the study. Quite often, once a first search has been carried out, the user will want to know how the data in the machine are related to his or her original question. Finally, the information gathered must be usable: it must be available for analysis (using appropriate programs) and must be represented, particularly in graphic form, in a way that enables the user to understand the state of knowledge on a perhaps unfamiliar subject, quickly, accurately, and without having to learn anything (or more than a bare minimum) about the software involved. To achieve this, the raw data and the sum of knowledge associated with it need to be brought together in an accessible place.

Data Banks and Databases

Not quite a library, but something akin to it, the idea of a place for storing knowledge about all sorts of domains has existed for many years now, and we have grown used to hearing the term *data bank* without having more than a vague idea of what it is. More recently the term *database* has appeared, and the two are quite often used interchangeably. Is there any difference between them, and if so, what does it imply? There is in fact a significant difference between the two concepts, and the two words ought to be distinguished, so that those who use them or read them know precisely what they mean.

The knowledge accumulated in any domain of learning can be brought together in a summarized fashion, often in the form of a filing system, where the knowledge is recorded on cards, in images (photographs for instance), or in sound recordings on magnetic tape. This record has its own intrinsic value and is similar to a record of the securities kept in the safe at a bank, which is why the corresponding collection of information is called a data bank. If the analogy with the text in a book is regarded as particularly apt, one can also speak of a data library. But anyone who has used a library knows that if it is anything over a minimum size, the classification system is crucial. If we are to find the information we want quickly, the books must be organized in a particular way in relation to one another, and there must be a search method designed to suit the type of information we want to find. This requires an abstract representation of the information, a model (technically called a *data schema*), which will subsequently be translated into a concrete form. It is the same with data banks: once they are over a certain size, it makes no sense to go through the entire contents to find the information we want. Indexing, using keywords, and searching only those files that contain the appropriate keywords is a basic way of improving access speed, but it is still too primitive. A minimum requirement might be links connecting related information. To achieve this, the data need to be structured around a model of what the user might want to know, and this model may be complex. The process of conceptualizing and organizing the knowledge adds considerable value to the "flat" data bank, made up of just a collection of files.

A bank structured around a model in this way is called a database. Usually, when the information collected comes from a very broad domain, the data must be divided up into *specialized databases*. In parallel with the International Nucleotide Sequence Database (which is strictly speaking a data library), there are a great many databases each specializing in a particular genome or a particular class of genes, and their number is growing all the time. In the main, databases exploit the existing links between the entries identified in terms of specific data fields such as "sequence name," "identification number," "sequence," "length," "gene name," and so on. The database designer makes a model by creating appropriate fields and setting up useful links between these fields in the different entries, so that users can carry out several searches at once on the basis of several criteria of their own choosing. The search is carried out by a server called a Database Management System (DBMS), which is transparent to the user. Normally it manages links or relationships, and for this reason the corresponding databases are called relational databases. At present, two commercially available relational database management systems, Oracle® and Sybase®, share most of the market between them, but there are many others, including public ones such as MySQL, designed for different hardware and especially for microcomputers.

Among the relational databases devoted to genomes is a generic server called GenoList, set up at the Pasteur Institute by Ivan Moszer, Claudine Médigue, and their colleagues. It is designed for databases that can evolve, and which contain appropriate biological commentary, and it includes some sequence analysis methods. GenoList is used to manage the sequence data on the genome of the model bacteria *Escherichia coli* (Colibri) and *Bacillus subtilis* (SubtiList), as well as pathogenic bacteria such as the tuberculosis agent *Mycobacterium tuberculosis* (TubercuList) and the bacterium that causes stomach ulcers, *Helicobacter pylori* (PyloriGene). These databases are divided into sections, and, using appropriate rules, they create secondary data in the form of gene products, which are translated using the genetic code and therefore hypothetical to begin with. They use commercial DBMSs and are available to the entire international community via the Internet. With the GenoList server, it will soon be possible to manage the knowledge associated with a large number of bacterial genomes.

Integrating Methods of Analysis and Sequence Information

In some cases, instead of managing links between fields, the entities need to be handled as a whole, as objects belonging to classes, each object inheriting the properties of the more general classes of objects it belongs to. Databases that use this management approach are called "object-oriented databases," and the corresponding management systems are object-oriented database management systems. Although these are fashionable, and they allow the user to handle text as well as images, they are still relatively little used on a large scale. However, one of them, ACeDB, is widely used in the biological community. Designed in 1990 and developed by Richard Durbin of the Sanger Centre and Jean Thierry-Mieg at France's National Center for Scientific Research (CNRS) at Montpellier, it was custom-designed for sequencing projects. Unlike commercial software, the source code is in the public domain, and Durbin and Thierry-Mieg are continuing to develop the system, with the help of contributions from scientists all over the world.

The name ACeDB is an acronym for "A *Caenorhabditis elegans* DataBase," and the name applies to the DBMS as well as to the database itself. ACeDB automatically sets up appropriate links between objects as soon as it is loaded, and uses a hypertext approach to move through the links created by the designer of each base. Several graphics features reflecting the requirements of geneticists have also been integrated into the software. Typical objects that the user might wish to handle are clones, genetic maps, genes, alleles, articles from scientific journals, and sequences. Every object is part of a hierarchical class corresponding to the model the designer is using.

However, those who consult these databases, even the most up-to-date ones, soon want to go much further. Not only do they want to carry out searches using multiple criteria simultaneously; they also want to run all kinds of algorithms to extract whatever information seems relevant from the sets of data they already possess. They need an environment capable of handling factual information about sequences, as well as the methodological knowledge associated with it; but organizing the construction of such an environment is difficult to achieve in practice, and is still essentially in the domain of informatics research. A good way to achieve this would be to create an interactive platform to assist with sequence analysis. Such a system, conveniently represented in a way similar to object-oriented DBMSs, would need to:

- run easily the analysis methods required by the user
- help the user to select the appropriate method or methods for a given task, and to string the methods together in the case of the most complex tasks
- memorize and handle both the analysis data and the results generated by the methods applied
- evolve, taking account of new methods as they are created (and new ones are appearing every day)

This last capability would require the system to integrate new methods and take account of the way they are used, without its original performance suffering.

Clearly, all this means enlarging the concept of a database to that of a "knowledge base" and, in the case of research on genome sequences, bringing together within one computational tool both biological objects (sequences and their products, keeping as close as possible to what is known about the information flow in the cell) and methodological objects (methods for analyzing sequences and biological objects in general). The system would need a modular structure, around a core or engine, to make it both extensible and flexible. "Extensible" means that neither the descriptions of the entities involved nor the list of methods it can offer should be fixed by the system; it must be possible to add new methods, which use and generate entities of a kind that may be completely new. To do this easily, the methods must be explicitly described. "Flexible" means that users can choose the methods they want to use, and in what order, and can refuse the advice of the system, either totally, or partly and temporarily.

Various research centers are developing integrated systems of this sort, some that are based on specific models of the knowledge of the domain in question, and some generalist systems. A central concept in their design is the concept of the object, not just that of the relationship. Knowledge bases developed in this way highlight aspects of the way system and user cooperate to resolve a problem. First the system finds ways of automating the tasks as much as possible, in order to handle the simplest, most frequently occurring problems quickly. But as some problems are better solved by the user than by the machine (particularly with graphic and image interpretation), the system knows when to hand over to the expert user, who will be able to spot what the machine cannot (at least not yet). Other tasks require a contribution from the user, for instance in setting the values of certain parameters required by the methods of analysis, so that a whole domain of possible solutions can be explored, rather than being limited by the parameters suggested in the automatic method. Geneticists are used to imagining possibilities on the basis of graphic representations of genomes and the genes within them; so if these knowledge bases are to be widely used it is essential that they give priority to the visual programming of new objects, methods, and tasks.

What is particularly original about these bases is that they incorporate a coordinated suite of sequence analysis methods, and are not just a library of programs to be called one after the other according to the level of analysis required. In this sense they are distinct from software libraries, in the same way as databases are distinct from data banks. The knowledge bases described above are designed around the objects they accept as input and generate as output. The designer creates models of knowledge to represent these objects, and the description of these classes of objects is an integral part of the knowledge base. The use of the task as a kind of methodological object makes it possible to describe adaptive strings of methods: breaking down a task into simpler subtasks depends on the entries the task has to deal with. The task itself thus contributes methodological knowledge, enabling the system to help the user select and string together the methods needed to solve a given problem, such as looking for protein coding zones in an unknown sequence.

In the knowledge base model developed by François Rechenmann, Alain Viari, and their colleagues at Grenoble and by Claudine Médigue at Evry, near Paris, a problem in analysis is resolved by breaking it down recursively into subproblems until the problems produced are simple enough to be resolved directly by one method, called automatically by the machine's exploration engine. At every level, breaking a problem down is opportunist, in the sense that it depends on the entities the problem applies to, entities that are themselves the product of previous solutions to other problems. A first step, in which the problem in hand is characterized by hierarchical classification, determines the most suitable way to break the problem down, and the problems and subproblems are represented by tasks and subtasks linked by operators that control the classification and breaking-down processes.

In this model, a cartographic interface has also been developed, to represent and then interact with sequence data and the results of analyses. Given the diversity of the objects involved and the possible ways of representing them, the specifications of this interface have been planned to be as flexible as possible. Users can thus create an on-screen layout showing several different maps, adjust their size, select the class of objects they wish to appear, specify the scale and units of measurement they wish to use, choose or even define the icons associated with the different categories of objects, and so on. As well as conventional functions such as zoom, the interface offers several advanced functions. These include setting up links between several maps, and synchronizing the rate at which objects appear in the viewing window, taking account of these links. It is also possible to show different zones from the same map in different windows. A typical use of this function might be to enable information associated with the three reading frames on each of the two strands of DNA to be very clearly represented. Finally, users can make particular classes of objects disappear and reappear on a map, as required.

As we can see, *in silico* genome analysis is still in its infancy. It will be complementary to *in vivo* and *in vitro* analysis, and as the third millennium begins it heralds an entirely new and fascinating kind of genetics, in which models of genetics will be developed in the computer. They will be tested there, too, before becoming a reality, *in vivo,* thanks to the techniques of genetic engineering, which have made it possible to modify a gene or any other DNA sequence by replacing it with an analog created *in vitro.*

Looking for Neighbors

Researchers involved in whole genome sequencing naturally want to exchange the knowledge they hold, and integrate it into the processes used in analysis, as quickly as possible. Initially this was done by express mail or slightly more complex procedures in which very large files were exchanged daily between the three centers of the international data library. But the real explosion, whose effects began to be felt from 1992 or 1993, came with the systematic use of the possibilities for dialogue and information exchange presented by the Internet and hypertext. Browsing on the Web, we can read a text "in depth" by following links that take us to other texts, images, or even sounds. It is easy to see why this way of presenting a text appeals to geneticists involved in genome sequencing programs. Once we know a portion of a genome, we will usually want to know what gene it belongs to, what other, similar genes are known, what signals control its expression, what articles have been written about it, and so on. Trying to answer these questions involves handling connections between different parts of a large area of knowledge. So it is a good idea for the knowledge to be organized according to a rational schema, as described earlier. This is something the Web is very good for, provided that one computer can be entirely dedicated to it (a well-designed database will be heavily consulted) and that the core of the exploration program (which will of course be totally invisible to the user) has been written with appropriate database management software. The number of Web sites devoted to genomes and their analysis is huge and growing constantly, so we can ask all sorts of questions about a gene and get a reply very quickly. The number of methods of analysis available is increasing, too, and it is out of the question to centralize everything we need to know about genomes from the biological and methodological points of view. A substantial research effort in computing is dedicated to *distributed* data management. This is a central issue, although handling the sharing of immense amounts of data and potentially incompatible methods or hardware can be difficult, especially if procedures need to be very fast.

The Web is particularly well suited to the conceptual analysis of information search methods. Most of the time, when we are searching for information, we are not absolutely sure what words we need to use in order to find it; only when we use a dictionary do we use the precise word whose definition we want to find. With an encyclopedia, we often need to work by a series of approximations until we find the information we are looking for. We use the idea of the relationship between terms or people's names and what we want to find. We do not use a fixed, precise object, but an idea of the *neighborhood* of that object. This simple idea gave David Lipman, of the NCBI, the inspiration for a very effective concept that would increase the efficiency of bibliographical research. Given a database of articles, containing the title, author, and abstract of each one, he considered the use of words as a further criterion. He analyzed this set of texts very simply, looking at the frequency of certain words other than the most basic ones, and the proximity of certain words, and then created an immense table, a veritable dictionary, in which every word was recorded together with its frequency in each article. So in the database, in addition to the keywords, authors, title, and abstract, each article also has a field or vector representing the frequency of use of the words in the article. An extremely simple mathematical technique (linear algebra) is used to calculate the scalar product of the vector associated with the article and the corresponding vector of all the articles in the database—a particularly simple operation for computer processing, and so, very fast. The articles with the highest scalar product are noted—the first ten, thirty, or fifty—and these are the nearest "neighbors" of the chosen article, in the sense of weighted word frequency. This makes it easy to find related articles that do not refer to each other (naturally it also makes it easy to spot plagiarism) and also to research forward in time. Taking an old article, we can look for its nearest neighbors, which may of course be very recent, as the date is not one of the criteria used.

There are many other ways of looking at neighborhood, and the concept is often now incorporated into software in one way or another. For instance, the nearness of two words in a text is exploited by the Advanced Search function of some Web-browser search engines, allowing us to use the concept of neighborhood via the term *near.* To find pages that deal with Mars on the one hand and life on the other, we can ask for Mars *and* life. But if we want to restrict our search to pages about the planet Mars and life together, we can find just these with the instruction Mars *near* life. Databases use this exploration by neighborhood in a particularly efficient way via hypertext. A simple link to a neighborhood search program at the end of a reference is enough to call up all the references that are near neighbors, and these in turn can bring up their own neighbors, and so on. One step at a time, the user can build an entire bibliography of the subject concerned. Of course this supposes that the texts have been "vectorized" (this can be done automatically or semiautomatically), but the result is so rewarding that the approach is sure to spread and be refined very quickly. The best current example of this kind of software is still the NCBI's PubMed library, which extends the neighborhood search method to all the references in the bibliographical data bank Medline, which collates all articles closely or loosely connected with medicine, amounting to over 15 million articles in 2001. We will see more facets and uses of the concept of neighborhood in the next chapter.

3

WHAT GENOMES CAN TEACH US

Why sequence genomes? There are several different ways of answering this question, all rooted in the ambiguous duality at the heart of our scientific culture. There are theoretical answers, related to the research undertaken in connection with sequencing programs and its contribution to knowledge, and there are practical answers, involving the technical application of the theoretical knowledge acquired. We have encountered this duality in a very explicit form in genome sequencing, because the starting point of these programs was in many ways much more political and technical than theoretical. I will come back to the practical side later, when I talk about genome engineering, but for now I want to analyze the content of the theoretical knowledge created by our understanding of genome texts (and which may later be useful in practice).

At the beginning of the third millennium, we are starting to get to know the whole genome texts of hundreds of living organisms. It has been announced that we know almost all the human genome itself, and that it will be "entirely" decoded by 2003. This was unthinkable twenty years ago. This entirely new knowledge will have immense consequences, which we cannot really appreciate. If it were no more than a demonstration of an amazing technical achievement, it would not have much real value, but as I hope to show, it is much more than that: knowing the genome texts enables us to understand a little better what life is, where it has come from, and where it is going. The most fascinating thing that we shall discover about life is that it is *not a mechanical process,* and that even if we do not deny its deterministic character, what we can know about it *does not enable us to predict its future.* Life is simply the one material process that has discovered that the only way to deal with an unpredictable future is to be able to produce the unexpected itself.

How can we read the story that the genome text has to tell? In the same way as paleographers do, decoding their ancient texts. To begin with, all we have to go on is what little we know about the mechanisms by which the genetic program is expressed. In particular we can use the genetic code, and luckily it is practically universal. The next step is to make *models* of life, which conform to the knowledge we have at the time we make them. We compare them against what we can expect from reality, then modify them, to make them reflect that reality more and more accurately, usually step by step but sometimes in great strides. This is when real discoveries are made.

The genesis and evolution of models thus resembles the genesis and evolution of species, being gradual and, occasionally, sudden. The very great difficulty we find ourselves facing is that throughout its long evolution—already more than three and a half billion years—life's exploration of reality has been based on *symbolic transposition.* Unlike physical or chemical objects, biological objects are more than just a site where actions occur; they *represent* functions. Very often they no longer correspond to them *directly.* As an illustration of this concept, a long way up the evolutionary scale, the use of words, the written words you are reading, takes the place of objects, relationships, and processes, and these words are not in any way directly involved in them. Philosophers have investigated this question in depth, and I shall not elaborate on it, but just stress that it is rooted at a level much lower than the human brain, at the level of the cell. The nucleic acids and the proteins, which are the very foundation of the objects, relationships, and processes that make up life, are made from completely different chemicals from each other, and the DNA of a gene that codes for the synthesis of an enzyme has absolutely no biochemical connection with that enzyme's function or shape. If life exists elsewhere in the universe, it is quite likely that the correspondence between the level of the memory passed on from one generation to another, and the objects that carry out the "biological" functions, will be based on entirely different chemical structures from the ones we know (even if it is not unreasonable to assume that they use carbon chemistry). In fact, one of the results of the genome programs might well be, in some not too distant future, the creation of other living organisms, whose basic chemistry would be different from the chemistry we know. Chemists have in fact already made peptide nucleic acids (PNAs), which hybridize or pair up with strands of normal DNA. If we were able to make machinery that could reproduce PNAs, we could create new organisms that had no nucleic acids as we know them.

To understand what life is, we will need to use all the resources of our imagination. We must be ready to reason by induction, from effects back to causes, rather than by deduction, from causes to effects, and to use all the information at our disposal, particularly the information in the genome text. It is this approach to discovery—which I will illustrate by exploring the way living things manage to determine, and select, the structures that enable them to carry out the control functions that are one of their most important characteristics—that will enable us to arrive at the most amazing, marvelous discovery the genome texts have to offer us: there is a map of the cell in the chromosome. It was already suspected that this was true of the body plan of animals, but it is true of the cell as well. We will be able to explore both the past, to look for the origin of the major biological functions, and the future, to ask what living organisms will become by themselves, and what we will make of them. We will end by taking a look at these questions, even though we know very well that this is a game we can never win, and that we can expect nothing but surprise—and awe.

Models and Reality

atcttttcg gcttttttta gtatccacag aggttatcga caacattttc acattaccaa
cccctgtgga caaggttttt tcaacaggtt gtccgctttg tggataagat tgtgacaacc
attgcaagct ctcgtttatt ttggtattat atttgtgttt taactcttga ttactaatcc
tacctttcct ctttatccac aaagtgtgga taagttgtgg attgatttca cacagcttgt
gtagaaggtt gtccacaagt tgtgaaattt gtcgaaaagc tatttatcta ctatattata . . .

This is how the text of the *Bacillus subtilis* genome begins. With its 4,214,630 base pairs, the text goes on and on for 2,000 pages, written in the four-letter alphabet. We know it has a biological meaning, because it enables bacteria to live and multiply in all sorts of environments, so we need to find a way to say something about the text, and find that biological meaning. This is clearly no simple task. However, we are certainly not starting from square one; far from it. First, this bacterium has already been studied for a long time, and a great many mutants are known (characterized by the modification of one or several genes, together with the corresponding text). Each of these mutants has an associated *phenotype,* a specific character (this might be resistance to a poison, or the inability to metabolize a particular sugar, or perhaps dependency on a vitamin in the growth medium). So there is a connection between the gene's function, as shown in the phenotype, and its structure, just as there is a connection between the instruction PRINT and the fact that this sentence appears in black and white in front of you.

We must beware, here, of the widespread belief that there is a straightforward link between the structure and function of gene products, and that the necessity for this link should be intelligible *a priori*. It is quite often very frustrating to find that, faced with a protein whose function is a mystery, knowing its architecture, even in detail, does not help us much in discovering the function. Very often the link between structure and function is only *historical,* and can be understood only *in retrospect:* a given gene product has become associated with a given function only because of the arbitrary conjunctions of events that have shaped the organism as we now observe it. This link is not logical and necessary; it is only symbolic and arbitrary. And it is precisely because it is symbolic that it can be arbitrary, in the sense that the atomist philosophers of ancient Greece meant, that is, fortuitous, unpredictable *a priori,* but obviously, because the function does exist, necessary *a posteriori*. The great discovery that brought material systems to life was that it is possible to make use of what is present, but not necessary, simply because it is there, and that this simple presence can provide a meaning. The whole problem of understanding genome texts rests on this enormous difficulty, which is similar to the difficulty in deciphering written languages. Written Chinese, with its ideograms, gives a clear illustration of this. Some characters are explicit representations, such as the one that represents a cart; it becomes a little more complicated and abstract when it represents a barn (a cart under an open shelter) or an army, where strength is linked with the cart; and to represent "cut," or "decapitate," the cart is linked with a knife.

In these variations on a theme, which originally represented a direct link between the structure and the function of the ideogram, the linkage is gradually weakened, and finally disappears altogether. This is exactly what happens as genes evolve. It is especially noticeable when moving from unicellular organisms, where the variations on the theme provided by each gene are relatively little exploited (although those variations that are used are very significant), to multicellular organisms, where the exploitation is intense. We should particularly bear in mind that things we might dismiss as redundant if our reasoning is too simplistic are usually not so at all. The evolution of living

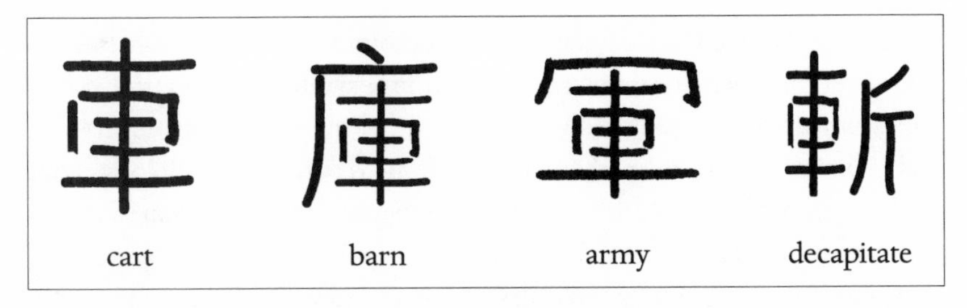

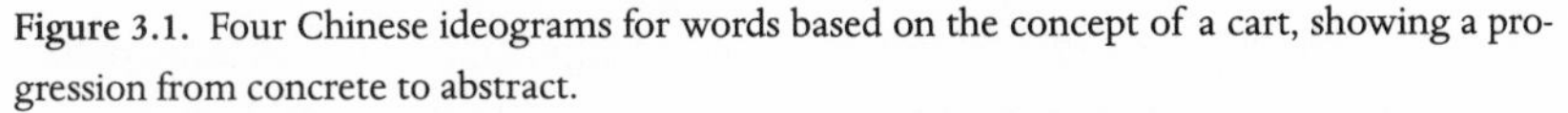

Figure 3.1. Four Chinese ideograms for words based on the concept of a cart, showing a progression from concrete to abstract.

organisms has discovered, in apparent redundancy, what turns out to be a very effective means of exploiting the symbolic power of biological objects, by finding new functions for them, revealed by the accidents of history.

The more biological knowledge we can bring in at the first identification stage, the better our chances of correctly identifying a gene's function. Biologists are thus in a similar situation to that faced by physicists when they want to account for a new phenomenon. Reality, in all its depth and richness of hidden information, must be compared with a model of reality that incorporates everything known about it so far. This is exactly what the NIH instructions for research funding require, when they state that they support "hypothesis-driven" research, or when the Wellcome Trust says of the projects it finances that "a cardinal feature of project grants is that they are hypothesis-led. They begin with an assertion, and the experiments comprising the project are designed to test whether the assertion holds; generally, attempts are also made to explain the association, in terms of the underlying mechanisms." What form does this correspondence between model and reality take?

Science is relatively recent as an autonomous discipline—it has existed for only two and a half thousand years, as against the two hundred thousand years of modern man's existence. During these two and a half millennia, science has been a process of improvisation and systematic exploration of models of the world, based on its own logic and steeped in symbolic thought and language. Though rooted in history, it has gradually had to detach itself from mythical images of the world, to replace them with something else, which would have a more effective relationship with its practical consequences and would be more universally communicable. From simple observation, science progressed to experimentation, in which reality is made to react to the hypotheses proposed. Finally, it has become profoundly anchored in abstraction (ultimately, mathematical abstraction) and in a symbolic theoretical framework, which is increasingly distant from the world of sensory experience, and in which the role of mathematics and logic is paramount. Biology is one of the most recent illustrations of this movement from observation to experimentation and then to abstract formalization—what could be more abstract than a genome text? But it is still very deeply anchored in observation, and as we are ourselves the first concerned, biology is particularly resistant to conceptualization, because the soul-searching this involves inevitably revives all the questions that have troubled generations of thinkers. It is important to understand how ways of producing knowledge evolve, from observation to experimentation and conceptualization.[1]

No human being is born a *tabula rasa*—a blank slate. We are the product of a long evolution, both biological and social. Our biological self is immensely important—even if we often forget it—because it fixes the limits of our perception and organizes our general plan of the world we live in, the way we see it and the way we act within it. Our history as animals means we inherit a perception of the world in which objects are organized according to two main sets of requirements: those of our metabolism, as creatures that have evolved for walking, within a particular environment; and those of the continuation of the species. From our history as social animals we inherit both an organized conception of interpersonal relationships and the use of language. As for the first questions we ask ourselves about the world—and asking questions is certainly one of the most important biological constraints, because this is how we build up our perception of the world—we initially answer them with *preconceived ideas.* In every society there is a preferred way of seeing the world, or rather the way it *appears,* based on stereotyped behavior or oral traditions, which serve as ways of explaining the world or of dealing with it. Francis Bacon remarked that human thought is often led into delusions, under the influence of "Idols." By Idols of the Tribe, he meant human nature's tendency to interpret phenomena according to preconceived ideas. The Idols of the Cave represent our judgment according to the limited viewpoint of our individual ego, with a reference to Plato's allegory of a cave-dweller who sees things only as shadows thrown by the light of a fire. The Idols of the Marketplace are fallacies arising out of interpersonal life and the traditions of the community. Finally, the Idols of the Theater are errors induced by our uncritical tendency to be swayed by the aesthetic power of traditional, established systems of philosophy.

Over the course of human evolution, the development and extension of language and tools have led to increasing interaction between human beings and their world, and many inaccessible places have become accessible. In parallel, ways of representing the world using signs and symbols have developed. It was probably the ancient Greeks who first made a consistent, operational distinction between the world and its appearance, between the truth of reality and the opinions held about it. This fundamental distinction, which can give an impression of absolute powerlessness, because controlling opinion about the world does not give us the power to control the world itself, is however quite different from *relativism.* It does not deny the reality of the world beyond man's perception of it. But it does show that it is possible to explore the world and predict its properties through the use of *models,* which we refine and try to bring closer and closer to reality. Rather than being an ineffectual doubt, this distinction involves the implementation of a *method* that enables us to produce, in a constant progression, the models of the world that are the basis of scientific knowledge and scientific reasoning.

Because we will have to use models of the genome text in order to understand it, I will describe this method briefly. This argument is intended as a preparation for genome analysis by computer, and is a personal synthesis of the work of many philosophers and logicians. Although it may look superficially similar to Karl Popper's presentation of the nature and construction of scientific thought,[2] it includes elements of the nature of models and theories which stem from the work of Rudolf Carnap and Charles S. Pierce,[3] combined in a structure that can be implemented in a bioinformatics approach. The main distinction from Popper's view is in the "interpretation" steps (abstraction and instantiation), which intrinsically separate the model from reality, making a demarcation between science and nonscience much more difficult than the simplistic criterion (refutability or "falsification") proposed by Popper. Of course this is a résumé of a process that has probably never worked exactly according to the steps I will describe. But as a whole, the set of processes used is similar to this description.

Even if we reject the impotence of solipsism (the view that the self is the only knowable reality) and accept that the truth of the world exists in its own right, the only way we can access it is through the mediation of a representation. We cannot get away from this. Being able to recognize and accept this separation is in fact a very difficult part of scientific reasoning, and the more we understand and apply this constraint, the better the quality of our thought will be. We must be content to make models of the world, since it is through knowing and understanding models of the world that we can explore and act on it, and that we can, ourselves, make reality behave in entirely new ways. Creating models means introducing a distance, a process of symbolic coding between ourselves and the world. But the particular consequence of the existence of symbolic coding processes is that they are in themselves *a source of creation.* This is how science draws its strength from the absolute weakness of knowing that it is by definition incapable of reaching The Truth.

The ideas we have inherited from our forefathers form the framework of our most basic representation of the world. But we have to ask ourselves how adequate this is: is it good enough to fulfill our expectations? Throughout history, many people have thought it is (or imposed this belief on others). But there have always been a few who wonder about what remains unexplained. How should we go about questioning accepted knowledge, and forming a new representation? A first step is to simplify the problem by separating out from our vague collection of preconceived ideas a set of fundamental principles which will not be questioned, at least for the time being. First we identify some *phenomenon* that stands out clearly from the surrounding confusion. Together with the phenomenon, we accept a set of unquestioned criteria, rules of scientific behavior, or *postulates,* which, temporarily, make up the founding dogma of the *theory,* an entity consisting of the phenomenon being studied, the postulates, and the model, together with the conditions and results of testing it. It is legitimate to call this attitude "dogmatic," because in order to test it, we treat the model *as if it were* the ultimate truth. This is why the expression "the central dogma of molecular biology" is used to refer to the rewriting of the genetic program, from DNA to RNA and then to proteins. This use of the word *dogma* (a religious word par excellence) to refer to a scientific theory should not be confused with its meaning in religion, and taken as indicating that the theory can never be challenged. We have to keep in the back of our mind that later it may have to be changed.

However, although establishing postulates is an important step, it does not allow us to explain anything or predict anything. To get any further, we need to begin a process of construction, in which the essential step will be putting together a model of the phenomenon. The postulates have to be *interpreted,* in the form of objects (either concrete or abstract) or relationships, so that we can handle them. At what is generally considered to be the ultimate stage of perfection, where abstract objects are manipulated, these objects will be *axi-*

oms and *definitions,* which can be manipulated using the laws of logic. Making a model thus involves, initially, a process of *abstraction,* which enables us to move from postulates to axioms. However, very often we will not be in a position to express the postulates as axioms. They can be represented only by an analogy with the basic elements of a *different phenomenon,* one that is better known. In this case the model will be a *simulation.* For instance, an electromagnetic phenomenon can be simulated using a hydromechanical phenomenon. We can also change the scale of a phenomenon (as is done when we use physical scale models or mock-ups). With genomes, we can represent a "theoretical" or "realistic" chromosome—the model chromosome—by writing a sequence that has the same statistical profile, preserving the relative frequency of the four letters of DNA, the observed frequency of the sixteen dinucleotides in the real chromosome, the same proportion of coding zones, and the same bias in the average use of the genetic code in the organism being studied. This model will then be compared with the sequence of the real chromosome.

But however useful and effective it is, simulation is clearly an admission of defeat. It represents, without explaining, and we want to do better than that. We must also understand and *explain,* reducing the representation to the simplest (and the fewest) basic principles. With genomes, this raises the question of their origin and evolution. We also need to understand the formal nature of control processes (feedback is a particularly important case). A simulation can be a good model for a particular genome, but it is a temporary model, and includes too few explanatory elements for it to be generalized and transferred to other organisms. As soon as possible, the postulates used to create it will need to be translated into well-formed statements, constituting axioms and definitions. We should note that even this axiomatization stage cannot get away from metaphysics (in the Aristotelian sense, of course). The rules of logic were themselves established through a thought process that necessarily draws on metaphysics, because the laws of logic cannot be established within logic itself. The next step is to bring the axioms and definitions into relationship with each other, to constitute a *demonstration,* which leads to a *theorem,* or more often to a simple *conjecture.*

The model—made up of its axioms and definitions, the demonstration and theorems—is then an entirely autonomous entity, which can be justified on the basis of its own rules. To be valid, it has to be true according to the rules of mathematical logic. This is one criterion of truth, an essential one, but it does not mean we know the truth of the phenomenon itself. *The truth of the model is not the truth of the phenomenon.* It is a common confusion between these two kinds of truth—the norm in magic—that sometimes sanctifies the model (which is regarded as part of the real world) and gives the scientist the role of priest.

To make the model, we had to start with the phenomenon being studied—in our case the genome text—and now we must go back from the model to the real world. This requires a process symmetrical with the one that gave us the basis of the model: an *instantiation,* a interpretation of the conclusions summarized in the theorem as instances. Instantiation involves making *predictions,* observations, or experiments, which are of two types, either *existential* (they predict the existence of an object, a process, or a relationship, which we must then discover) or *phenomenological* (they describe something), and they are subject to verification, and therefore *refutable.* We must set up an experimental procedure able to explore, and confirm or disprove, what has been predicted. With hypotheses about genes, this can be done using *reverse genetics,* where a gene is replaced by a modified counterpart or moved to another position in the chromosome.

The response of reality, either through passive observation or by provoking a reaction, enables us to validate the model and to measure how suitable it is. The only constructive way to do this is to identify where the model is *not* appropriate, which means that the most important element is disproving some of the instances predicted as consequences of the model. This essential part of the process avoids a knock-on effect, in which one finding confirms another in an infinite succession. Put another way, it is obviously the responsibility of the model's creators to show that it fits reality. Besides being totally ineffective, and contrary to the laws of logic (nonexistence can be proven only in a *finite* universe), it would be counterproductive to make an irrefutable model. Of course this is an easy way to spot charlatans. They stand out, and make themselves easy to recognize, by saying, "Repeat my experiments and prove me wrong!"

When some of the model's predictions are not borne out, we must question how well it reflects reality, and go back along the path that led to the incorrect predictions. In the course of this return journey, we will come back to the postulates that were used to create the model, to redefine and modify them, and if we can *reduce* the number of postulates, this will increase the model's power to explain the phenomenon. In practice, this return journey is long and difficult. Resistance in favor of the model organizes itself very quickly, if only because its designers cannot help believing that, rather than *representing* the truth of the world, it *is* the truth. It is quite natural to think that the reason for the incorrect prediction is, first and foremost, not that the model itself is inadequate, but that its conclusions were inappropriately interpreted at the instantiation stage. Obviously this is a step in which formal terms are interpreted into real behavior, involving a great many adjustments. Because it is hard to distinguish between the effects of the model's inadequacy, and the effects of an incorrect interpretation of the connections between the model and the phenomenon studied, what usually happens is that the model survives, often for a long time, despite a good deal of evidence of its imperfection.

During this critical process, the very nature of the model is called into question. The contradictions encountered serve to make the model's constructions, the meaning it represents, more precise and specific. The terms of the instantiation of its predictions, or of the abstraction of the founding postulates, are themselves refined and made more specific. This is why this "dogmatic" stage has a positive, essential role. A model that proves very inadequate will be quickly rejected, without contributing much to the genesis and progression of knowledge, while a succession of adjustments to a model that is useful, though not perfect, will lead to an increasingly detailed representation of the phenomenon. Finally there comes a time when we may have to reassess the founding axioms themselves, as renewed abstraction from the initial postulates calls them into question. This is obviously very rare and difficult, and it is the source of real scientific revolutions, and paradigm shifts (to use Thomas Kuhn's word),[4] from which new models, founded on hypotheses that are very different from those of their predecessors, will be born, develop, and die.

Finally, we should note that the approach I have just described is the one that is easiest to formalize; it corresponds to the hypothetico-deductive approach, the most conventional approach to science (it is what Americans call "hypothesis-driven"). We will also have to deal with the enormous quantities of facts that must be organized in inductive ("data-driven") approaches. Of course context plays an essential role in the guessing games called "abductive" or "context-driven" approaches by the advocates of artificial intelligence, and the study of neighborhood, which we have already mentioned, is the simplest example.

The Lactose Operon Model

To see how the hypothetico-deductive scenario works in terms of the biology of genomes and their expression, we shall look at a famous case, the discovery of the way adaptation to a particular food source is controlled in *E. coli*. If it is grown in the presence of the sugar lactose, the bacillus synthesizes a particular protein, β−galactosidase, which enables it to convert this sugar into a form it can assimilate in order to survive and multiply, but if lactose is not present, this activity is not detectable. When production of β−galactosidase appears in the presence of lactose, we say that it is *induced*. Explaining this regulatory phenomenon would be very interesting and useful, and it might be possible to generalize from this to other, similar adaptive behavior. The process of constructing a model of the phenomenon draws on a series of postulates.

In the earliest model, dating back to 1945–1955, it was suggested that the protein (β−galactosidase) already exists in the bacterium, but in a catalytically inactive form, which becomes active only through interaction with lactose. This model is the simplest one that comes to mind, because it makes a direct connection between the structure of the protein and the function studied, the induction of its activity. It is illustrative of an attitude still met with everywhere, that of systematically looking for an obvious relationship between the structure and function of biological objects. Research institutes and grand, often expensive international projects are still being set up on the basis of this simplistic idea, but which at least has the merit of being easy to understand (although in many cases the relationship between a protein's structure and its catalytic activity is in essence very enigmatic).

Transforming the postulate into axioms involves representing the protein theoretically, making models of it, and studying what happens to it, especially changes in shape, via the behavior of these models. To do this, the formal characteristics of the protein must be defined, taking into account our assumptions about the way it can change shape and become active when in contact with its substrate. If the model matches reality, we ought to be able to deduce a certain number of behaviors for β−galactosidase from these properties. This is the demonstration phase, still using the formal model of the protein. As it is an enzyme, it must recognize not only its substrate, lactose, but also analogs of it (poisons work on this principle). Some of these must convert it into its active form, while others must prevent it from becoming active. The logical combination of these formal properties leads to the conclusion that the observed effects must be detectable on the protein itself, *out of its context in the cell.* This means that the effects of various analogs of lactose that are recognized and hydrolyzed by β−galactosidase in its pure form can be explored in the test tube, *in vitro.* If the model is correct, all these analogs induce the protein to convert to its active form. They must therefore have this same effect on the protein *in vitro.* It follows that if we discover a lactose analog that, while intact in the cell, does not induce the production of active β−galactosidase *in vivo,* but on the other hand is hydrolyzed by the *pure* protein *in vitro,* the model is contradicted. In fact such an analog does exist: phenyl-galactoside is hydrolyzed by pure β−galactosidase, but does not induce its activity *in vivo.* Of course, suitable control experiments show that phenyl-galactoside does penetrate into the cell (this would have been a possible objection to the rejection of the hypothesis). The model suggesting that β−galactosidase is prompted to change shape by its inducer must therefore be rejected. Other arguments could have been added to reject this first model, in particular the need to find a source of energy for the change of shape.

A quite different model proposed for the same phenomenon suggests that the protein does not exist initially, but that it is synthesized *de novo* when the environment contains lactose. A new concept appears in this model: the *control* of gene expression. This would *separate* the protein's catalytic activity from the induction of that activity. In fact there seems to be no real reason why these two processes should be combined. One involves chemical transformation (one simple transformation in fact: hydrolysis, a reaction with water that breaks down a product, lactose). The other involves control, a process that is formally abstract but is typical of the kind of processes found in living organisms. It is clear at once that a formal representation of this situation will be completely different from the first one. There is no reason to postulate architectural properties in the protein that would allow it to change shape. The protein can remain an undeformable object, or it may undergo only local changes in shape, related to the energy variations that occur as the catalytic activity takes place.

On the other hand, the synthesis mechanism, which was not considered in the first model, now becomes the focus of attention. A gene codes for the protein: what kind of relationship is there between the gene and the protein? Is it direct or indirect? If there is an intermediary, how is it synthesized, what can we say about its chemistry, its life span, and so on? As in the previous case, formalizing these hypotheses produces a well-defined model, precise enough to be used to devise experimental tests. However, the logical relationships between the interpretations of the different starting postulates need to be both well constructed and exhaustive. Otherwise, it will not be possible to make a precise enough prediction, deduced from the logical chaining of axioms. A basic aspect of the instantiation of this second model of bacterial adaptation to growth in the presence of lactose is that it makes an existential prediction. This is a very powerful constraint in every type of model, because existential predictions are qualitative: they cannot be adjusted when one of the parameters is changed. There is a significant difference between this and the situation with the deformable β−galactosidase model, which is more difficult to refute, because it is always possible to imagine some quantitative parameter connected with a change induced in the enzyme by the substrate, which would explain why what has been observed *in vitro* cannot be reproduced *in vivo.* The new model predicts either that the enzyme is unstable and disappears after a while (but this is contrary to experimental results, which show that β−galactosidase is an extremely stable enzyme), or that an intermediary with a limited life span intervenes between the gene and the protein. If it were not limited, the protein, derived in some way from this intermediary, would be permanently present in the cell, bringing us back to the first model, which has already been refuted. The second model also predicts the existence of an intermediary responsible for controlling the synthesis of β−galactosidase. If this model adequately reflects reality, experiments should be able to demonstrate the existence of two new classes of biological objects, an intermediary between the gene and the protein, and a molecule responsible for controlling its synthesis.

There is another crucially important aspect of this second model. It was particularly well illustrated by Jacques Monod, but strangely it is often forgotten forty years later. The separation between the catalytic and regulatory functions indicates that the function of *substrate* of the catalytic activity and the function of *inducer* are two different things. The first can and indeed must have some connection with the structure of the enzyme, β−galactosidase, which must recognize lactose. For the second, in principle this connection is totally arbitrary. Other molecules, or even a physical phenomenon such as light, or a change in temperature, might control β−galactosidase synthesis, playing the role of inducer. This does sometimes happen in other systems. In fact, although there is no point in giving the details of the corresponding chemistry, it is not lactose itself, but allolactose, a similar molecule derived from lactose by the action of β−galactosidase, that actually induces the enzyme's synthesis.

At this point in the description, I would like to reemphasize the arbitrary character of the association between a function and the control of its expression. This is a first level of an aspect we normally call "symbolic," when we are talking about human communication. This arbitrary, symbolic character allows the cell to manipulate associations situated at a high hierarchical level, between apparently unrelated functions. Life has made systematic use of this remarkable phenomenon. This is what makes it possible to introduce relationships between physical parameters, as well as chemical ones, into gene expression. It is found in butterflies, for instance, where the control of a plan for the development of the wings has been used in another context to create the ocellus or false eye patterns with which they frighten off predators. This symbolic aspect is typical of the most important biological functions. The arbitrary association between different phenomena is so universal that it can even be used by scientists to trick a host cell, and to control the expression of numerous genes for commercial ends (including making it express foreign genes).

Both models have been explored, but experimentation shows that it is the second, later model that better represents reality. The two intermediaries it predicts were discovered: the carrier of the gene's meaning is a kind of RNA, which was named *messenger RNA* for this reason, and the object responsible for control is called the *repressor;* because its presence in the absence of allolactose represses the synthesis of the mRNA needed for β−galactosidase to be produced. This model, which Jacques Monod and François Jacob called the *operon* model, paved the way toward understanding the relationships between genes and the products they define as an algorithmic program. The deep meaning of this particularly powerful and creative model is still far from being understood or even explored more than superficially. It is the basis of the genesis of shapes in plants and animals. It explains how bacteriophages (viruses that infect bacteria) can become integrated into the bacterial genome as *lysogenic phages* or *prophages,* where they remain hidden, setting off the cycle of virus production (and killing the host) only when the circumstances are right. All sorts of variations have been discovered. In particular, situations have been found in which the control is positive, and where an *acti-*

vator, rather than a repressor, triggers transcription in the presence of an inducer; or others such as the control of the premature ending of transcription, called *attenuation,* rather than its start. This model evokes a way of representing the world that is profoundly different from the way we usually account for the physical world. It adds abstract symbolic relationships to the objects of chemistry and physics. The difficulty of understanding this symbolic aspect explains why biology in general and what we call "molecular" biology in particular are the subject of so much misinterpretation and misunderstanding.

Phenomenology

It is difficult to connect the text of genomes with biological functions. Knowing the text of a gene, predicting the sequence of the protein it specifies, visualizing its architecture, does not directly give us its function. The best we can do is to modify the gene or inactivate it and to study the genetically modified organism. But then we are faced with the difficult situation of studying phenomena (in the etymological sense, not with the Husserlian connotation of transcendental phenomenology). What is the best way forward? How should we interpret what we observe, and avoid taking our wishes for reality? Unlike in a number of domains of physics, where phenomenology is already well established and the theoretical, *a priori* approach is highly developed, we are not in a position to make a model of what we want to observe according to the criteria I have outlined. First we must observe and account for a phenomenon: growth under certain conditions, use of a particular molecule, sensitivity or resistance to a particular variation of a physical parameter. Simple phenomenology, because of its approach in which observation is only very loosely connected to a well-defined and delimited theoretical corpus, is on the borderline between science and an unstable form of thought, often close to a kind of primitive magic. This is not often recognized, but it explains why a large part of scientific work, even work that is institutionally recognized, is in fact of very little value in advancing scientific knowledge. It also explains the existence of many activities in the field of biology that are close to ignorance or even fraud.

In the first place, the phenomenon must be worthy of interest. Its boundaries must be fixed, so that it stands out from the preconceived image of the universe and the usual world. At this stage the postulates taken into account are often very coarse and vague. An important element in the Galilean practice of experimentation, and its attempt to capture reality in mathematical terms, is that it begins by allocating quantitative values to the observations made about the phenomenon. Implicit in this is the hope that it will then be possible to relate it to the various worlds of numbers, where variations are slow or sudden, monotonous or cyclical, continuous or discontinuous, or, to take up a modern theme, regular or chaotic (bracketed together with the universal truism "nonlinear"). The model that is perhaps most frequently used today (often without people's realizing it) is what we may call the *black box* model. In this model, an attempt is made to study how an object (which must first be isolated, an initial experimental stage in itself) reacts when placed in conditions that are varied at will, and a table is drawn up setting out correspondences between these variations, in accordance with the conditions imposed. The phenomenology consists in cataloguing the observed behavior of the object. Correlations will be drawn from this catalogue—a list of covariations. Only then is it possible to start making a model of the phenomenon. We can see that this leaves room for all sorts of misinterpretations. The observed behavior may in fact be connected to all kinds of constraints, some of which are caused by the equipment used in the experiment, rather than to the true nature of the object studied. Moving from phenomenology to representation by a model is a difficult step, but it is crucial to scientific progress.

At the turn of the century, a novel technique for studying living organisms, using DNA chips, gives us a perfect example of a new phenomenology. A DNA chip is a physical support for holding either whole genes, or oligonucleotides carrying the sequence of gene fragments, in the form of a grid of ordered spots, known to the user (this can be done because the genome sequence is known). To carry out an experiment, these chips are hybridized with RNA that has been extracted from cells isolated in a particular environment and labeled radioactively or with fluorescent nucleotides (in fact usually complementary DNA created *in vitro* is used, because DNA is much more stable than RNA). The chip is then explored by a machine that measures the intensity of the radioactivity or fluorescence of the spots. The result is a table of spots, some very brightly fluorescent, some very dark, and others in between. A different set of experimental conditions will give a different image. This produces an enormous table of numbers associated with each of the thousands of genes of the organism, obtained in different conditions, and giving an image of the phenomenon under analysis. The phenomenon can be used in diagnosis (which is never explanatory). But how can we get from the phenomenological image to a model of cell function? A large number of researchers worldwide are working on this, especially on the basis of advanced statistical analysis, associated with approaches in which biological knowledge is brought into play.

We have all heard the story of the boy who pulled the legs off a flea and showed that it could no longer jump when he told it to, thus proving that fleas hear with their legs (which might even be true, but is obviously not proven by this experiment, even if it is perfectly repeatable!). It would be a mistake to laugh at this example, which unfortunately could stand for a large proportion of the phenomenological experiments undertaken in biology (and which, sadly, often lead to medical applications). Another common example of this kind of mistake is *instructive* interpretation, usually known as Lamarckist, which often occurs in experiments connected with the evolution of species. We know for instance that August Weismann cut the tails off dozens of mice over many generations, to see whether this characteristic could be inherited, and that he found no change in the offspring, which were always born with long tails. But this was a matter of luck! There is in fact a series of mutations that produce offspring with short tails. Weismann could easily have accidentally bred mice like this. One shudders to think of the consequences this would have had for the development of genetics, so prevalent is the "finalist" interpretation in biology. We also know of course that the phenomenological approach is at the heart of behaviorism. Of course all these experiments can perfectly well be reproduced. It follows that *the fact that an experiment can be reproduced is not a sufficient criterion of validity for the interpretations of a phenomenon.* It is obviously a *necessary* criterion, but it is neither sufficient nor even an indication of the intellectual value to be attached to a phenomenon. For the validation to be effective, the parameters describing the phenomenon have to be changed, by linking them to a first attempt at interpretation, which will be used temporarily as a postulate, allowing an initial model of the phenomenon to be constructed. It is this model that we will then attempt to validate, preferably by looking for its flaws.

How Statistical Models Should Be Used

Very often it is a probabilistic or statistical approach that enables us to discover whether a phenomenon is worth studying or not. We may not know much about the phenomenon, except that it produces variations in certain observable characteristics. We highlight them, measure them, reproduce them, and attempt to find how consistent they are when the same conditions are repeated. There is a theory of chance in this approach, usually unacknowledged, and always heavily burdened with *implicit* postulates. In particular, the idea of chance itself, if it really existed, would prevent any scientific approach, because the chance inherent in experiments would, by definition, make reproducing them pointless. As Heraclitus said, the water in the river is never the same. On the other hand, we can simply identify chance with the unknown rather than with the unknowable. When I talk about statistics it is this second approach that I am referring to (which is why I prefer to use the word *contingency* in its original sense, or *arbitrariness,* rather than *chance*). The statistical approach, which is taken for granted in quantitative experimentation (the response of any system subject to testing necessarily varies from one occasion to another), has two main aspects, one descriptive or exploratory, the other inferential (logical), which enables the models of the phenomenon to be explored more extensively. (In certain cases this may even lead right outside science and into politics, via the financial decisions made about the corresponding research project, as has happened with many experiments in astrophysics and high-energy physics, and genome programs in biology.) When a phenomenon is first revealed, we are in the exploratory phase of statistics. The main objective at this stage is to designate correlations. This is a particularly dangerous stage, as there is a tendency to *find correlations we wish to find,* because we believe in the explanatory power of the preferred model, whereas methodologically speaking we should try to avoid having to prove that we are right (which is by definition impossible). In any case, of course, *correlation is not causality.*

Another particular feature of the statistical approach also has to be taken into account. In the case of genomes, we are studying a *finite* text, not a collection of experimental results about one biological phenomenon. This means that statistics have to be used with great care, because there are a great many pitfalls in the statistics of finite sets (especially because *any finite set contains an element of regularity* merely because it is finite, and this regularity has nothing to do with the functions attributed to the set). For example, if we study the frequency of the words in the genome text, we must always remember that the statistical observations are meaningless unless they can be correlated with a biological phenomenon that can be studied objectively, *independently of the analysis undertaken.* Studying correlations is a job that is difficult and often misunderstood: it is hard, and sometimes impossible, to prove that a correlation is significant. First of all the distribution of the measurements that revealed the correlation should have a particular form; it must show a certain regularity. Then of course there must not be any other, hidden variables involved; but this is a major difficulty, because in the early stages we necessarily form as *few* hypotheses as possible about the parameters involved (statistical analysis begins with Ockham's razor). Finally we must take into account the fact that the statistical justification for a correlation is normally tautological, because the results are explained with the help of the data that were used to obtain them. So to have a more convincing idea of the relevance of the correlation, we need to identify *independently* some characteristic of the studied phenomenon that is associated with the statistical correlations obtained.

If we want to explore the genome text further, the statistical approach involves using *models* of the DNA in the chromosomes, and of the biological processes underlying its sequence, its function, and its evolution. We must be able to compare the real text with a model text about which we know something *a priori,* in order to identify what is original and unknown in the studied genome. If we decide to begin without forming any hypothesis specific to biology, in order to avoid introducing too much bias into the early stages of the observation, the only models used will be those that derive from statistical theory or the calculation of probability. So at this stage it appears that what is observed will depend on the theoretical model the real genome is compared against. Once a certain number of specific traits have been characterized, biological knowledge then has to be introduced, to deepen our understanding of the genome text.

A sequence is a text, which we want to describe, for instance in order to look for relevant elements in it—those that are statistically significant—or simply in order to manipulate it, to identify what are considered to be regions of interest. To carry out experiments, these regions can later be physically associated with other regions, using "cut and paste" techniques as in word processing. Doing this requires a model of what DNA is. Why not introduce biology into the model, instead of the laws of chance? The most important characteristic to be introduced is the fact that DNA contains coding sequences, and these must be defined in the model. Without going as far as the nature of the genes and the fact that they code for proteins, if we remain at the level of the physicochemical nature of the DNA molecule itself, DNA polymerase does not know anything about the function of the text it is copying, and it is sensitive to the context of the text at only a very local level. However, if we think in terms of *concrete* biological processes, we realize that the polymerase is subject to numerous constraints. For instance it must take account of the cell metabolism. When replication takes place, the new DNA molecule is constructed using the four nucleotides as building blocks (these are produced by the cell metabolism, using a number of enzymes). So from this stage onward there is a connection between the nature of the genome text and the stock of nucleotides available at the time of replication. There are other, similar connections, particularly because in a certain context, some DNA bases are modified after replication, to label them so as to mark the site of certain cell functions, especially control functions.

This necessarily introduces a constraint and a first element of meaning into the genome text, in that there is a certain accord between the general metabolism of the cell during its existence and the balance of bases in the DNA sequence of its genome. When genomes are compared one with another, it would be absurd to consider their percentage of certain nucleotides as being independent of this constraint. Strangely, even scientists who are famous for their work as statisticians do make this kind of risky assumption. This is why the scientific literature is full of misleading examples of overall DNA sequence analysis.

Here is an example showing that the contribution of biological knowledge supports the statistical result, and leads to the discovery of biological functions connected to genome sequences. We will take the case of the *E. coli* genome. The codon sequence of each gene unambiguously specifies the distinctive sequence of letters in the gene product, the protein's amino acids. With 61 codons out of the possible 64 triplets, and 20 amino acids, on average an amino acid can be specified by 3 different codons. This means that each gene has its own *bias in the way it uses codons;* in other words the frequency of use of each type of codon for a given amino acid varies according to the gene. For instance, GGC, GGT, GGA, and GGG all code for glycine, but some genes use GGC and GGT more often, and others are unusually rich in GGA. Each gene can be represented in a 61-dimensional space, appropriately normalized, with each axis representing the frequency of use of the corresponding codon. The set of genes is represented by a cloud of points in this space (more than 4,000 points in the case of *E. coli* or *B. subtilis*). The form of the cloud can be studied easily thanks to very simple mathematical techniques, for instance by projecting it onto a plane. We can then ask: does the shape of the cloud of points enable us to spot any rules in the way the genes of an organism like *E. coli* use codons? If any rules exist, do they enable us to learn anything, perhaps not about the gene product itself, but about its origin during the course of evolution, or about the circumstances in which it is expressed in the cell? With several thousand genes, it is possible to draw up a (very large!) table of codon use in all the genes of this bacterium. The next step is to discover whether there are any correlations between the way the genes use codons and their biological function. Numerous statistical approaches can be used to perform this analysis. Some trace their origins back to the classes established by Vicq d'Azyr in the eighteenth century, and were popularized in France by the mathematician Jean-Paul Benzécri at the end of the 1960s.[5]

This analysis has revealed that *E. coli* genes fall into three classes, distinguished by the way they use codons. The same thing—a division into three classes—can be seen with *B. subtilis*. Is the distribution significant? Only if we observe a correlation that is *totally independent* of the statistical analysis. Gene function was not involved anywhere in the production or the analysis of the shape of the cloud. But obviously it is possible, *a posteriori*, to label the genes in each of the classes discovered, and to give each point in the cloud a name. This makes a further correlation appear, and in fact we find that functions that have comprehensible relationships with each other are spread out in a way that correlates with the class the gene belongs to. It is this, and only this, that enables us to justify the distribution discovered by this "data-driven" approach and to give it a biological meaning. We now have something much better than the simple observation of a phenomenon: it appears that the way the genetic code is used depends on the functional nature of the gene in question.

There are a number of other methods of analysis that enable unexpected functions to be suggested for genes, so that gradually the genome text reveals its unexpected meaning, giving a remarkable idea of what life is.

Preludes to Discovery: Exploring the Neighborhood

Most of the time, we explore the paths of knowledge by moving from the known to the unknown, and extrapolating from what we already know. We very often study the *logical consequences* of what we know, by successive deductions, making the most novel hypotheses possible, and exploring their consequences as profoundly as possible. Effective as it is, this hypothetico-deductive method has the drawback of being able to refine only knowledge that we already have, without giving us a way of forming hypotheses that are both new and pertinent. How can we find original ideas, but with an originality that is not alien to what we are studying? This general question is relevant to all approaches to discovery, whatever domain is explored. How can we advance inductively, how can we explore *upstream,* and not downstream as with deduction?

This is a profound epistemological question, which I cannot really go into here. We will consider only one approach, because it is particularly effective in the case of genomes: that of induction by exploring the *neighborhood* of the objects we want to consider. The idea behind this approach is that each object exists in relationship with other objects. John Donne said:

> No man is an Island, entire of itself; every man is a piece of the continent, a part of the main; if a clod be washed away by the Sea, Europe is the less, as well as if a promontory were, as well as if a manor of thy friend's or of thine own were; any man's death diminishes me, because I am involved in mankind, and therefore never send to know for whom the bell tolls; it tolls for thee.[6]

The same is true in biology. No object exists in isolation—or if such objects do exist, it is less important to know them, because their isolation means that they have little to contribute to the phenomenon being studied. It is precisely relationships between objects that are at the heart of life. So we know in advance that, among the things we need to discover, there are relationships that have a particular form, whose implementation enables vital functions to be expressed, such as the regulation of gene expression. Of course we do not know exactly what these relationships are *a priori,* but we know that they do exist. We do not know what form they take, but we know that they demand a certain proximity between objects, whether in terms of space or time or other forms of mediation.

Genes are the most important objects in genetics, and give living organisms their character and phenotype. These fragments of DNA include, first of all, control elements (such as transcription promoters and the regions where transcription ends). They also include a string of nucleotides that are expressed via their transcription into RNA, and indeed there are some genes whose expression ends here, as RNA is the gene's final product: transfer RNA, for instance, and ribosomal RNA, which forms the core of the machinery that translates mRNA. Other genes are finally translated into a protein. So it is natural to take genes as the central object of *in silico* analysis. Inductive exploration consists in finding all the *neighbors* of each given gene, as a starting point.

"Neighbor" is to be understood here in the broadest possible sense. It is not only a geometrical or structural notion. Each neighborhood will have its own particular light to throw on the gene of interest, and will provide clues for researching its function. The scientist behaves rather like Sherlock Holmes, combining the clues contributed by all sorts of relevant, neighboring objects, in order to understand the crime. One natural kind of neighborhood is proximity on the chromosome. With the genome represented by a text, it is natural to look at those elements that are neighbors to one another in the text itself. Earlier, we saw that in bacteria such as *E. coli,* transcription is controlled by a repressor, which expresses a messenger RNA. This mRNA directs the synthesis of an enzyme, β−galactosidase, which breaks down lactose. But Jacob and Monod discovered that the messenger does more than this. It is also responsible for the synthesis of another protein, permease, which belongs in the cytoplasmic membrane. A particular feature of this protein is that it enables lactose to be concentrated within the cell. So transcription coordinates two processes, the concentration of lactose, even if it is very dilute in the cell's environment, and also its degradation. Lactose is used as a source of carbon skeleton molecules, to build the cell and manage the energy it needs to function. A control structure coordinated in this way—an operon—can contain many more than two genes. But this organization shows that genes located close to one another can often be functionally connected. Here, then, with the operon, we have a first example of neighborhood, which can contribute useful hypotheses, provided we know something about the function of one of the operon genes. Unfortunately there are numerous transcription units that consist of only one gene, and eukaryotes do not organize the transcription of their genes into operons, which makes things more difficult. We must therefore look for other kinds of neighborhood besides simple geographical proximity.

The evolution of species proceeds by variation on ancestral themes. Consequently, many genes are descended from common ancestors, and just as children look like their parents, so genes, or more often their products, have points of resemblance. This is a rewarding kind of neighborhood to consider. In the case of the genes themselves, because of the redundancy in the genetic code the resemblance in the DNA text soon fades, even if their products remain very similar (for instance because they are responsible for the same function over successive generations). It is mainly because their proteins resemble each other that two genes are considered to be neighbors from the evolutionary point of view. Often their structure or architecture will be very similar, and quite often there will be a functional similarity, too (although it is dangerous to generalize on this point). So it will be useful to draw up a table for each gene product considered, showing all the similar products found in the sequence data banks, and to establish their genealogy, while retaining the annotations summarizing what is known about these different products.

There are many other ways of finding neighborhood. In particular, a gene may have been studied by researchers in laboratories all over the world. For one reason or another, the gene may have properties that have made these researchers associate them with other genes, so it is worth looking for a gene's neighbors in the sense that it is mentioned in their company in the scientific literature. This kind of *in libro* analysis can help to produce an image of what we are studying, which is enriched with the knowledge of a great many researchers, in a way that accumulates over time.

A gene's similarity with others can also come from similar physicochemical characteristics of their products (such as their amino acid composition, their electrical charge, or the fact that a gene product has an address label that indicates that it belongs in a specific cell compartment). Similarities can be local rather than global, indicating that two proteins recognize the same cofactor or the same substrate. Similarity might also be a matter of the absence, rather than the presence, of certain motifs (for instance, certain proteins that come into contact with oxygen lack the amino acid cysteine, because this reacts very strongly with oxygen, so a protein containing cysteine would be systematically modified). Giving free rein to the imagination can help us discover other kinds of neighborhood, which are added to the list to broaden the scope of what we can say about a gene and its product. Neighborhood can be structural, if the products of different genes share the same cell compartment, for instance the cytoplasm, a membrane, the nucleus, or a particular organelle. But there are also kinds of functional neighborhood. As the molecules involved in metabolism undergo interconversions, there are enzymes that are neighbors because they use the same substrate, produce the same product, or follow one another in a metabolic pathway.

Finally there are more complex kinds of neighborhood, and studying these can bring particularly rewarding results. To take up the example I outlined earlier, that of bias in the use of the genetic code, we find, for instance, that two genes can be neighbors because they use the code in the same way. It is interesting to study all the genes surrounding a given gene, in the cloud of points that describes the use of the genetic code in all the genes in that organism. When this is done, we begin to discover some very unexpected properties of genome texts.

A Closer Look at the Way E. coli *and* B. subtilis *Use the Genetic Code*

We will be drawing on the analysis of codon use in *E. coli* and *B. subtilis* several times in this chapter, because it has prompted some profound thinking about the way cells are physically organized, while keeping a firm grasp on the abstract nature of the genome program. Work undertaken a long time ago indicated that genes could be classed into two main families, which could be distinguished by their metabolic activity. Most genes belonged to the first class, which showed very little bias in the use of the genetic code. All the codons were used in more or less the expected proportions, depending on the amino acids they specified. Codon use in the genes in the second class showed a very strong bias. However, very little had been said about the reasons that might have led to the formation and preservation of these classes. Then, at the beginning of the 1990s, when large sections of genomes were beginning to be known, Alain Hénaut and I thought it would be a good idea to give this project to two young researchers who had taken the risk of showing an interest in *in silico* genome analysis, first Claudine Médigue and then Ivan Moszer. To our great astonishment, we found that in these model bacteria, the genes grouped themselves unambiguously into three classes rather than two.

When we labeled the genes in each of the classes, we were surprised to discover that the biological meaning of the classes was as follows. The first group corresponded to genes responsible for most of the biosyntheses of small molecules. All these genes share the property of being normally expressed at relatively low levels, but either continuously or very frequently. The second class represented the central genes in cell metabolism: those of translation and transcription, the core of intermediary metabolism (the part of metabolism that distributes atoms and energy to all the molecules used to build the cell), and the control of protein folding. These genes are expressed at a very high level when bacteria are growing exponentially. The third class was different for *E. coli* and *B. subtilis,* but it grouped together the miscellaneous genes that play various roles in the lateral exchange of genes between organisms (horizontal genetic transfer), a phenomenon that particularly concerns those who worry about the spread of genes in the environment, from genetically modified organisms.

In the *E. coli* genome, there are genes corresponding to a variety of processes which indicate the versatility of the mechanisms of horizontal genetic transfer in bacteria with double membranes (including receptors for bacteriophage viruses; genes that control the integration of their genome into the host chromosome—lysogeny; and genes for the transposition of particular sequences, called insertion sequences). In fact bacteria of this type, with an envelope made of two membranes (called Gram negative bacteria, because they do not take up a stain in the method invented by the Danish bacteriologist Gram), are usually able to exchange genes very easily via a process called conjugation, in which plasmids, pieces of chromosomes, or even whole chromosomes can be injected from one cell into another. This property also exists in Gram positive bacteria, but it is much less common. In both cases there is also another type of exchange, carried out by bacteriophages. Within class 3 in *E. coli* we found gene regions that allow the transposition of DNA (called insertion sequences), genes that allow conjugation, others associated with recombination, and bacteriophage genes, or receptors on the membrane that allow the bacteriophage to enter the cell. In the equivalent class in *B. subtilis* we found only the last two groups.

Two remarkable observations complete this first image of the organization of bacterial genes into functional classes. The first is a beneficial example of natural genetic modification: accuracy in replication can be transferred from one organism to another. In *E. coli,* among the genes involved in horizontal genetic transfer, there is also a special family of genes that code for the proteins needed to correct replication errors. Why should this be? DNA replication cannot be totally accurate—errors cannot be avoided when several hundred thousand pages of text have to be copied. DNA polymerase makes mistakes, at a rate set by the constraints of chemistry at normal temperature at between one error per thousand and ten thousand bases. It is complemented by a correction stage, which corrects a large proportion of these mistakes on the spot, but once again one out of every thousand to ten thousand slips through. The result is an error rate of about one in a million. This is very low, but it is still too much. In normal conditions *E. coli* actually makes less than one mistake per billion bases (which means that only one bacterium in a thousand has a mutation compared to its parent). To achieve this, it uses a family of "antimutator" genes, whose products have the specific role of correcting certain types of errors as well as they can—oxygen damage to guanine (G), for example, or the action of ultraviolet light, which affects thymine when two T's occur one after the other. But what is quite remarkable is that these genes belong to the class of horizontally transmitted genes.

This prompts the intriguing hypothesis that in nature, most microbes like *E. coli* are not genetically stable species, but collections of individuals that can mutate extremely easily (a thousand to ten thousand times more easily than the laboratory species). These individuals produce an immense variety of descendants, similar but not identical, and this variability of their own enables them to explore the highly variable characteristics of their environment. In general, precisely because of this environmental variation, there is no disadvantage to the microbes in being variable themselves (and above all, being randomly variable, *in no particular direction,* as the future of their environment is unpredictable). But it may happen that the environment does not change for long periods. Then, if a bacterium is especially well adapted to growth in this particular environment, it has little chance of evolving into a better-adapted form through mutations. On the contrary, any mutation can be only neutral or harmful. If, because of the structure of its environment (via the presence of other bacteria that possess the required genes), it can capture genes that will prevent its genetic program from changing, the individual that remains fixed will thus have an advantage over all the others, and after a few generations will take over from them.

A second observation is that viruses can transmit pathogenicity from one bacterium to another. This is useful to bacteria but harmful to their hosts, and it suggests some interesting hypotheses. The *B. subtilis* genome contains a large number of bacteriophages in the lysogenic state, in other words hidden in the chromosome, inactive but ready to become active and multiply if circumstances permit. It is noticeable that close to or even within these regions, we often find genes that confer on the bacterium the ability to resist toxins such as heavy metals, arsenic, or certain antibiotics. We also often notice the presence of genes whose product is similar to proteins annotated as conferring pathogenic virulence. The inference that immediately comes to mind is that these genes spread from one host to another, after infection by bacteriophages, because of the selective advantage they confer on their host. Provided they belong to the same class (the one containing the genes needed for the horizontal transfer itself, as well as other genes showing the specific codon usage bias noted), these virulence genes can even spread between species. It is easy to understand in the case of toxin resistance. But it is also comprehensible in the case of virulence: for a bacterium, becoming pathogenic means overcoming a host, which will be a substantial source of food. However, this conclusion is rather worrying. It explains how new diseases can suddenly appear, caused by previously harmless bacteria. As I have said, *Bacillus subtilis* is much used in industry, especially the agricultural and food industries. It is what the Food and Drug Administration calls "generally recognized as safe." But there is nothing to prevent it from suddenly becoming pathogenic. The stage is set!

A Look inside the Cell

The only effective way to move on from here is to proceed systematically, not by looking for an answer suggested by formal reasoning, but by trying to visualize what is physically possible in the cell. In fact we have kept relatively distant from most of the physical and chemical processes in the life of the cell. But the genome text and its meaning are closely connected with an architecture, which is real even if it is minuscule. One consequence of the domination of biology by biochemistry, which favors the study of objects in isolation, has been to encourage an image of the cell as a miniature test tube. In this view, the concentration of molecules is seen as uniform, and the standard thermodynamic approach is normally used to measure the course of biochemical reactions, as if that were what happened in the cell. But this is very misleading. To avoid the fallacies of Bacon's Idols, we need to look afresh at what are taken to be commonsense ideas or well-established facts. It is essential to have a realistic idea of what happens inside cells, even the smallest ones. We need to imagine ourselves smaller than the Lilliputians, and try to describe the internal landscape of a cell.

A typical animal or plant cell is made up of a *membrane* that separates the inside from the outside, a *nucleus,* which contains the chromosomes, and the *cytoplasm,* where the main stages of metabolism take place. The chromosomes are made up of DNA and of proteins that enable the very long strand of DNA to be folded up, but at the same time to maintain its role as a template for gene expression. Bacteria, which are single-celled organisms, are organized in a similar way, except that there is no membrane separating the nucleus from the cytoplasm. To keep things if not as simple as possible, at least with as few compartments as possible, let us try to visualize a bacterium on the molecular scale. Bacteria are barely visible with an optical microscope. Most of them are minute, although some larger species do exist—there is a bacillus called *Bacillus megaterium* (meaning "large animal") because it can be seen in a fair amount of detail using this microscope, and the "Sulfur pearl of Namibia," *Thiomargarita namibiensis,* is almost visible to the naked eye. Many studies involve just one cell, and this fact has allowed particular features of the internal organization of the cell to be highlighted, not just the general properties of life. With the electron microscope, details can be analyzed in a specimen that has been colored and fixed, or frozen and etched after evaporation in a vacuum (and thus necessarily dead).

Every cell is surrounded by a membrane, which is often part of a more complex structure surrounding the cell, given the generic name of *cell envelope.* The envelope adds a whole series of functional and architectural characteristics, which make the cell rigid and resistant to all kinds of attack. In particular they give cells their individual shape. If cell membranes were completely flexible, the difference in the concentration of ions and molecules inside and outside the cell would induce osmotic pressure, which would act like the air in a balloon, making them spherical. Despite several decades of work by tens of thousands of scientists around the world, we still know very little about the fine detail of how cell membranes are organized, apart from the fact that they are made of lipid molecules with a hydrophilic head and a hydrophobic tail. These molecules form a double layer, with proteins between them. Among these are the *receptors* or *sensors,* structures that play a vital role in the cell's recognition of its environment by monitoring the state of that environment. Another important class of proteins in the membrane has the specific role of carrying metabolites between the inside and the outside of the cell. These proteins are called *permeases* or more generally *transporters,* to refer to their "vectorial" action (to transport a molecule in or out of the cell).

The double, fatty layer, which separates the interior from the exterior of the cell, prevents matter from getting in and out, except for molecules that are chemically similar to the ones the membranes are made of (molecules soluble in lipids) or molecules recognized by the permeases. At the same time as it ensures the selective separation between the interior and the exterior, this structure also creates a strong electric field in the membrane. This is due to the electrochemical gradient caused by the difference in the concentration of electrically charged molecules or *ions* inside and outside the cell. We don't think about it very often, but a cell membrane generally represents a huge difference in electric potential between the interior and the exterior, in the order of 100,000 volts per centimeter (a difference of a few thousand volts per centimeter is enough to produce a spark in air, as with a piezo-electric gas lighter for instance).

The organization of the cytoplasm is still not very well understood. In bacteria it is almost unknown, because the detail of living cells, as distinct from fixed ones, is too minute to be seen with the tools currently available. This is why many biochemists often tend to think of the cytoplasm as a microscopic test tube, as if it were a homogeneous medium, whereas in fact it is not homogeneous at all. For instance, the DNA molecule must be folded up very tightly, because it is a thousand times longer than the bacterium itself. In cells with a nucleus, there are large numbers of membranous structures inside the cytoplasm, as well as organelles that manage the use of energy (*mitochondria*) or the production of chemical energy from light energy (*chloroplasts,* found in the cells of the green parts of plants). In reality we are increasingly convinced—we are almost certain, conceptually speaking, because of the geometrical program which is revealed in the genome—that the cytoplasm is highly organized, even in bacteria. The implication is that the way the objects in the cytoplasm interact with each other is harmoniously controlled. This is a powerful evolutionary constraint which needs to be taken into account in considering the way genomes evolve: besides the selective pressure of the exterior environment, the environment within the cell must exercise a significant and powerful selective pressure of its own.

As we go down to a slightly smaller scale than that of the cell as a whole, we can see from a simple calculation that our unfortunate tendency to extrapolate from the biochemical image of the cell, in which the different components are both numerous and homogeneously distributed, produces a very inaccurate picture. The cytoplasm is far from being a simple test tube. Let's consider the acidity of the medium. The acidity of a solution is determined by the number of "free" protons.[7] The solution is an *acid* if the concentration of protons is higher than that in water, and a *base* if it is lower. It is neutral if the concentration is of the order of 10^{-7} mole per liter. When the proton count is used as a measure of acidity, using a logarithmic scale, 10^{-7} mole per liter is expressed as pH7. A concentration of 1 mole per liter corresponds to Avogadro's number, the gigantic number of 6×10^{23} individual molecules to the liter. Neutrality, at 10^{-7} mole per liter, is one ten-millionth of this concentration, but in a liter of water this still indicates a huge number of protons, 60 million billion. This can legitimately be treated as an average, from a statistical point of view.

Things are quite different inside a bacterial cell. For example, a colibacillus has a volume of about 1 μm^3 or 10^{-15} liter, or, if this is any easier to visualize, 1 trillion bacilli in a *milli*liter. At neutrality (pH7), this means $6 \times 10^{16} \times 10^{-15}$, which works out at 60 protons, but as the pH in the cell is estimated to be 7.6, there are only about fifteen protons ($10^{-15} \times 10^{-7.6} \times 6 \times 10^{23}$). This number is much too small to be treated as an average, so it is not legitimate to talk about an intracellular pH.

It is quite clear, then, that it does not make sense to extrapolate from *in vitro* biochemical data, as is unfortunately so often done, if we want to understand what the situation really is within the cell, *in vivo.* There must be some organization that is both structural and dynamic, which guides the flow of protons within the cell and couples it to the generation and recycling of energy. And because the concentration of the macromolecules within the cell, taken all together, is at a high level, they must play an important role as a reservoir and channeling system for the protons (so the usual laws of diffusion are generally not applicable inside the cell). This is just what was observed when the complete genome of baker's yeast was sequenced, and we will come back to it later.

These simple facts remind us that we are still a long way from understanding how a cell functions, despite the work of millions of scientists over half a century. But we do know that the space in the *E. coli* cell is very crowded. The same reasoning we have just applied to protons also applies to the translation components of the cell. Consider the ribosomes, the organelles that decipher the text of messenger RNA, using the genetic code, and are at the heart of gene expression. The volume of a ribosome is the equivalent of a cube measuring 200 Å (20 nanometers) on each side. In bacteria reproducing exponentially in a rich medium there are 15,000 to 20,000 ribosomes, taking up over 15 percent of the cell volume. The volume actually free of ribosomes is in fact much smaller if we take into account the space occupied by the chromosome together with its transcription machinery. And photographs taken with an electron microscope do show the cytoplasm as a ribosome lattice, in which the local diffusion rate of small molecules as well as macromolecules is relatively slow. Along the same lines, the concentration of proteins in the cell is calculated at about 100–200mg/ml, which is very high. Taken together, these observations show that the interior of a cell is more like a stiff gel than a solution.

Water takes up no more than three-fourths of the cell volume, and it is possible to count the water molecules displaced when a DNA-binding protein interacts with a portion of the chromosome at the site it normally recognizes. (A typical case studied experimentally showed that during this interaction 300 molecules were moved, which is not very many.) Similarly, if we count the molecules that cluster around the ribosome, we end up with limited numbers. It becomes clear that we cannot really talk about the "concentration" of this or that molecule. Doing so would suggest a uniform image of the cell environment and imply large numbers of each type of molecule, whereas in fact we need to consider each type of molecule individually, because there are only a few of them around the ribosome. We should regard each ribosome as a "magnet" for a limited pool of tRNA molecules. In this context, the cytoplasm becomes a kind of network of ribosomes, moving slowly in relation to each other. Only the small molecules around them move quickly enough, and are numerous enough, to be described by the normal laws of diffusion. This situation creates a form of selective pressure, resulting in the adaptation of the tRNA pool used by the translated message, depending on its position within the cytoplasm. Looked at from this angle, it is no longer a question of diffusion as in a continuous process, but of the movement of individual molecules around potential targets. The problem has to be considered locally, by trying to visualize what is taking place, rather as if we were in a factory. We need to consider first of all the relative inertia of the various components of the processes in this biological factory if we are to discover how the molecules are organized within the cell. It is important to stress that in this way of looking at things the *macroscopic* laws of thermodynamics are simply not applicable. This undermines all the theories and analogical models inspired by thermodynamics and wrongly applied to biology—as should have been plain enough.

Threads and Knots: The Cell as a Textile Factory

If we look closely, one of the most surprising things about the cell is the DNA molecule folded up inside it. In bacteria this molecule is about one millimeter long, in other words a thousand times longer than the cell itself. Standard polymer theory tells us that, at physiological salt concentration, a random polymer of this length with the same intrinsic rigidity (called its *persistence length*) would spontaneously fold up into a sphere with a diameter of 10μm—ten times that of the cell. So it must be terribly tightly rolled up on itself to fit into a cylinder 0.5 to 0.8 micron in diameter and only one micron long! There must be superordered DNA structures to account for the DNA packaging in the cell, including supercoiling, organization into domains, and attachment to specific sites. Are these physical constraints reflected in the genome sequence? Preliminary studies with the yeast genome suggested that such structures do indeed exist in this organism. We should note here that the need to pack DNA into a small compartment is a strong selective pressure, which explains the existence of a structure such as a nucleus: this limits considerably the number of states available to the molecule, and enables its behavior to be organized. Thus the number of degrees of freedom offered to DNA increases when the compartment grows (the cell or the nucleus). As a consequence, replicating DNA has a spontaneous tendency to occupy the new space offered by cell growth, creating a natural process for DNA segregation into the two daughter cells.

DNA is certainly a highly organized structure, but it is hard to see this with electron microscopy, because the methods used to make the molecule visible against background noise interfere with its structure. Work by Abraham Worcel and colleagues at the beginning of the 1970s showed that the chromosome of *E. coli* is made up of between 50 and 80 very large loops associated with a network of proteins and perhaps with the cell membrane. In cells with a nucleus, the chromosome is coiled up around a core of proteins called histones, like a string of pearls carefully rolled around itself, creating a superordinate structure. The double helix of DNA appears to be subject to precise architectural constraints, which make it fold in a way that allows transcription and replication to take place without creating too many contradictory forces. During replication, the replication fork, where the enzyme system that copies each of the strands of the double helix is located, has to deal with this preexisting, superordinate DNA structure, and has to replicate this at the same time as it replicates the two DNA strands. It also has to work alongside a correction mechanism that deals with the inevitable replication errors, so as to minimize variations in the text as it is copied. Bearing in mind that the strand of DNA must copy itself with each generation—about twenty minutes in a culture of *E. coli* growing exponentially—replication, transcription, and translation have to be meticulously organized.

The DNA molecule, which is the repository of the text of several thousand genes, continuously transcribes local regions in the form of messenger RNA molecules (more than a thousand of them, normally, and this represents only a part of gene expression). If we measure these molecules, it is clear that each messenger is about the same length as the cell. Of course these may be folded up, but their very function—to be translated by ribosomes—implies that they must be fairly well spread out while translation is going on. So the cell has to deal with a strand folded back on itself a thousand times, producing thousands more strands each as long as the cell. How do they manage not to form an inextricable tangle of knots within the cell? In the vast majority of cases, the mRNA and proteins produced by transcription and translation are correctly formed. The main folding problem with long polymers such as DNA or RNA is that they have a great many possible states, and if they were free to diffuse, this would be incompatible with any organization of the cell architecture. In fact, even if they diffused through an organized lattice (such as the ribosome lattice), a set of freely moving long polymers would still rapidly tangle into an unsortable bulk of knotted structures. There is a way out, however. Anchoring points provide a very efficient way of drastically lowering the number of states available for polymer conformations. A single anchoring point, as is assumed in the standard models of transcription, would already severely restrict the number of explored states. This would limit the formation of knots, but as can be seen with what happens to long hair that is not combed, it still might not be enough to reduce sufficiently the number of states that transcripts would be able to explore. It is well established that *two* anchoring points, instead of only one, would strictly limit the exploration of possible states, reducing it to a manageable number. This might be achieved by the mRNA molecule folding back as a loop with its 5′ end binding to the RNA polymerase complex, as we shall see in a moment.

Several models of DNA transcription, supported by experiment, indicate that under the direction of the network of ribosomes, peristaltic waves run through the folded-up DNA strand, bringing the region being transcribed to the surface of the chromoid (the chromosome and its associated proteins). The transcription enzyme RNA polymerase advances locally, like an inchworm contracting and extending itself by turns, instead of having to go around the DNA. This allows the transcribed strands of mRNA, which are themselves often folded into specific structures (especially near their ends), to insert themselves into the lattice of ribosomes. Electron microscopy images of the translating machinery show the ribosomes spaced out along a messenger RNA molecule in a remarkably regular order. Nascent RNA coming off the DNA is pulled by a first ribosome, which scans for the start codon and begins to translate, and then by the next ribosome. The translation machinery actually pulls it in, mechanically, using a great deal of energy, moving it from one ribosome to the next like a thread in a spinning machine. Most of the cell's inertia and the majority of its energy are in this machinery: although it is energy costly to synthesize mRNA, one mRNA molecule is translated at least twenty times, and energy is used for loading amino acids onto tRNA and elongating the polypeptide chain in the ribosome, as well as for moving the RNA strand. So it is the structure of the ribosome network that organizes the mechanics of gene expression (with a further energy cost involved in proofreading), and it is the coupling between translation and transcription that makes the DNA move, bringing new genes to its surface ready for transcription. The physical organization of the cell has no reason to follow the genetic information flow, which goes from DNA to RNA to protein.

Messenger RNA thus passes from one ribosome to the next, controlling the synthesis of the protein it specifies at each ribosome. During this process one new protein is synthesized on each ribosome of the sequential set, so it is the *linear* diffusion of the mRNA molecule that dictates the distribution of new proteins within the cell. This avoids the difficulties that would arise if diffusion took place in three dimensions—which would be much too slow to ensure the proper organization of the cell (except very locally or for very small molecules). It should be noted that this realistic representation of what happens in the cell is exactly the opposite of the widespread—but wrong—textbook image, in which the ribosomes are represented as moving along an immobile strand of messenger RNA, and not the other way round.

Finally, as soon as an appropriate signal reaches the ribosome at the same time as the translated messenger, a (yet unknown) degradation process is triggered, breaking down the mRNA from its 5′ end and thus ending its expression. A refinement of this model assumes that ribosomes start to translate loop structures, rather than that nascent mRNA molecules are translated from their 5′ end. This alternative view supposes that the 5′ end of the message folds back and remains linked to the RNA polymerase until a specific signal, which can be located way downstream, tells it to detach (and to terminate transcription). Transcription "antitermination," which has been thoroughly investigated in the case of the protein N of bacteriophage lambda, is readily compatible with a scanning process that enables the 5′ end of the RNA to explore what happens in 3′-downstream sequences. However, no clear-cut picture has yet emerged of the events that control these processes. Since it is certainly very difficult at this time to observe the ongoing transcription process in living cells, it will be interesting to look for 5′–3′ correlations in the nucleotide sequences of operons. This hypothesis thus sends us back to *in silico* study of the genome text, once again demonstrating that, in order to understand genomes, it is necessary to go back and forth between the study of the physicochemical aspects of gene expression and the formal study of the genome text. We would expect two distinct fates for transcripts: either they form loops, with the 5′ end scanning the 3′ end until it encounters some termination signal, or the 5′ end folds and forms an RNA-protein complex, with specific binding proteins, shifting away from the RNA polymerase transcribing complex. This would be the case with ribosomal RNA, which associates with ribosomal proteins, but also with complexes such as the 5′ terminal regulator of the transcription control of tRNA synthetase genes in *B. subtilis.*

This brief description, sketchy and partial as it is, is valid for bacteria (or at least for bacteria without a complex network of internal membrane structures, as some seem to have). The description would have to be much more complex in eukaryotes, because of the greater compartmentalization. These cells, which include animal cells, have not only a nucleus, in which transcription takes place, but also a very complicated arrangement of membranes within the cell, which guides the proteins to the precise place where they must carry out their function. Such cells also contain mitochondria, which manage the cell's energy, and chloroplasts in the green parts of plants. Finally, the nucleus is the site of a large-scale transcription process that synthesizes very long precursors of messenger RNA. These have to be shortened by cutting out some internal sections and splicing together the matching cut ends. We must therefore suppose that there are appropriate mechanical structures here, too, to prevent knots from being formed or the end of one messenger from being spliced onto a different messenger. This suggests some structure similar to the distaff on a spinning wheel, but we do not yet know much about it.

Is there any strong argument in support of these scenarios? As they are based on the idea that the messengers translated by the ribosomes are a driving force in cellular dynamics (via translation, which is a process that consumes a great deal of energy, making this idea plausible), we should predict that the messengers are subject to a certain mechanical tension, which would place a strain on the connection between translation and transcription. This leads to a simple prediction: it must sometimes happen that *the messenger strand gets detached from the RNA polymerase, and is not completely finished.* If this is true, what happens? A truncated messenger arriving at the ribosome will begin to be translated as its codons pass through. But if the message is interrupted, the situation is very different from normal, because the ribosome does not find a recognizable stop codon. If this situation really exists, and happens often enough, evolution is sure to have selected an appropriate mechanism that can take this into account. This is in fact all the more important because, as many proteins form part of complex structures, there is a risk that incomplete pieces of protein—exactly what would be synthesized by a truncated messenger—will take the place of the whole protein and prevent the complex from functioning, if only for architectural reasons. What good would a window frame be, with a lintel of half the proper length?

In fact, the cell does have a remarkably clever way of coping with this particular situation. In bacteria, there is a particular RNA, tmRNA, much longer than a tRNA, which can fold itself up like the tRNA that carries the amino acid alanine, so as to be recognized by the appropriate enzyme, which then gives it an alanine just as if it were the normal, specific tRNA for alanine. When a messenger arrives at the point where it is truncated—and thus lacks a stop codon—this tmRNA, loaded with alanine, substitutes itself for the missing part of the messenger, adding an alanine residue to the end of the protein being synthesized. Then, after having behaved as a transfer RNA, it *goes on to behave as a messenger RNA* (this is why it is called tm) and presents ten successive codons to the ribosome. The ribosome translates these in the normal way, adding to the end of the truncated protein ten further amino acids, as specified by the sequence of the tmRNA. At the end of the series of codons there is a stop codon, which enables the ribosome to complete the sequence normally and to free the new protein into the cytoplasm. The consequence of this process is immediately clear: proteins constructed from truncated mRNAs *all end with the same tail of eleven amino acids.* The system is complemented by another cell structure: a protein degradation device that recognizes this special tail and completely breaks down any protein that carries it. The cell is thus freed of these useless, toxic broken pieces.

We can thus see that, as we predicted, it must quite often happen that genes are incompletely transcribed into messenger RNA. This supports the hypothesis that it is the organization of the cell's cytoplasm and its network of ribosomes that provides the mechanical force essential to DNA transcription. One clear consequence of this interpretation is that the position of the genes on the chromosome is important.

Transcription itself also introduces constraints into the DNA structure. Remember that this molecule is closed in on itself and formed of two intertwined strands. There must be a way of releasing this constraint (and there are in fact *DNA gyrases* and *topoisomerases* whose function is to regulate the degree of twist in this intertwining). But it is certain that the frequency of certain modes of transcription must have an effect on the neighboring genes. As DNA is folded back on itself thousands of times it forms a kind of web, folded up into a more compact structure with an inside and an outside. Some genes will thus be more easily transcribed because they are found on the surface, or else there is a mechanical process of waves running through the folded web formed by the DNA, to bring different parts of it to the surface one after another. In particular, for the genes situated on the inside of this folded web, access to the translation machinery must be rare, except at the moment when the replication machinery goes past. We should therefore expect to find that weakly expressed genes are expressed in a cyclical fashion, at the same time as the DNA polymerase goes past. In each case it seems that transcription and translation are closely connected to the structure of DNA and the distribution of genes in the sequence. Of course replication must be added to this, as with each generation the DNA must be copied into two identical molecules.

Despite its simplicity, this view is not very common, and biochemists in particular have been more interested in the *local* dynamics and constraints of the processes involved in gene expression. Similarly, they have almost always preferred experimental approaches in which proteins can be expressed as a *single,* pure product, and they have only rarely taken into account the fact that in reality, the different objects needed to ensure the functioning of the cell are much more likely to form complex structures involving *several* proteins.

In bacteria, genes are organized into operons, bringing together functions that have some points in common. But can we find some more profound organization? The fact that the chromosome is folded in on itself at the heart of the cell rules out finding any obvious local geometrical organization of the genes, apart from those located immediately next to one another. In an operon, as soon as the first protein has been translated, and while it is still close to the ribosome, the second one follows. So it is easy to understand that enzymes made of multiple subunits are often coded in operons, where the genes specifying the various subunits are contiguous. They are distributed in the cell when the message passes from one ribosome to the next, in alignment with each other. We must also consider the ribosomes located at right angles to this line. The products synthesized on these ribosomes are also close to those in the alignment. It is thus quite likely that messengers that specify related functions or are structurally connected must be read by neighboring ribosomes. This could happen if the transcription machinery was organized like a team of draft animals in harness, working together. One prediction of this hypothesis is that, if this is the case, there must be factors whose function would be not to control transcription directly, but to associate several RNA polymerases in a transcription unit, in which several messengers, transcribed from nonadjacent regions of the chromosome, would be transcribed simultaneously. The group of messengers would then be organized as in a textile. The weft threads would correspond to operons, and the position of the warp threads would correspond to functions with features in common, in terms of the cell's structure and dynamics.

There Is a Plan of the Cell in the Chromosome

If this organization exists within the chromosome, how can we find evidence of it? In pathogenic bacteria, it is illustrated by the "pathogenicity islands," where genes related to the function of virulence are grouped together. Eduardo Rocha and Agnieszka Sekowska have also recently demonstrated that genes involved in the metabolism of sulfur also form clusters,[8] suggesting a higher-level organization of the corresponding gene products, probably due to the fact that, sulfur being a highly reactive atom, the gene products that deal with it must be compartmentalized in the cell to protect it from the environment.

The observation that there are several classes of genes that use the codons differently is further evidence of architectural constraints in the cell. It suggests that there has been a permanent, systematic bias over the course of the evolution of the species considered. If codon usage were undifferentiated, the spontaneous mutations that inevitably arise over time would tend to even it out, so that the pattern of use would be proportional to the number of codons that correspond to each amino acid. A strong selective pressure must be necessary for this bias to be maintained. The process of translating mRNA into proteins requires one amino acid to correspond to each codon. Often, thanks to a certain amount of "wobble" in the match between codons and anticodons, a given tRNA can decode several of the codons that specify the same amino acid. For instance, the tRNA for phenylalanine can recognize both UUU and UUC. This capability reduces the number of different tRNAs required to decode the 61 possible codons, but not in all cases. The tRNA that reads the codon CUA does not read the codon CUG very well. So there must be a certain harmonization between the relative concentration of the various tRNAs and the distribution of codons.

Let us consider the ribosomes that translate a messenger whose codon use is very biased. Because they frequently call on the corresponding tRNAs, they act as "attractors" and tend to keep the appropriate tRNAs nearby. Messengers will then be translated quickly, so this raised local concentration will make a bias toward these particular codons an advantage, if the messengers translated have the same bias. We can therefore expect the great majority of the mRNAs translated by these ribosomes to have the same characteristic bias. This selectively reinforces and stabilizes the particular bias, provided that these mRNAs are always *translated at the same place* in the cell. There is thus an architectural connection between the codon usage bias and the site where the messenger RNA is synthesized. Consequently, for distant ribosomes, the availability of tRNAs that correspond to the codon bias will decrease, and there will be a selective advantage if the messengers that are decoded on these ribosomes have a codon bias appropriate to the lower concentrations. Similarly, because of this initial bias, selection will favor other messengers that come to be translated by the attractor ribosomes, if they show the same codon usage. We thus have a local set of ribosomes, which translate messengers that all have the same strong bias in the way they use the genetic code.

One conclusion of this reasoning is that the ribosomes in the cell are not all equivalent. But what is a ribosome anyway? In 2000, their structure was determined—a remarkable technical feat, which was strangely little noticed—and, as Luigi Gorini had guessed in the 1970s (but contrary to current opinion), it was suddenly discovered that they are essentially an enzyme factory, made not from proteins but from RNA. This finding justifies their name, in retrospect. It all looks as if these objects we call ribosomes are the core of a much more complex machine, which includes elongation and initiation factors and the enzymes that fix the amino acids to the tRNAs. In fact the method biochemists use to prepare them is the same as that used to stone cherries: it is done by centrifuge. The stone is forced out and the flesh stays behind. The structures we call ribosomes are the stone, the RNA core of much larger objects that are probably in contact with each other (this would explain the fact that with electron microscopy the ribosomes are seen to be regularly distributed along the length of the messenger, although they do not appear to be in physical contact with each other). During translation, the messenger would thus be like the thread through a necklace of beads. Each bead or ribosome, wherever it is found in the cell, would have the same core, whereas the surrounding body could vary, and correspond to the bias found in the use of the genetic code, which varies with the nature of the genes. And as helical structures are the easiest to make, it is likely that necklaces of ribosomes, organized in helices, form under the membranes and distribute the products of translation through the cell.

The cell is made of multimolecular structures, composed of different proteins, produced from different genes, often assembled around a framework made from RNA. At first the ribosomes were the only known structure of this kind, mainly because they are visible with an electron microscope; then the polymerases, DNA polymerase and RNA polymerase, were found. But numerous other structures are being discovered. The "degradosome" breaks down mRNAs, but there are also complexes made of the chaperone proteins that play a role in protein folding, and all sorts of complexes responsible for electron transfer, nucleotide synthesis, the secretion of proteins across the cell membrane, or energy management. The "proteasome," for instance, is responsible for breaking down proteins that are foreign to the cell or those that have been denatured by heat or by some other phenomenon.

Pascale Guerdoux-Jamet made a striking observation: when a particular function requires the participation of proteins that are found in different cell compartments, from the exterior membrane to the cytoplasm via the periplasm, in *E. coli* for instance, the codon utilization gradient appears to follow the spatial distribution in the cell. Similarly, when the genes correspond to identical enzymatic activities expressed in different environmental contexts (for instance at different temperatures, as is the case with *E. coli,* which can live either inside a mammal or outside), we note that the genes expressed in each of these contexts are grouped together precisely by genetic code usage.

All this can be summed up in one particularly striking statement: *there is a map of the cell in the chromosome.* Genes are not randomly distributed in the genome text; their position relates to their mode of expression, depending on the nature of the environment, and to the location of their products in the different cell compartments. In retrospect it explains certain contextual effects that had been experimentally observed but were still not understood. For example, it had quite often been observed that a gene's expression changed if it was moved to another position in the chromosome, although the elements that regulated its expression—especially its promoter—had been carefully preserved. It also accounts for the difficulty often faced by commercial laboratories when they want to make an organism express a heterologous gene, only to find that its product is expressed only weakly or in an incorrectly folded form.

Analyzing the genome text thus has a lot more to teach us than just the most basic function of each of the genes within it. It also tells us where their products belong. The position of the genes for ribosomal RNA in particular, generally near to the origin of replication, becomes perfectly understandable once we realize that merely because of the space it takes up, the formation of the ribosomes produces a pressure that pushes the daughter chromosomes apart during replication and stretches the cell envelope. There is a strong evolutionary argument supporting an architectural basis for the role of ribosomes in cell organization. Analysis of the genes of the tuberculosis bacillus, *Mycobacterium tuberculosis,* shows that they have a slight bias in their genetic code usage. This is a bacterium that develops extraordinarily slowly (one generation every 24 hours), and, unlike *E. coli* or *B. subtilis,* the genes of its ribosomal RNA are situated at the opposite end of its chromosome from the origin of replication; in fact they are close to the replication terminus.

Clearly, if these semantic constraints on codon usage have some meaning, it is that they create a preferential pathway for the positioning and shaping of proteins. Not only do scaffolding proteins (called molecular chaperones) take part in this, but also the protein is born into a particular environment, so it does not interact with just anything. This implies that the entire local environment has a role to play in its folding and its correct positioning. It also explains the fact that each genome has its own particular style, because the long evolution that has produced its sequence means that each protein "knows" *where it is born.* In a manner of speaking, all its rough edges (as they would be perceived in its normal environment) have been smoothed off during the course of evolution, so that in this environment it goes straight off to interact with the other proteins of the complex to which it belongs, without interacting with any others. An organism's proteins are thus (in formal terms, not in reality) either "spherical," without rough edges, or "sticky," and recognize particular targets.

This raises an important question for genomics: Is it possible to find out, just knowing the genome text, whether a gene product will form a protein complex? This is unlikely, although alignment with model proteins whose structure is known will certainly help to predict a structure, because of the need to take into account the selective forces that have, during the evolution of species (or phylogeny) led to the actual fold found in the models. Combined with the comparison of the many genome texts we now have at hand, this approach (called "threading" because it involves using the "thread" of a polypeptide chain in the mold of the fold of a known protein, and testing whether it fits in terms of physicochemical constraints) is often used, but it should be extended to the study of protein complexes, by taking into consideration the contacts between subunits in the models. In fact, the future of structural biology does not lie in collecting three-dimensional structures of all the proteins of a genome, as is often proposed, but in the identification of protein complexes, another example of the "neighborhood" approach we advocate as a prelude to discovery.

Some Simple Architectural Principles: Helices, Honeycombs, and How Structures Accumulate

Genes are not randomly dispersed along the chromosome of a bacterium such as *E. coli*. This fact is clearly connected to the function of the proteins they code for, and to the cell architecture. But this is all very mysterious: how can we understand what connection there might be between a symbolic text (that of the genes), its products, and an architecture? For a correspondence to exist, there must necessarily have been a *physical* link between these different aspects of reality at some time or other. Somewhere between the gene and its product, and its location in the cell, there must be an addressing process. And if we want to respect the simplicity of Ockham's razor, and avoid multiple hypotheses, we must look for *simple* physicochemical principles behind this. I propose to tackle this by taking the approach of the pre-Socratic philosophers, and by looking at universal constraints—accepting the risk of being too general and too imprecise, but in the hope of finding some new directions to explore. We shall reason by symmetry.

We will look at what happens to the product of a single gene, a protein synthesized on a ribosome. There are two possible situations. Either it does not interact with itself, and the absence of any constraint (we will see that this is what entropy means) can equally well take it anywhere in the cell (it may become stuck to other structures with which it has a certain affinity). Alternatively, it is attracted to itself (repulsion, which is theoretically possible, is in fact rare in biological objects, at least at the molecular level if not at the cellular level). A site on its surface, A, interacts with a site B on a second molecule of the same protein, and binds to it. But this two-molecule unit or dimer now has a free site B on the first subunit and a free site A on the second. As the ribosome continues to synthesize more copies of the same protein, these will bind to the ones already there. This agglomeration is not random; it always occurs at the same particular spot on the protein's surface, which is complementary to the other specific site on the following protein. But these sites are unlikely to be in exactly symmetrical positions on either side of the protein molecule, so generally as A binds to B a slight twist is introduced. A succession of identical twists will form a helical structure. *The helix is thus the most basic shape in biology,* the most frequent and the most commonplace. So far, this is fairly straightforward, and our reasoning has led us to recognize an essential asymmetrically constructed form, which is the basis of all living things.

Evolution has explored all kinds of helical forms. This immediately has one remarkable consequence. Combining different helices can lead to unexpected properties: in particular it enables nature to *measure lengths* very precisely. This has happened in the genesis of the tails of certain viruses (which they use to inject their genetic material into their hosts). A calibrating device or "vernier" is constructed, using the product of just two genes, like this: a first product makes one helix, leaving space on the inside for a second helix *with a different pitch* (the distance between one turn and the next). However, these two pitches, that of the inside helix and that of the outside helix, are still proportionate to each other, so after a certain number of turns they will meet at the same radius on the base circle. This brings construction to an end. Then, via a mechanism connected to the continuing construction of the virus (making its head), the inside helix that served as scaffolding is taken down, releasing the tail, which was formed by the agglomeration of a number of subunits fixed by the ratio between the pitch of the outside helix and that of the scaffolding helix. Similarly, the pitch of the helices of the nucleic acids RNA and DNA is a powerful constraint on other helical structures in objects associated with it. This is very probably how the telomeres, structures found at the ends of linear chromosomes (such as the human chromosome), are reconstructed with a given length by an important enzyme, telomerase. At the end of 2000 at Oxford, Jeffrey Errington discovered a helical structure in bacteria that seems to be directly associated with their general shape.

But evolution is exploration, and through the genetic variation produced by mutations it explores all sorts of different pitches and diameters, which leave internal spaces of different sizes. Among the pitches explored will be a zero pitch, which places all the subunits on the same plane, producing cyclical structures. The geometrical properties of such structures have been studied in detail since antiquity. In particular, they form the faces of the polyhedra studied by Plato, which he describes in his book *Timaeus.* And these polyhedra do indeed exist in a large number of biological structures. Viral capsids often take such forms. A construction of this type (such as an icosahedral virus) has a simple property, demonstrated by the reasoning we have just followed (it is made of a single subunit): if it does not interact with anything in particular, it will tend to explore the entire cell by diffusion. Eventually it will find the cell membrane and get out.

Among all these structures, there is one with a remarkable property: the hexagon. Regular hexagons necessarily fit together, like tiles, to form a plane like a flat honeycomb. Although they can also fit together to make tubes, they cannot make any other shape. Imagine what happens to a piece of hexagonal tiling being synthesized on a ribosome. Unless there is some specific ad hoc constraint, this flat piece of tiling cannot just stay where it is. It will tend to drift away from the place where it was made, following the internal movements of the cell, or an electrostatic gradient, or any other form of diffusive movement, and generally speaking will keep moving until it meets an obstacle. The first obstacle it meets is a structure that is itself flat, locally at least, and is thus ideally suited to interact with a piece of tiling—the cell membrane. Will the tiling bounce off again into the cell interior? In most cases, definitely not, because in water, a flat structure such as a piece of tiling limits the number of positions and orientations that the water molecules around it can take up. Its presence in the middle of the solution is thus a powerful constraint, from the point of view of entropy. If, moving around by diffusion, it meets another flat surface, the water molecules lying across the interface between the two planes will be pushed away, into the surrounding solution, and will thus take up a very large number of new positions and states. The Second Law of Thermodynamics will thus *encourage* the accumulation of these structures. We can expect that all hexagonal structures, or any other more complex structure that forms a tiled plane, will quickly become stuck against the cell membrane. Furthermore, while a subunit of the piece of hexagonal tiling is still being synthesized, and is thus still attached to the ribosome, the tendency of the completed part to move toward the membrane will draw the ribosome itself gradually after it. Despite the short-term, local tension between the ribosome and the DNA, mediated by the translated messenger RNA, the long-term effect of this cumulative, global, entropy-driven process is to bring the network of ribosomes into association with the membrane.

This happens without the need for any specific characteristic of the electrical charge in the cell (although there may be a mild electrostatic effect, which channels the pieces of tiling toward the membrane more quickly), or any interaction with the lipid membrane, to account for this original architectural property. The principle of the interaction is simply that flat surfaces accumulate easily, merely because they are flat. In fact, in comparison with other geometrical structures, a plane has the greatest surface-to-volume ratio, and makes the greatest entropic contribution when it sticks to another plane. We can thus easily imagine that this extremely simple physical principle—which uses the propensity of things, the Chinese *shi,* to go along with an increase in entropy—could have the power to guide the construction of the cell's architecture. It is still too early to be sure that it works like this, but the first results obtained from studying the structure of the *E. coli* genome indicate that genes whose products form hexagonally tiled planes are distributed at regular intervals along the chromosome.

The ubiquitous presence of membrane structures is thus a distinctive feature of living organisms. In fact a general "strategy" of evolution has been either to compartmentalize the cell with a single—albeit sometimes very complex—envelope, made of a lipid bilayer, or to multiply membranes and skins. *A posteriori,* a membrane corresponds to an efficient way of using the natural tendency of things to increase their entropy. Liquid water is a highly organized fluid, with a natural, built-in tendency for its molecules to occupy as many spatial and energy states as possible. This causes other molecules in contact with it to separate out, according to their properties. Hydrophilic molecules interact easily and blend with the water, whereas hydrophobic molecules, which require energy to interact with water, are "squeezed out" of the mixture, leaving room for water to occupy as many places and states as possible (this is what happens to oil and vinegar in a salad dressing: shaking the mixture provides energy, enabling mixing to occur, but after a while they separate out again). Entropy increase, the Second Law of Thermodynamics, is consequently the driving force for the construction of many biological structures, by grouping together objects with similar properties. This physical parameter not only is at the root of the universal formation of helices: it also drives the folding of proteins and the formation of viral capsids, it organizes membranes into bilayers and creates more complex biological structures.

The Role of Time in the Life of the Cell

Up to now I have spoken only of the spatial organization of the cell, and of its very probable strong connection with the spatial organization of the genome. But of course we must add the time dimension to this. It goes without saying that the way the genome text is organized allows time to be divided into periods, which can (and must) have some meaning. It takes a certain amount of time to transcribe or to translate a gene. Roughly speaking, in bacteria at room temperature, this happens at a rate of about fifty nucleotides a second. This represents a transcription time for a typical gene of between about twenty seconds and a few minutes. But for a long gene (particularly with genes that have long introns) it can take ten or twenty minutes, or even hours. Translation follows transcription (in bacteria they are usually coupled together) and works at about a third of this rate, implying a minimum of about a minute to produce the amino acid chain in a protein. To this must be added the time required for it to fold and to move into the appropriate compartment, and generally for it to associate with other proteins. Once again the minimum time required is counted in minutes. Clearly, adding a section to be transcribed introduces a timing element, which can have an important effect on the cell's dynamics, *simply because of its length,* without the corresponding nucleotides' necessarily having any particular meaning. This is another reason why the distribution of genes along the chromosome is not random, but well organized. It is probably a function of introns, too, especially if they are long: they enable the gene's products to appear at specific times. Calculation shows that genes one megabase long (these do exist, especially in the immune system) will take about twenty-four hours to be transcribed. Given that such timer genes exist, we can predict that it is essentially their length that is important, not the sequence of their nucleotides. In that case, comparison of related genomes should reveal regions where *the length is preserved,* although the sequence is not.

It also throws doubt on the idea that certain DNA sequences for which no function has been found are totally inactive. It is often stated, particularly in respect of eukaryotic genomes, that a large part of the DNA has no meaning, and is just "junk" DNA. There are reasons to doubt this. Imagine the following scenario, for instance: on a circular flat surface sit two cubes, a small one and a large one, and the small cube has a ball on top. Then answer the question:

> What does the smaller cube the round support supports support?
> The answer is of course *a ball.*

This shows that even repeated words can have a meaning! So even when DNA segments correspond to repeated sequences or to mobile elements that can cross from one genome to another, this does not imply that they are meaningless. In other cases, they may function as synchronization or timing devices (they may also be *spacers,* with a role in the architecture of the genes within the chromosome). Of course the invasion of a genome by a repeating sequence may have a purely arbitrary, inactive character *a priori,* and may be the result of an accident at the time of its arrival in the chromosome. But *a posteriori,* it is unlikely that the distribution of these sequences in the genome text will *remain* random. It must respect the constraints that preserve the harmony of the cell's architecture and dynamics. Only sequences that are not incompatible with the life of the cell are preserved, and this introduces a considerable selective factor, although it is not easy to assess how strong this is, knowing only the genome text. But, over the course of evolution, any new sequence that respects these constraints is ready and waiting to adopt a new function, merely because it is there. Furthermore, precisely because the original invasion was arbitrary, there is a good chance that when it appears, the new function will have the symbolic character that is essential to the definition of life. *A priori,* the sequence has no connection with the function it will come to code for. This implies that it will be a hidden, very high-level function, one which it will be hard to discover. This is a task for the future!

Comparing Genomes: Back to the Origins of Life

Just as with the Rosetta Stone, comparing genomes with each other has enabled us to understand the meaning of many genes and their products. But the similarities we observe are more than just that: they have a great deal to teach us about the evolution of genes and genomes. Finding similarities enables us to go back toward the past and discover a little about the origins of life. We can easily tell which genes in a genome make the heart of life: the "housekeeping" genes, which every cell needs, and the specific genes that give it a preferred environment. To identify the first of these, we will go back to the smallest known genomes, so as to discover the functions needed for a minimal definition of life.

Mycoplasms and Minimal Cells

Mycoplasma genitalium has (at this time) the smallest known genome of any autonomous organism. It contains genes for the essential functions of replicating the genome, transcribing it into messenger RNA, and translating it into proteins. Besides these, there are functions that enable all the metabolites required for making genomic DNA, RNA, and proteins to be transported through the membrane, and others responsible for some of the interconversions of intermediary metabolism. In fact this bacterium has the simplest possible metabolism. Apart from macromolecular synthesis, there are a very few intermediary metabolic syntheses. Control elements appear to be practically nonexistent. We seem to be dealing with a set of genes with no hierarchical coordination apart from the flow of information from DNA to proteins. All this represents about 500 kb of the genome. The other 80 kb code for proteins the bacterium needs in order to occupy its habitat, the genital epithelium. It would seem difficult to reduce the genome much beyond this, if the organism is still to be autonomous rather than a parasite, so we can consider that a cell needs 300 to 400 kb of DNA to be viable. This is so small that it is almost the size of a virus genome, such as the smallpox virus or the cytomegalovirus (and a virus is a pure parasite, rather than a true living organism).

Reasoning brings us to a figure of around the same size. However, contrary to what clever marketing would have us believe, it is extremely unlikely that this is a model of ancestral life. Indeed we have to admire the persuasiveness of the propaganda that has led some people to believe there are *ethical* problems in wanting to try to create even smaller genomes from this one, or even in recreating life.[9] This is obviously not like creating Frankenstein's monster, as some have claimed; it is simply a continuation of a normal evolutionary process, toward a cell that becomes extremely vulnerable through the loss of functions. Indeed, if absence of control is to be viable, genes and their products need to be perfectly suited to each other, and for this to have arranged itself *de novo,* the organism would need to be clairvoyant. *Mycoplasma genitalium* must therefore be a *degenerate* organism, and therefore very highly evolved, and certainly a descendant of a much more complex organism. It is in no way a minimal model of what life was when it first began, and even less is it an organism that would escape from its maker's control!

Once life exists, there is simple catalogue of functions that have to be provided. Most of these are involved in the machinery for replication and gene expression. Next, interactions with the environment and the general flow of metabolism have to be taken care of. Replication requires a DNA polymerase, and associated proteins that recognize the origin of replication and join the interlaced strands together again once it has finished. As replication is directional, it must be discontinuous on one strand if it is continuous on the other, and this requires a system, a ligase, to enable the disconnected segments to be joined together on a strand. All this takes at least 20 proteins (and thus 20 genes). Transcription requires an RNA polymerase and at least one factor to recognize promoters in the DNA. It probably also needs a factor to indicate the end of transcription. Finally, appropriate factors enable the geometrical constraints produced by transcription to be resolved. This represents at least 10 proteins. Translation is more complex. Ribosomes are made from three kinds of RNA and at least 50 proteins, to which we must add the 20 enzymes that load the transfer RNA (and the redundancy of the genetic code means there are unlikely to be fewer than 24 of these). Factors are also needed to start translation, to elongate the peptide chain, and to terminate translation. It would also seem useful, if not absolutely necessary, to have a factor to couple transcription with translation, so as to prevent the messenger RNA from breaking too often during transcription as it is pulled by the ribosome reading it. This means more than 100 proteins, plus about 30 RNA genes. In all, the information transfer circuit alone requires more than 150 genes.

But the import and export of essential metabolites also needs to be managed, as does the energy flow. At least 10 fairly nonspecific permeases are needed to manage the import and export of the molecules used to build the cell (as in our hypothesis they are not made inside the cell). As for energy management, the nucleotides (4 for RNA and 4 for DNA) have to be converted from their low-energy form to their high-energy form. This takes at least 15 different proteins. Messenger RNA and truncated or incorrectly folded proteins also have to be degraded. Molecular chaperones are needed to help proteins fold correctly. This takes a couple of dozen genes. These processes have to be linked to the transport of the nucleotides and the maintenance of the cell's electrical balance. Finally, the membrane must be constructed and its integrity maintained; another 10 proteins at least. All this can hardly come to less than 200 genes. So we cannot expect to discover autonomous organisms with genomes much smaller than that of the first mycoplasm.

Other bacteria with small genomes, such as *Rickettsia prowasekii* (the typhus agent) and several *Chlamydiae* (which cause many important diseases), cannot survive outside cells, and the typology of their general functions shows that this is more or less the borderline between an autonomous (though fussy) organism, and an organism that needs the life of another cell to support its own life. *Rickettsiae* look very much like the bacterial ancestors of mitochondria, on their way to becoming cell organelles (at which point the genome size of these degenerate bacteria is reduced to that of small viruses; 16 kb in the case of human mitochondria).

Minimal Life: A Look at Some Bacteria that Transform Minerals into Living Matter

Fewer than 500 kilobases seem to be enough to define the minimal genome required to sustain life in an environment containing everything needed for cell construction—amino acids, nucleic bases, polyamines—in a manner of speaking an environment that has already been enriched by the presence of life. Is much more needed in an entirely mineral environment? One would have thought so. After all, now not only does the machinery for general reproduction need to be put together, and we have seen that this requires several hundred thousand bases, but also machinery capable of building all the molecules the organism needs, starting from bare minerals. This means making not just amino acids and nucleotides, but also lipids for the membranes, and especially the cofactors responsible for all the catalyses required for life.

The organism must build everything we have described as found in the general cell metabolism, some 700 or 800 small molecules. In each case this implies the interconversion of one molecule into another, and therefore one enzyme each time, so 700 to 800 genes. Together with the 400 kb needed for the minimal cell, this gives a total of about 1.2 Mb. But would it not need much more? Results obtained from the various genome programs dealing with organisms that live in purely mineral environments do seem to show that this minimal limit can be reached. Apart from a set of genes needed to control gene expression (about 10 percent of the genes is enough for this purpose), not much more is needed. Indeed both archaea and bacteria living in entirely mineral environments (and able to survive and multiply using just a few metal ions and some phosphate, water, and gases: nitrogen, carbon dioxide, hydrogen sulfide, hydrogen, or methane) have a very compact genome, not much longer than 1.2 Mb.

This is obviously very small—only three times the minimum for life—and we shall see in the next chapter why this is not a paradox. In fact there is no direct connection between the quantitative and the qualitative. It is not the number of genes that is important, despite the endless empty speculation on the number of human genes (whether in amazement that there are so few or, on the contrary, saying that there is no difference between humans and other animals); it is the relationships between them that count. Humans and chimpanzees cannot be reduced to each other, but there is no doubt that they have (almost) the same genes. The simplest way to imagine the beginning of life is that organisms were formed by adding genes on, building up to a large population in which horizontal genetic transfer was common. There would still have been no special relationships and no control. Then, genes began to make increasingly specific products and optimized their relationships, allowing genomes to become smaller, and true species to form. Since then, an endless conflict between increasing and contracting the size of genomes has produced organisms as we now know them. Today's DNA is the result of large numbers of multiplications of the same set of genes, followed by variations and reductions. Above all, we must not forget that this has taken three and a half billion years of evolution. Despite everything, this shows that it does make sense to explore the functions of life by beginning with sets of genes as small as this. We now have increasing numbers of genome texts available, and by comparing them we can improve our understanding of the way they are organized. But once again, because of the accumulated corpus of knowledge about them, it will be the model genomes that have the most to teach us.

Some Model Bacteria: *Escherichia coli, Bacillus subtilis, Synechocystis* PCC 6803, and Others

After the whole genome sequence of *Synechocystis* was known in 1996, those of *B. subtilis* and *E. coli* were published almost at the same time, in the late summer and fall of 1997. This was an important source of comparison between genomes, and it enabled scientists to begin to explore their functions, first *in silico* and then *in vivo.* To begin with, these bacteria seem to have little more in common than just over a thousand genes (which corresponds to the minimum genome we have just seen). Beyond this, they have become specialized to suit the particular environment in which they live, and this enables them to be clearly distinguished: *E. coli* lives in the intestines of mammals, and *B. subtilis* on the surface of leaves. To judge by the genes found in its genome, and some experiments suggested by these findings, the *B. subtilis* genome text seems to indicate that, in line with its name "hay bacterium" in Japanese and Polish, it is adapted to life on the leaves of a few low-growing plants. As it is also found on the ground, *B. subtilis* is often described primarily as a soil bacterium. Comparing the genomes demonstrates that there is a clear separation between a core genome, which corresponds to the everyday life of the cell, and the set of genes needed for its survival in a particular habitat (which ecologists call a *biotope*). Through comparative studies, we can discover in ever-increasing detail both what makes the core of life, and what makes it specifically adapted to a particular biotope. In the latter case, it reveals functions that are original and, more often than not, quite unexpected.

In the general organization of their genes, these model bacteria have very little in common. But analysis of the distribution of related functions reinforces the broad outlines of the idea that there is a map of the cell in the chromosome. We have also already seen that, each in its own way, both these bacteria have been heavily influenced by horizontal genetic transfer, indicating that the exchange of the genes in the environment plays an important role. Careful study of the words in the DNA shows that *E. coli* and *B. subtilis* react differently to the presence of foreign DNA. With *E. coli,* large sections of DNA are introduced into the chromosome and can often remain where they were inserted, even when they are almost identical. In contrast, with

B. subtilis, as with *H. influenzae* (both of which are suitable for transformation by exogenous DNA), it seems that the influence of a recombination process is important in building and maintaining the genome, more so than the errors of DNA polymerase. There are very few exact repetitions, and those that do exist are close to each other. Comparison with *Synechocystis* shows an even greater divergence: apart from the operons responsible for the synthesis of the transcription machinery and the ribosomes, and the operon of eleven genes that specify the synthesis of the enzyme ATP synthase, which regenerates the cell's energy (and in *Synechocystis* this operon is separated into two pieces), hardly anything in the genetic organization of these organisms is the same. Understanding what is so different in the cell structure itself in this bacterium, whose ancestors led to the chloroplasts that fix carbon dioxide and produce oxygen in plants, will be very rewarding. The management of these two gases is almost certainly the main organizing force that has led to the cell architecture we know. It is noticeable that the genes whose organization is best preserved from one bacterial species to another are those for the cell's motors. Situated within the membrane's double lipid layer, ATP synthase is in fact the smallest motor known, with a rotor and a stator. It is about 10 nanometers across, and if a fluorescent marker is placed on one blade of the rotor, it can actually be seen under the microscope to rotate in one direction when protons are provided on one side of the membrane, and in the other direction when protons are supplied on the other side. Is it because inventing the wheel is so exceptional, and requires such great precision, that its parts are still coded for by neighboring genes today?

Model Genomes and Unknown Genomes: The Case of the Human Genome

The first eukaryotic genome known was that of brewer's yeast. This reference organism is studied by more than 5,000 scientists working worldwide, and the functions of a large proportion of its genes are well understood. It is a typical eukaryote, with its nucleus, its mitochondria to manage energy, and its network of reticulum membranes, vacuoles, and lysosomes. Mitochondria have their own (small) genome, but only a minute proportion of the genes they need are coded there, most of the mitochondrial proteins and RNAs being synthesized in the nucleus and transported through the organelle's complex double membrane to their proper location. The yeast model is therefore particularly valuable in investigating the interaction between the nucleus and the mitochondria, a significant problem in animals, where a great many genetic diseases involve a mitochondrial partner. Finally, the complicated network of intracellular membranes, typical of eukaryotes, is quite similar in yeast and in mammals, and studying yeast could well teach us a great deal about their roles and functions. Yeast multiplies easily, grows fast, and is harmless (except for the large quantities of alcohol it secretes, tempting to those who find it hard to resist drinking too much beer . . .).

Even leaving aside the obvious ethical questions that would be bound to arise, managed genetic studies are clearly impossible with an organism that has one generation every twenty-five years. A great many human genetic mutations are known, but this is only because the human population is very large, and above all because many individuals who could not survive the "natural" conditions in which other animals live are spared, thanks to our social organization, our use of tools and language, and our very significant mutual support behavior. However, it is very difficult to study these mutants, especially as no gene functions in isolation, because, as we have seen, gene products are usually members of complexes that bring together several proteins. Any geneticist would begin by creating *isogenic* models of the mutation, in other words a set of reference individuals, with the same unmodified, wild-type gene, and individuals with the mutated gene, but with all the rest of the genome *identical* with that of the genetic reference group. Clearly, this is impossible with humans. Except in the simplest cases (metabolic diseases, where one particular enzyme is missing, as with phenylketonuria, which is detected at birth in Western countries and treated with a special diet), models have to be used before reasonable hypotheses can be made, and then these can be tested directly. Remarkably, however, despite what is generally thought, it has often been easier to experiment with humans than with animals, especially in the field of reproduction, and in many areas connected with health. This is indicative of a recent global shift in ethical principles, in that what is now considered to be ethical is simply whatever makes some kind of profit.

Two or three hundred human genes had been characterized at the turn of this century as corresponding to hereditary diseases (around five thousand of these are known, linked to an appreciable percentage of human genes). If we compare them with yeast genes, we find that about a quarter of them have a definite and very similar analog in yeast. As far as we can tell, there are probably far fewer human genes than was first imagined on the basis of the early sequences. There are probably between 30,000 and 40,000 human genes, only five or six times more than in yeast, making the study of this model even more interesting. With genes that code for some enzymatic activity, comparison with yeast genes often gives an initial idea of their function, but it rarely tells us the conditions under which they are expressed (in terms of time, and in the roughly 250 different mammalian cell types). In other cases it is more difficult to understand the function, which is often a regulatory one, and difficult to extrapolate from one organism to another. As more genes for human diseases are discovered, this proportion will certainly be confirmed, and the study of yeast will make a substantial contribution to human genetics. But we will need many other models—and there are even bacterial genes that have a human counterpart implicated in various diseases—not only to help discover the function of the genes involved, but also to propose treatments to reestablish the normal role. This ought to be taken for granted, but unfortunately the opposite is often stated, giving false hopes to many patients by suggesting that just knowing the sequence of the human genome will in itself enable us not only to understand but to cure diseases.

It was already known by the mid-1980s that the gene for cystic fibrosis, a common genetic disease that causes seriously disabling respiratory and digestive troubles, resembled a gene in yeast. These two genes were not functionally similar, but they belonged to the same family of proteins, called ABC permeases, which selectively transport certain substrates across membranes, using energy from ATP, the nucleotide that is the universal energy donor in all living organisms (ABC stands for "ATP binding cassette"). This meant that yeast could be used as a laboratory model to study the functioning of the human gene. Its product enables the cell to set up its electric potential and to control the transport of all sorts of other electrically charged metabolites. Ten years later, another relationship was discovered, between the gene (ATM) for a hereditary ataxia *(ataxia telangiectasia)* and a yeast gene (TEL1) that is important in the control of the *cell cycle* (the reproductive cycle of the cell) and enables the cell to multiply in optimal conditions. This was particularly important because this genetic disease predisposes to cancer. But of course these discoveries do not mean that we are close to finding a cure! Finally, more recently, the yeast protein analog to PrP has been discovered (this is the protein that causes prions to form). As in mammals, in yeast it exists in two forms, a soluble form, and an aggregating form that acts as a catalyst in just the same way, converting the soluble form into prion form. Here again is a model that enables the mysterious behavior of this protein to be explored, and perhaps this will help to find a way to prevent or cure the disease.

However, there are two interrelated levels of meaning in plant and animal genomes. In this book, I concentrate on the cellular level, at which there is a very direct link with what life is (and where yeast can be a useful model). Beyond this, there is the more integrated level of the organism, at which the base elements are no longer molecules or atoms, but the cell itself. The genetic program is integrated with these two levels on several different scales, making the exploration of gene function in plants and animals much more difficult than in an isolated cell. In particular the semiotic or symbolic character of genes and their products, the absence of a straightforward relationship between structure and function, becomes increasingly common at the highest levels of integration. We can no longer simply substitute the study of yeast for the study of man, whether it is brewer's yeast, baker's yeast, or another yeast, *Schizosaccharomyces pombe,* which is used in tropical countries in similar roles, and which in many ways resembles human cells much more closely than *S. cerevisiae* (especially in the control of the cell cycle). We can learn a great deal about the human *cell* through the use of these models—and we should add certain bacteria, which are good models of mitochondria, not only because of genes similar to those in the mitochondrial chromosome, but also because of the genes in the nucleus that code for mitochondrial proteins—but this will not be enough if we want to learn anything about humans themselves. We will need models of multicellular organisms if we want to go any further. We have seen the unexpected kinship between vertebrates and *Drosophila,* but we will soon need to get closer to man than that, and the mouse genome sequencing program is an essential part of the Human Genome Project.

Of course the mouse is not man (nor the cow, nor the sheep), and it is not even a very good model, as rodent evolution is not typical of mammalian evolution; but with mice it is possible to replace genes *in situ* with genes modified *in vitro,* enabling us to understand how the move from the cell level to the level of organization of a mammal operates. The essential concept of function will become more closely defined, and gradually better understood, with the help of these *transgenic* mice. This *reverse genetics,* to give it its usual name, will give us a means of getting close to the way genes interact with one another within a complex organism.

Here is an example that shows the contribution that human genome sequencing can make to medicine. At the time when Craig Venter and TIGR were publishing tens of thousands of human cDNA sequences, most of the pharmaceutical companies decided to analyze these as quickly as possible. Their hope was that this would enable them to discover potential targets for therapeutic agents or for diagnosis (especially for prognostic study of the possible effects of drugs, including undesirable side effects, on given individuals—what is now called pharmacogenomics). While the debate about patenting raged—and I will discuss this briefly in the epilogue—Merck decided to publish large quantities of its sequences every day, so as to make as many as possible available to the combined wisdom of the world's scientists, and to limit the possibilities for taking out untimely patents. Because of the way cDNA is obtained by recopying RNA messengers extracted from different cell types, their length is heavily biased (it is limited to a few hundred bases), as is their frequency, because the rarest messengers often slip through the analysis, despite the use of clever enrichment techniques. Nevertheless, by 2001, millions of sequences were in the data banks and available for study.

The sequences contain errors, because more often than not they were obtained from a single experiment. However, using the genetic code and a few correction rules, they can be translated into protein fragments. Searching the data banks finds their known counterparts, with a greater or lesser degree of similarity. This is how the company Amgen became interested in a gene, TNFβ, which codes for the receptor for a molecule that is important in modulating the activity of the immune system. Amgen explored everything in the data banks that resembled this gene, and found that a fragment of one particular hitherto-unknown protein had a significant similarity with TNFβ. The next step was to "fish" out the messenger RNA with an appropriate probe, and then make the complement of the entire messenger, sequence it, express the protein in a heterologous system, make antibodies for it, in other words to try all imaginable and possible techniques for studying it, up to and including using the fishing technique to find the homologous gene in the mouse, so as to study the function of its product in transgenic mice.

Up to this point, the work is relatively easy and the path well trodden. The first results of *in silico* analysis of the gene's complete product show that although the protein is partly similar to TNFβ, it cannot be a receptor as TNFβ is, because it has no site that would enable it to attach itself to a cell membrane or to couple any signal it might be capable of receiving to a probe present in the cell. Nevertheless it does belong to a family with several known members, most of which are indeed receptors of external signals that allow the cell to trigger processes specific to the expression of its genes. So the analysis is both interesting and disappointing. This product probably is connected to a function worth investigating, but there is no way of knowing what that function is, or even if it is something like the reception of a signal (as is the case with its homologue TNFβ). These preliminary studies are relatively inexpensive, but once transgenic mice are used as models, the cost of the associated experiments climbs rapidly (at the turn of the century, creating a strain of transgenic mice cost around $40,000), especially since many genes important in mice living in the wild do not seem to be critical when the gene is entirely deleted. There is thus a strategic choice to be made at this stage.

The most productive approach is certainly to "construct" transgenic animals, putting them together out of genetic "spare parts." But several different types have to be made: first, mice in which the gene has been totally inactivated, and second, mice in which the gene is overexpressed. Then it is important to have animals in which the gene can be conditionally expressed, by adding an appropriate supplement to the feed. It is also useful to have the gene linked to a reporter gene, which allows its expression pattern in the animal to be traced. With these there is a good chance of observing a phenotype (but once again, this is by no means a certainty). In general, if modifying the gene interferes with the normal development of the embryo, this will show up very early, as it would lead to poorly viable or even nonviable offspring. However, if no problem is encountered at this stage, the gene's expression must be explored in detail in the tissues of the young animal as well as in the adult, and all possible anatomical or cellular alterations identified. One way to do this is to use hybrid genes between the gene of interest and the *E. coli* gene coding for β−galactosidase, because the activity of this enzyme can be monitored in exquisite detail using appropriate substrates that generate colors, such as the deep blue seen in many pictures of genetically modified organisms. Alternatively, the green fluorescent protein (GFP) gene, from the jellyfish *Aequorea victoria,* makes places where the gene is expressed fluoresce when illuminated with ultraviolet light. This allows researchers to find out where in the animal, and at what stage in development, the relevant gene is expressed. In the case of the gene similar to TNFβ, inactivating the gene did not produce any remarkable phenotype. In contrast, its overexpression clearly modified the skeleton of the transgenic mice, which had much stronger bones than those of a normal mouse. Detailed study of the bone metabolism enabled the scientists to show that this gene's product controls the development of osteoclasts, the cells that recycle bone matter. In fact, it has even been possible to show that injecting the pure protein locally inhibits the development of osteoclasts, which are much less numerous after injection. This accounts for the thicker bones, because the bone matter is not reabsorbed as quickly. Amgen naturally explored the properties of this protein further (as well as related ones), as it may be very valuable in the treatment of osteoporosis, for example (a common condition that affects most women after the menopause, sometimes seriously). Of course, development into a pharmaceutical drug involves at least ten years of testing before the product can be safely used, and accepted by the legal authorities. This shows the importance of strategic choices, because of the high cost of the research involved.

This example is enough to show the directions that applications of research into the human genome can take. Although this kind of study is not guaranteed to find interesting new concepts (man is well known to be an ill-suited subject for genetic research), it remains true that it will lead to useful new applications.

4

INFORMATION AND CREATION

Standing as the epigraph to this book is the question asked by the Oracle of Delphi: What is it that makes the boat a boat? The answer was that the boat *is the relationship* between its planks. The physical nature of the planks themselves is fairly incidental, and what counts is the way they relate to one another. We have seen that the physicochemical nature of the nucleotides that make up the genes is also fairly incidental, apart from the fact that it allows them to join together. Again, what counts is the way they relate to one another. On the most basic level, it is the *relationship* between the three nucleotides that gives each codon its meaning and allows it to represent an amino acid. On a higher level, it is the *relationship* the genes have with one another, and with the signals that control them, that gives an organism life. Among these relationships, those to do with heredity are obviously of prime importance. Within every cell, something is passed on from generation to generation, transmitting what common sense calls *information,* by a process that current models of heredity compare to the running of a program, in a sense that is close to the way this word is used in computer science.

This is the heart of what I want to discuss. Understanding life demands more than just being familiar with a finite (though substantial) catalogue of original objects, all of which have their own individuality. When we consider function, a whole set of relationships is added to the object being considered. We need to know what these relationships are, what particular connections link these objects to one another. The information in the genome text describes all of this. The links that can be discovered in the genome text itself are not unlike the association between the text of a book and its meaning. We shall use this to help find a way toward understanding the deep meaning of the genome text, justifying the huge projects that aim at characterizing hundreds and presently thousands of them.

Although a metaphor can be very valuable as an illustration, it does not explain anything. We cannot understand life or the meaning of the genome text by relying on anecdotes and simple images alone, as is often done in superficial accounts. More often than not, the pathways they create in the mind lead in the wrong direction, and rather than helping, they prevent real understanding. So I will now have to ask you to be prepared to make a real conceptual effort. The ground we will have to cross is a minefield. For at least a century, it has been dominated by discussion of the nature of the *order* underlying physical phenomena (and, by extension, biological phenomena), on the basis of various conceptions of the macroscopic world. This approach usually draws directly on intuition, and is consequently shaped by the established ideas of the time. In parallel, the preoccupation with formalization, which has stamped our era since the time of Newton and Leibniz, and which was especially marked during the first half of the twentieth century, culminated in the formal representation of phenomena in terms of mathematical equations. Perhaps because mathematics has played a leading role in scientific thought, taking the place of the arcane texts of the alchemists, these equations have almost become a creed. They have often been used to compel developing ideas, which are inevitably vulnerable to pressure, to submit to general principles, effectively those of an almost political vision of the world. I will therefore have to ask you not just to construct new concepts, but to question established ideas as well.

The effort of abstract thought we will have to make is all the more difficult because, under the influence of established ideas from our history, our language and civilization, we have a natural tendency to *overlay* reality with all sorts of models of what we think it is, without even being aware that we are doing so. This is particularly true with biology, which deals with a kind of knowledge that seems familiar, so we feel that we know it already. We are used to hearing many of these concepts mentioned, just as we are familiar with the maxims in the dictionary of quotations, without necessarily understanding either. In the same unthinking way, both can be used to justify the privileged social position of powerful or respected public figures. Talking about life involves speaking of the order, the complexity of things, and knowing information that seeks to predict the future on the basis of the past. The genetic program is more than just what determines an organism's birth and development; it is the result of its history and that of its ancestors just as much as it is the story of its future. There is nothing innocuous about this; it affects our deepest beliefs and our most profound motives. Think about the feelings that the words *genetic* and *genome* conjure up spontaneously. For some people, the connotations of the first may be negative, no doubt because of the legacy of the monstrous Nazi view of the world. The second has more positive connotations, perhaps because the word *genome* has appeared more recently, in specialized literature, and because it is associated with the idea of "high technology." In this chapter, we will try to start afresh, and explore some established ideas from the beginning.

We will be exploring the formal, general properties of what genome texts tell us.[1] We will consider them in only one light, as a chain of a limited set of chemical motifs, represented by the alphabetic metaphor of a text written in the four symbols A, G, C, and T. This is not just one of those convenient images we must beware of; it is a precise and appropriate metaphor, because it is by manipulating these texts written in four letters that it is possible to move back and forth between living things and the genetically modified organisms (GMOs) that can be created using them as a pattern. We must be clear that this process concerns only a *fraction* of what genomes are. However, it is absolutely not a process of reduction, and if this simplification reveals interesting properties in them, it shows that the biological reality must be even more interesting.

As a basic philosophical rule, we accept here that life stands among normal physicochemical processes. But this does not mean—and we must be clear about this—that we *reduce* life to just these processes. We recognize that life has its own properties, laws which are constructed in a way that is not predictable from knowing just the laws of physics or chemistry, and which thus appear arbitrary, accidental, not inevitable *a priori*, but are obviously necessary *a posteriori*, because everything that followed could not have happened without them. The physicist Niels Bohr remarked on this in an address delivered in 1932 at the opening meeting of the international congress on light therapy in Copenhagen, and published in *Nature* in 1933:

> if we were able to push the analysis of the mechanism of living organisms as far as that of atomic phenomena, we should scarcely expect to find any features differing from the properties of inorganic matter. With this dilemma before us, we must keep in mind,

> however, that the conditions holding for biological and physical researches are not directly comparable, since the necessity of keeping the object of investigation alive imposes a restriction on the former, which finds no counterpart in the latter. Thus, we should doubtless kill an animal if we tried to carry the investigation of its organs so far that we could describe the rôle played by single atoms in vital functions. In every experiment on living organisms, there must remain an uncertainty as regards the physical conditions to which they are subjected, and the idea suggests itself that the minimal freedom we must allow an organism in this respect is just large enough to permit it, so to say, to hide its ultimate secrets from us. On this view, the existence of life must be considered as an elementary fact that cannot be explained, but must be taken as a starting point in biology, in a similar way as the quantum of action, which appears as an irrational element from the point of view of classical mechanical physics, taken together with the existence of the elementary particles, forms the foundation of atomic physics.[2]

The most striking example of what Bohr refers to as "an elementary fact that cannot be explained, but must be taken as a starting point in biology" (and this is much more profound than seeing life as a "vital" principle), is the law usually known as the "genetic code," which some have called the "central dogma" of molecular biology,[3] a surprising name which reveals the almost religious tone that creeps in as soon as we start to think about life. It is absolutely impossible to imagine this law *a priori,* knowing only Schrödinger's equation, the equation that rules the lives of atoms. The genetic code law is *symbolic*: it is similar to the link between a fact and its representation by a symbol. The correspondence between an amino acid in a protein, and the codon that specifies it, is entirely arbitrary. There is nothing in the equation that allows us to *predict* that, among all the chemical structures that are possible, some should specify each other, as one strand of DNA specifies its complementary strand, and even less to predict that, thanks to a complex set of chemical machinery, a correspondence should exist between the sequence of bases in DNA and the sequence of amino acids in a protein. (In order to avoid the temptation of vitalism, we should also note that this apparent powerlessness to predict applies not just to biology but also to much more fundamental things: it has not yet been possible to use Schrödinger's equation to predict the lattice structure of a simple crystal of salt either.)[4] Nor is there anything to say that, somewhere in the universe, there cannot be other relationships, just as compatible with the laws of physics as those we know of, and which might have created organisms that are *formally* similar to living organisms as we know them, but with a completely different chemistry. *A posteriori,* the law of the genetic code nowhere contradicts Schrödinger's equation; indeed, it *scrupulously* obeys all its constraints. But scientists need to know both the law and the equation, to understand what life is. What governs life is therefore absolutely not outside physics—it respects all its laws; but a law such as the genetic code cannot be simply an *automatic* consequence of the laws of physics. This is what I am summarizing when I say that it cannot be *reduced* to physics.

What I have just said has enormous consequences. In fact, many modern thinkers refuse to accept this originality that distinguishes life from other material processes. They want life to be in itself an unavoidable consequence of things. This creates a very strong tendency to attempt to represent life not just as a possible and predictable result, but as an inevitable, logically derived consequence of the laws of physics. This reduction of life to the physicochemical world has culminated in studies which postulate more or less elaborate connections between various dynamics of simple physical systems, and which are summed up by an expression that is as fashionable as it is vague and inappropriate, *self-organization.*[5] By sheer tricks of language and abuse of metaphor, the authors of these studies seek to "explain" life in terms of the complex behavior of oscillating chemical reactions, or the spontaneous appearance of organized structures on different levels.[6] This painfully reductionist attitude completely fails to recognize what is the basis of life, *symbolic abstraction.* The objects that make biological functions happen often have no mechanical relationship with them; they are only their mediator, their *symbol.* What makes the Delphic boat float is the nature of the relationship between its planks, not their physicochemical nature. Whether they are made of oak or pine or aluminum or steel is irrelevant to their function. In the same way, think of two eggs of the same age, one fertilized and the other not. After a while, if we were able to measure the physical and chemical composition of the atmosphere surrounding the two eggs, we might observe a few differences, but nothing significant. The real difference becomes clear when suddenly one egg cracks and out comes a very strangely organized structure, chirping loudly! The entropy of both eggs has certainly increased, and nothing in the laws of physics has been violated, but it is clear that there is something special about the process that has resulted in the birth of a chick.

This "something special" corresponds to concepts such as *control, coding,* or *information,* which appear as the semantic dimension appears, when the genetic program is put into operation. They cannot be *reduced* to the concepts of physics, but obviously that does not imply that they are *incompatible* with them! *Matter, energy,* and *time* account for the mechanics of change in the normal world of physics. Despite the dearest wishes of reductionist thinkers, it remains impossible to explain life in terms of physics alone, be it the physics of chaos, dissipative structures, or complexity. Control and coding are the result of a long history that has selected a set of material systems whose stability over time is demonstrated by the fact that they produce (almost) identical systems, endowed with the properties I describe elsewhere. This history, evolution, is simply an ordering of the world on the basis of a trio of functions, variation, selection, and amplification, which do not normally come into physics and chemistry.

This shows why understanding the principle and orientation of the evolution of living material systems requires a fundamentally new approach, one able to explain the necessary variation associated with physicochemical processes, as well as how evolution selects the most stable ones among them and ensures that they multiply. This is what Charles Darwin understood, with remarkable intuition. Of course, selective theories had in fact existed for a long time. Aristotle discussed them, and in the Middle Ages his critics often referred to selective theories in order to refute them as contradicting the idea of the creation of the natural world by God. During the Renaissance and the centuries that followed, more emphasis was placed on Plato, and the selective line of thought was almost forgotten, although it was sometimes discussed by scientists, and Samuel Pepys refers to it in his famous diaries. Rediscovered in the middle of the nineteenth century, it was formulated in modern terms by Darwin.

The complementarity that exists between the material world produced by physics and the symbolic world produced by natural selection can be explained by the logic underlying the self-reference or recursiveness produced by the genetic code. The laws of physics and natural selection operate as complementary constraints: the laws of physics describe the unchanging part of phenomena, those properties that living organisms cannot in principle dominate or control. The theory of evolution through natural selection seeks to explain the way in which living organisms do, however, progressively improve their control over those laws. The artifacts produced by mankind, whether in order to create or to destroy, are the most striking sign of the degree of control achieved, as well as the limits on that control reached by the evolution of species. In this sense, we rightly think of ourselves as the ultimate product of evolution, because of our social evolution, and because of our language, which is a classic illustration of the symbolic world in which living things go about their business—but even if this is true today, it will certainly not be tomorrow. Prometheus, who was punished by the gods for daring to steal their fire, represents the ultimate fate of all forms of life.

The laws specific to biology are able to exist because of a particular aspect of their role: they do not affect the nature of physical and chemical objects, but govern the relationships that exist between certain objects. These objects have a *meaning,* which is connected to their *function* in the physicochemical processes of life. This gives them an original order of abstraction, quite distinct from what physics tells us: metaphorically speaking it is the type of order found in the plan of a building. However, there is much more to life than this architectural metaphor suggests. In living beings, the plan is not static, frozen in time, unaltered. It is permanently oriented toward the future, because it implements the processes that *express* life, through metabolism, exchange between the inside and the outside, and the use of information in duplicating the hereditary memory and in managing the functioning of the objects that make up cells, by appropriate, specific manipulation. This is what we are summing up when we speak of the *expression of the genetic program.* This space-time plan, this program that links together the material objects of physics in order to compose a living organism, is an abstraction. However, it cannot be regarded as arbitrary or as existing in itself, without the material support of the physicochemical objects of life. The links in question are not just any links; they have original properties which we must try to understand. They are the result of a continuing *selection,* in the normal course of an evolutionary process that can be measured by the survival and existence of the organisms in question. It is these links that we will have to find and account for when we attempt to understand the meaning of genome texts.

There is a mathematical approach that deals with the properties of sequences: *theories of information,* and I use the plural "theories" deliberately, rather than the singular we are more used to seeing. Here we meet what will be the central concept of this chapter, the concept of *information*. As we look at it in greater detail, I will try to bring out some of the particular characteristics of the laws of life, which are prominently displayed in genomes.

A Few Fashionable Words: Information, Order, Complexity, Chaos . . .

At the turn of the millennium, biology is the science of the moment. Like any fashionable subject, it is promoted through the use of vague, woolly words. We need to try to clear away some of the semantic fog surrounding the terms most often used to talk about life. I would like to start with the word *information,* and show how it is connected to the metaphor of the boat, still bearing in mind that this is an illustration, not an explanation. I have already used this word frequently, but without specifying its meaning beyond the everyday, commonsense one. If we look at it closely we soon find that *information* is a difficult word to define, an indefinite concept, and we should beware of assuming that we know precisely what it means. It expresses ideas in everyday life, such as the information we get from the newspaper or the television, and also an idea of value associated with a given piece of knowledge, as in "privileged information" or "insider information," which can give rise to accusations of "insider trading." Then there is the recent meaning associated with "information technology" and "informatics," linking on the one hand the communication of a signal, and on the other hand the meaning that signal carries.

The Many Faces of Information

An entire book could be written about this very rich concept. It is worth looking at some of its different aspects, reflected in various dictionary definitions. Each of these represents one *view* of information, relevant in a given context; none of them gives us the "whole truth" of the word. *The Delphic Boat* was originally written in French, and there are some important differences in the way French and English use the word *information*. Most of the definitions and examples here are drawn from the *Random House College Dictionary* (1991), *Webster's Third International* (1961), or the *Shorter Oxford English Dictionary*.[7] However, where examples in Alain Rey's great French dictionary *Le Grand Robert* throw an extra light on some aspect of the meaning, these have been included as well.[8]

The everyday, intuitive meaning links the concept of information with that of the news we read in the newspaper or see on the television, in terms of both the news itself and the act of communicating it. *Random House* gives "knowledge communicated or received concerning a particular fact or circumstance" and "the act or fact of informing." We also speak about "information services" or an "information bureau," and can say that "the function of a public library is information" (*Webster's*). Generally, information is something we can communicate; it has a source and a destination. Between the two, something can happen to it, and we can speak of suppressing information, or censorship, and of misinformation. In terms of our boat, this view of information is represented by its construction plan—the instructions for building it. It is this sense of information that is relevant when we consider the genetic information that is transmitted from one generation to another—the instructions for building an organism.

In an important, complementary meaning, derived directly from the Latin verb *informare,* and still current in French, but very little used in English, the verb *to inform* also means to *give form* to something. Earlier editions of the *Shorter Oxford English Dictionary* defined this original sense of *to inform* as "to give 'form' or formative principle to," but by 1996 (in the CD-ROM edition) the definition had been changed to "give a formative principle or vital quality to." This change underlines how our perception of the link with "form" is weakening, and in fact the corresponding sense of information as an action ("an endowing with form": *Webster's*) is regarded as obsolete. Aristotle studied this action in depth, and it was often taken up again by medieval commentators. The origin of the concept of information thus goes back at least to Aristotle's discussions on an individual's Being, perceived and described as a connection, a fortuitous relationship between substance, matter and form. Substance is the pure Being, in which we can distinguish the determinable, matter, and the determinant, form. Beings are therefore classified not by their substance but according to the degree of kinship that can be established when studying their form. Ultimately, individual (id)entities are usually distinguished by their specific characteristics of shape or form. This generic aspect of information has slowly faded from our consciousness, but in fact it is not merely of historical interest. In the context of interpreting genomes, it can be seen as the vital aspect of the word, the sense that is closest to that of genetic information when it is *expressed*—when the organism (or the boat) is actually built. The information contained in the genes and in the relationships between them is what "informs" the organism, expressed in the *action of giving the organism its form,* not just in the sense of the *Oxford Dictionary*'s abstract "formative principle or vital quality," but in the very concrete sense of shaping its body. We will see this illustrated at the end of this chapter.

In a very different context, since the invention of the telegraph and the first computers it has been possible to accumulate and transmit information very easily and quickly, whether this is the information in the daily news or simpler facts. *Random House* gives the example of "computer data at any stage of processing, as output, input, storage or transmission." This more recent view of information, in which the meaning of a message is no longer considered, was used in the theory of communication devised by the late Claude Shannon at the end of the 1940s.[9] His theory was a synthesis of the problems of communication and of different interpretations of what information is. Immediately after the appearance of Shannon and Weaver's book, *The Mathematical Theory of Communication,* in 1949, this view of information, shorn of its semantic dimension, became a scientific concept with a very restrictive definition. It was further developed by the physicist Léon Brillouin, and later became known as *the* Theory of Information (the singular implicitly conferring universality). *Webster's* describes it as "a theory that utilizes statistical techniques in dealing with the effect of encoding on the efficiency of processes of signal transmission and of communication between men (as in telecommunication or the printed word) or between men and machines or between machines and machines (as in computing machines)." The French dictionary defines this aspect of information as an "element or system which can be transmitted by a signal or a combination of signals belonging to a common structure," expressed in the form of a linear sequence of symbols. This interpretation has come to be seen by many as the only "scientific" meaning of the word *information*. But reducing information to a sequence of signals, with no meaning other than the order of the succession of signals, gradually empties it of its content. Above all it contradicts the sense normally associated with the word, in which *meaning* is an essential component of information. It is understandable that these two things—on the one hand a theory very strictly limited to a precise objective, the communication of messages through a channel subject to noise or interference; and on the other, how intuition sees the concept of information, as expressed in the everyday use of the word—have become confused; and as Shannon's theory is still very commonly cited when information is discussed, despite its necessarily limited and reductionist character, we will look at it in detail a little later. It is superficially tempting to see genetic information in this way, since the genome text is presented as a series of symbols, but we will see that this tells only a small part of the story, and that this narrow view is far from capable of characterizing what genetic information really is.

The development of electronic computers added further connotations to the word *information.* Familiar as computers are, it is not easy to understand how a machine can connect arithmetical calculation and logic with the manipulation of symbols, illustrated by the use of a word processor to write this book. What obvious connection is there, after all, between writing software for graphics, accounts, or word processing, and the laws of arithmetic and logic? We like to joke about "stupid" computers, but if we think about what they can do these days, especially when they manipulate objects on the screen, remember our habits, and present us with familiar objects in the order we expect them without our even realizing it, all things we take completely for granted—is that really so stupid? We need to analyze the concepts behind this connection, by going back to the broader meaning of information. There is a lot more to this word than our intuitive understanding of it would suggest, so it is not surprising that it can have different meanings, depending on the context in which it is used.

Several other words with imprecise meanings are used in parallel with information in similar contexts, and in order to clarify the conceptual domain we will be exploring, it is worth also considering them briefly. We will need to keep an essential critical principle in mind, that of restricting vocabulary to a minimum and avoiding the use of several different words to say the same thing. We shall look at two frequently used words that are complementary to *information,* and that are important when analyzing a text, such as the genome text: *order* (often associated with the scientific term *entropy*) and *complexity.* And since this is often associated with *chaos* by those who want to sound more "scientific," we shall analyze this term briefly, too.

Order and Disorder

Like the concept of information, the concept of order is imprecise (or rather, to avoid the negative connotations of this adjective, it is a *prospective,* inexhaustible concept, one that refuses to allow itself to be pinned down when we try to define it). Generally, *order* expresses an "intelligible relationship which can be understood between a plurality of terms" *(Robert). Webster's* definition is less abstract: "A condition in which each thing is properly disposed with reference to other things and to its purpose; method or harmonious arrangement." We should note straight away that as with *information,* because of its relationship with the way we understand things, *order* contains an element linked to that special capacity of the human brain we call conscience (illustrated by *Webster's* use of "properly," which is not a neutral word). Even if we try to keep value judgments out of the definition of *order,* to define it we still need to compare two domains, an observable domain and an observed domain. Order can be expressed only through an identifiable relationship between distinct systems: what we see, and what we are expecting to see, whether consciously or unconsciously. Finally, order is also the order of the world: "Conformity or obedience to law or established authority—*to maintain law and order" (Random House).* Our understanding of the concept is consequently heavily influenced by the religious or ethical thinking of the time.

One reason why it is difficult to use these terms scientifically is that they do not relate just to the physical nature of things—despite what people often pretend to believe. They are also affected by the implicit duality that results from the interrelationship between mind and matter, or, if we try to leave this dialogue out of it, between the physical nature of things and what I call their *symbolic* character, for lack of a more appropriate word (one without ethical or aesthetic connotations); that character which (physical) objects sometimes adopt when they establish arbitrary but reproducible relationships between well-identified systems. Discussing such subjects usually implies an underlying reflection (which may or may not be conscious) on the existence or relevance of this duality. The fact that this domain is close to philosophy, and also often to religious beliefs, explains why scientific logic is so easily forgotten in the discussion. It also explains the tendency to make peremptory, dogmatic statements, and the enthusiasm for an approach that is more often than not irrational.

It would be as well, then, to limit ourselves to the most elementary level of the definition of order. The commonsense interpretation of this word is essentially the notion of *regularity* in the arrangement of the individual objects identified. Both scientific thought and the experimental method are based on the search for an "eternal return," first the return of day and night, and of the seasons, and then, once experimentation succeeded simple observation, that of experimental results (an experiment must be "reproducible"). Without regularity it is not possible even to describe phenomena, let alone to explain them. Regularity is not enough in itself to make a phenomenon accessible to science (regularity can be seen in any finite set, merely because it is finite), but it is a necessary condition. Anything that appears irregular appears disordered, and thus evades (scientific) explanation: all we can do is observe it. Fads and fallacies masquerading as science (which are always with us, but seem particularly common in times of crisis, such as between the two world wars, or at the turn of the millennium) always turn out to be based on one-off observations, and to rely on some extraneous factor as an excuse when they cannot be reproduced.

If the word *order* is ambiguous, and hides assumptions alien to its use in science, this is because regularity can be difficult to spot, or because it may be connected to the systematic repetition of some specific feature of the experimental protocol, and not to the phenomenon itself. Besides, we often see regularity where we hoped to see it, projecting our own mental schema onto the phenomenon itself. We have to be particularly sensitive to this aspect, since meaning will have a major role in associating a function with a gene, for example. This state of affairs arises because order has a moral or aesthetic dimension, which we are usually unaware of but can certainly recognize when we consider some of its opposites: disorder, irregularity, chaos, complexity (which we will look at shortly), confusion, tangle, disturbance, anarchy . . . no doubt this explains the way superficial, ambiguous, or cynical thinkers (and also journalists, often) use it to mean anything and everything.

We will come back later to those strangely successful twins, order and entropy, and relate them to the concept of information, because this has important consequences for the way we access the meaning of genome texts. But we can already see that most of these words are so ambiguous that we should insist on their being used only where the context is precisely specified, and where there is a formally established definition of what they mean in that context.

Complexity and Chaos

A particularly striking example of this ambiguity is the word *complexity*, whose mathematical definition is a long way from the commonsense meaning, as we will see. Fittingly, *Webster's* defines *complex* as "so complicated or intricate as to be hard to understand or deal with." In fact this word means two diametrically opposite things. It can mean either that a phenomenon is very rich and complicated because it is *highly organized* but includes multiple elements (so many as to be generally reckoned beyond comprehension); or, on the contrary, that it is very tangled, very *disorganized.*

The second meaning is the one to which the word *chaos* usually refers (although it comes from the name of the Greek god who presided over the birth of the world out of an utter void, not out of a confused state of matter, as the modern sense would seem to suggest). But many of our contemporaries use "chaos" to refer to phenomena such as turbulence, whose order can be clearly seen, even if it is complex. We have all marveled at the harmony of the eddies that form and re-form in a river current. So the concept of chaos is itself unstable: from the obvious disorder of the chaotic piles of fallen rock after a cliff has collapsed (in French such a pile of rocks is actually called a *chaos*), it can become synonymous with a highly organized phenomenon, such as the beautiful fractal patterns similar to the figures described by Benoît Mandelbrot, which are supposed to be the result of chaotic behavior.

The danger of using these words carelessly—and they are not hard to find in studies analyzing genomes[10]—is that as the connotations associated with them are often precisely the opposite of each other, they are frequently used to say diametrically opposed things, often within the same discussion. But perhaps that is the secret of their success: confusion becomes persuasion, when influential people impose their own interpretation, to prevent others from trying to understand things independently, and what is more, they do so in the name of science! Just to complicate everything, mathematicians generally use concrete words to name their concepts, without worrying too much about the connotations associated with them, perhaps because they are frustrated by spending so much time in the realm of the abstract (or perhaps just because they typically have a mischievous sense of humor). Remember the "groups," "rings," and "bodies" of set theory, or the "catastrophes" of differential geometry. As Ivar Ekeland recalls, Hector Sussmann said: "In mathematics names are chosen freely. We can call a self-adjunct operator an elephant. We can call a spectral resolution a trunk. We can then prove that all elephants have trunks. But we cannot claim that the result has anything to do with big gray animals."[11]

We might think it is enough to note this with regret: after all, concepts are misused in lots of ways! But it is intellectually very dangerous to link together entropy, information, order, chaos, and complexity in this way, because under the guise of science, it promotes a pernicious ideology of degradation, and more generally a very negative conception of the way things develop, without making the necessary distinction between the simple *description* of basic physical phenomena and objects, and the *explanation* of the nature of the relationships (often abstract or symbolic) through which life is expressed. I will present a precise, specific use of the word *complexity* later, and will try to lessen the confusion surrounding the word *information* by using it in a much wider sense than the meaning Shannon and his successors gave to it, getting as close as possible to the richness of the intuitive sense that we use when we say, for example, that heredity transmits information from generation to generation. This will allow us to define a little more precisely what we have seen so far, in our discussion of the nature of life.

A Common Confusion: Disorder and Entropy

Science is an aspect of culture which claims to be universal, but it is not entirely free of ideology. This would not matter so much if the vagaries of history had not invested science with an almost religious power. In a world in which science is the new religion, often no proper distinction is made between the way a precise scientific theory uses normal words, and the connotations associated with the everyday use of the same words. This is complicated by the fact that today authority is judged in terms of social recognition (defined in terms of television ratings) rather than on genuine knowledge. This shift has brought with it an unstated, underlying discourse, which has the effect of spreading the consequences of the dominant ideologies in the name of science. These are the ideologies of disorder, the primacy of the individual over the group, dignified by the name "personal freedom," but which in truth is the right of the strongest; chance and indeterminism, and an aesthetics that is defined not in terms of the intrinsic value of a work of art, but merely by the fact that it calls itself a work of art.

A Philosophy of Degradation

In our unsettled time, values that were once certainties are now often challenged, and the idea of disorder, summed up in the metaphor of *noise,* causes (or reawakens) all kinds of evil, by proposing universal degradation as a natural explanation for the behavior of all sorts of things we do not understand. In genetics, the underlying idea of degradation surfaces in eugenics, where the aim is to discriminate among human beings as early as possible, even before conception, to keep it well out of sight. In the context explored in this chapter, where it ought to be possible to represent disorder and noise in a scientific fashion, it is *entropy*—a difficult concept from thermophysics—that has become the success story. It has taken over the esoteric position once occupied by *energy*, which Mesmer exploited so profitably in the nineteenth century with the magic of animal magnetism, and which is still regarded by some as being invested with magical power, for example through physical exercise such as the famous Chinese *qi gong*. All sorts of flourishing coffeeshop philosophies come to the aid of this apparent victory for imprecision and woolly thought, using *the* theory of information to shore up the doctrine of disorder and noise. It has even been said that things necessarily evolve toward increased disorder, as the entropy of an isolated system cannot but increase—unless of course some new principle (always the last resort of mysterious explanations) should oppose it, for instance minimizing the entropy produced when a system changes from one state to another. Or alternatively another simplistic principle, of "order from noise," is suggested as the root and explanatory principle of life. This is so common that it has become a kind of mantra, enabling complacent thinkers to justify their vague ideas without making any effort.[12] But is it not simply ridiculous to reduce life to these vague processes? It is easy to forget the reality of the eggs and the chick we considered earlier. As we shall presently see, entropy is more a measure of exploration than of anything else.

Of course this situation is not the result of a single line of thought. In fact it has its roots in the way our Western identity was constructed during the twentieth century, going back to the creation of industrial society in the preceding century. This is illustrated by the fact that, in the context of the birth of industry, energy was called *work*. The use of this term shows clearly that only the part of energy that could be used to drive machines was then considered meaningful. The rest was regarded as no more than wasted energy, degraded, worthless, even harmful, and it was essential to be able to distinguish explicitly between these two forms. It also demonstrates that it was the human, macroscopic scale that was thought to be an appropriate field of study. We urgently need a new Copernican revolution to make us broaden our interest to other scales!

Clausius invented the concept of entropy in the middle of the nineteenth century, to differentiate between the usable element in energy and the "degraded" element. He established the quantitative relationship that exists between entropy and the temperature of a thermodynamic system, when heat is exchanged between a hot and a cold source. This enabled the total energy in a system to be divided into work, which was usable, and entropy, which was not. One thing leading to another, this concept, which was relevant to the macroscopic world, quickly slid toward the idea of disorder, because it seemed obvious that the degradation of energy could produce only disorder. This approach was implicitly supported by the fact that a new physical principle had been demonstrated, once again in respect of heat engines. This was the *Second Law of Thermodynamics,* which defined the direction of the spatiotemporal evolution of material systems, stating that this evolution could happen in only one direction. Above a minimal level of complexity, a system cannot return to a previous state without consuming energy, and the increase in entropy is the measure of this irreversibility. It is easy to see how the idea of degradation, associated with the idea of irreversibility, soon came to be interpreted as the domination of disorder over order, especially in biological systems, where thermodynamics is particularly difficult to cope with.

Interestingly, this is not the vision we might expect from a world then still dominated by Christianity. Genesis refers to the creation by God of a magnificent order, from a nature that was seen as formless. We cannot go into the consequences of the slow fading of the biblical vision of the world in the West from the end of the eighteenth century onward, but we can note the parallel growth of a pessimistic vision that culminated in the Germanic countries with the rise of an ideology of the degradation of mankind, which was raised to the status of a policy under Nazism, and expressed in attempts to "purify" the human race by controlled inbreeding. In *The Descent of Man,* Darwin reminds us that earlier, in Prussia, humans were domesticated in order to breed the famous Prussian grenadiers:

> In another and much more important respect, man differs widely from any strictly domesticated animal; for his breeding has not been controlled, either through methodical or unconscious selection. No race or body of men has been so completely subjugated by other men, that certain individuals have been preserved, and thus unconsciously selected, from being in some way more useful to their masters. Nor have certain male and female individuals been intentionally picked out and matched, except in the well-known case of the Prussian grenadiers; and in this case man obeyed, as might have been expected, the law of methodical selec-

> tion; for it is asserted that many tall men were reared in the villages inhabited by the grenadiers with their tall wives.[13]

This feeling that degradation is inevitable is prevalent even among intellectuals who are far removed from the political ideology of that dark time in our history, and is particularly well illustrated in Erwin Schrödinger's *What Is Life?*[14] Written in 1944 by a physicist who had fled Nazism, it deals with the definition of life in terms of physics and chemistry. Schrödinger talks repeatedly of the degradation that lies in wait for living organisms, and discusses the concept of entropy at length (he even talks of molecules needing *Lebensraum,* room to live, which was hardly a neutral term in 1944!),[15] concluding that life is a kind of battle against the inevitable degradation of the world, and seeing living organisms as having the power to transform entropy, reestablishing the order of the world within themselves, and expelling disorder. But as Schrödinger is an excellent physicist, he takes care to say that these are not *proven* consequences of the laws of physics, but are his own ideas, and that his fellow physicists do not agree with him on this point.

The concept of entropy was thus first used to account for phenomena connected with the exchange of heat between macroscopic systems, with no concern for the underlying microscopic level. It was Ludwig Boltzmann, at the end of the nineteenth century, who made the link between the macroscopic and microscopic levels, and he was the first to make an explicit analogy between entropy and disorder. He was studying the special case of an ideal or "perfect" gas, and put into equations the movement and equilibrium of this model gas in a closed container. The perfect gas is formed of individual atoms, represented as tiny spheres that bounce off one another and off the walls of the container totally smoothly (this is an extreme simplification, as it is well known that atoms of gas cannot be spherical, impenetrable points). Boltzmann's intention was to show that it is possible to use *statistical* reasoning to go from individual gas molecules to the general behavior of a large number of atoms.

If the number of gas atoms is large enough, it can be shown that although individual atoms move at very different speeds, their *average* speeds are very similar, and this average can be precisely calculated. The corresponding entropy can then also be calculated, and shown to have a simple relationship with the logarithm of a function directly linked to the number of possible states of the system. (This is the famous law $S = logW$, which was written on

Boltzmann's gravestone.) Boltzmann's approach also associates this entropy with the concept of information, by saying that what is being measured is the "missing information," which can be evaluated as the number of possibilities still open to the system, once the information on its *macroscopic* state has been established. This demonstration is essentially based on a postulate of macroscopic, statistical analysis: the number of atoms in the system must be very large for the demonstration to hold good. In this model, it is not the behavior of individual moving atoms that is being described (thus excluding the explicit use of the laws of mechanics). Nor is it concerned with whether the states whose average is being calculated can *actually exist:* it postulates that all possible states must be considered, as if they could and did occur. It is therefore a thought experiment.

Until very recently, considering an infinite number of objects, processes, and so on was regarded as valid, and was taken to guarantee that the outcome of a theory could be regarded as a true representation of reality. A major advance in mathematical thought in the twentieth century showed that this position is not tenable. As far as very large numbers are concerned, we can infer only very general properties, using axioms that are much weaker than those applicable to large numbers that can be conceived in practical terms. In the same way, it is no longer regarded as valid to consider physical states of matter that cannot actually be explored during the time thought to be available to that part of the universe that is, in some way, accessible to us.

This thought experiment calculates the number of configurations and the state of each configuration, adds these configurations together, and studies their average properties. All the detail of the structure of the system under consideration is ignored. This means that what is being calculated in this average state is nothing other than the result of a *principle of symmetry.* This elimination of detail, this deliberate decision to consider the system in its *totality,* has an important consequence: if, in a system that has been considered only as a whole, details of a structure later appear, this will obviously seem mysterious, because these details were eliminated *a priori.* More precisely, the objective of modeling reality will then be a systematic attempt to find newly emerged properties of the kind found in living organisms, while reducing reality to no more than the laws of the dynamics of physical objects, represented statistically. Irreversibility is a natural consequence of this statistical approach, so it should not come as a surprise. The equations of mechanics that describe the dynamics of the behavior of atoms are certainly reversible, but this property can be used only on condition that each *single instance* is considered independently and *explicitly,* not the statistical average of a number of possibilities so large that only a minute fraction of them can be (and are) realized. In fact it is becomingly increasingly recognized, even in the most deterministic of worlds, that of integers, that what cannot be realized (the effective reality of very large numbers for instance) cannot be described in the same way as what can. It is not surprising, then, if we find statistical behavior confusing.

The common confusion between statistical effects and the laws of mechanics explains the huge success of analyses in terms of auto-this and self-that (*self-organization* being a particularly prominent case), which do no more than reveal underlying microscopic properties, carefully excluded at the beginning of the analysis and then surreptitiously reintroduced into the statistical formulation. We should ask ourselves: What does calling the way a pile of sand moves "self-organized" actually contribute to the description of the phenomenon (which is very interesting in itself, incidentally)? How does adding this adjective enable us to account for it? Does this organization have anything at all in common with the autonomous behavior of living organisms, as the prefix "self" attempts to suggest? What is really interesting is to calculate the stability of the pile, and the formation of avalanches, as a function of the size of the grains and how wet they are. Anyone who has ever built a sandcastle will remember thinking about things like this when trying to make pointed shapes or to delay the effects of the tide on the structure for as long as possible. We should focus on understanding how the local organization of the grains is connected to the shape of the pile, rather than hiding the explanation of the phenomenon by saying that it is self-organized! When such a vocabulary is inadequate for simple physical questions, it is easy to see that its implications are even more woolly when it is applied to life. Why is this prefix so successful? Unfortunately the answer is simple: it relies on a principle that gives the illusion of an explanation by implicitly referring to *something else,* not normally present in physical systems (here, the organizing power of living organisms is implied). This reliance avoids the need to ask questions about the nature of the phenomenon (especially the "symbolic" character of living things), and in fact frustrates any further curiosity. It is not possible to go into the social and historical reasons for this sorry state of affairs here.

In order to understand the evolution of the simple system of a perfect gas, it is visualized as being in a closed container, with two compartments separated by a wall with a small opening, and initially containing either two different gases or a gas in two different physical states. The system is seen to evolve inevitably toward a mixture in which the properties of the two gases become homogeneous. Boltzmann concludes that this is an evolution toward disorder. In this approach, it is taken for granted that *what is homogeneous is considered to be more disordered than what is heterogeneous,* without further discussion. But in what way is homogeneity *disordered?* On the contrary, is it not a very clear way of creating order, as we can see if we compare a trampled sandbox with what it looks like when it has been raked smooth? To say that entropy and disorder are the same is rather like restating the perfect gas situation in two dimensions and in color, with a square of blue gas next to a square of yellow gas, which mix to become a green rectangle, and then asking if a green rectangle is more disordered than a blue square next to a yellow square. Besides, even whether a mixture is seen as homogeneous or heterogeneous depends on the scale at which you consider it. This is the problem with the elimination of detail, just as we saw with the molecules in the cell. So this assimilation of entropy and disorder is not just debatable; it is also misleading. Although we can see that there is certainly a connection between what we perceive as order, and the way things evolve (this is partly the objective of the study of thermodynamics), we must recognize that order is a function not only of the observed reality but also of the observer. A pebble lying among others on a shingle beach may look exactly like a human skull, but it has the order or the information that corresponds to this resemblance only because a human brain sees it that way. More generally, what governs the order of a material system is the set of interactions it can create or capture at any given time in its evolution. We shall see in the next chapter that the creation of a new biological function is exactly this kind of capture of a preexisting structure, which suddenly acquires a function through the discovery of appropriate links.

As with the words *complexity* and *chaos,* so *order,* here associated with *entropy,* is used with entirely contradictory meanings, which may explain its success.

Entropy as Exploration

What *is* entropy, then, and is it an important principle for the construction and functioning of biological systems? One way to understand it is to try to be more realistic. If we look closely, we can see that an increase in entropy, in accordance with the Second Law of Thermodynamics, simply means that objects will spontaneously *explore all the environment accessible to them* (according to an entirely natural law of symmetry: as all positions or energy states that are possible in the experimental conditions are equivalent, it would require an intelligent god to break the symmetry and determine their path). In this context, irreversibility, characteristic of the Second Law, becomes easy to understand and is no longer a pretext for despair or a sign of the loss of a Golden Age. It is simply the expression of the fact that the total "space" of states and positions available to the objects considered can only increase, in the absence of any ad hoc constraint (and we would have to wonder where such a constraint could originate). If this physical space and range of states increases, there is nothing to prevent the objects to which they are offered from occupying everything they can possibly occupy. This is a simple way of looking at the Second Law of Thermodynamics, with its associated notion of irreversibility: the irreversibility of a system is not an intrinsic "defect," a lack of perfection, a misfortune; it is the natural consequence of the absence of constraint. There is no morality in physics; this is simply a principle of economy, a kind of Ockham's razor, which implies that in the absence of an explicit restraining force, everything that is possible within a finite time will actually occur.

However, we must insist that *not everything is possible,* because *time* is also a crucial consideration. It is meaningless to consider states that are theoretically possible but are inaccessible for lack of time. Once the number of objects considered is over a certain minimum, the number of possible states is so vast that they cannot all be explored. This is a fundamental flaw in the statistical model, right from the outset, and it is important to bear this in mind when considering what happens in real cases, but unfortunately this is almost never done. Entropy is therefore nothing but a measure of the extent to which everything that can be occupied is actually occupied, and an increase in entropy only accounts for the exploration of all this new space (perhaps we should say its *creation,* to mark the fact that it represents a transition from a virtual state to a real state, since the nature of the initial space was different from the nature it acquires when explored, because of the possibility of new interactions). The increase in entropy is nothing else but the *power of creation and exploration of a virgin territory of possibilities.*

The Second Law of Thermodynamics thus defines the tendency of any physical system to evolve naturally toward a state of equilibrium, in other words a state characterized by a certain degree of uniformity, in which every local variation tends to be quickly compensated by a variation in the opposite direction so as to maintain an overall state of symmetry. We should also note, as the late physicist Rolf Landauer (1927–1999) pointed out, that there is no reason why this evolution should follow a path where the production of entropy would be minimal. His 1975 article, which describes not only a theoretical model but also a practical example in the form of a simple electrical circuit, is remarkable for its date of publication, at a time when it was fashionable (and, strangely, had the blessing of the Nobel Prize committee) to regard the principle of the minimization of entropy production as a law for physical systems (which was used to justify all kinds of eccentric speculation in biology).[16] Of course a physical system can evolve to a situation that appears disordered, or more precisely, undifferentiated (because it has symmetry). But it can just as well produce other kinds of order besides simple homogeneity, particularly if during the exploration, objects that are capable of interacting but have not yet had the opportunity to do so come into contact with each other. This aspect of the exploration of the possible was noted by the atomists of ancient Greece. They distinguished phenomena whose existence was arbitrary or accidental *a priori* but which, because of their existence, became necessary *a posteriori*. This resulted from the way eddying atoms encountered one another and created new forms. The encounter, even if it were the result of a determined chain of events, could not know *in advance* that it would produce a new form. But once the form existed, it became explicable because of the encounter, and thus was the necessary result of that encounter. This is not the same thing as *chance, a priori,* because the encounters can be the result of real, independent, causal (and thus not random) chains of events, and yet *a posteriori* it is necessity.[17]

The evolution of any material system tends to bring it closer to a state that, *if it could exist,* would be a state of (dynamic) equilibrium. But again this natural evolution toward equilibrium has no reason (short of invoking an intelligent external constraint, as before) to happen at the same *rate* for all the subsystems considered. This has enormous consequences. Over time, the objects that are present at the same moment are generally different from those that were together at the beginning of the evolutionary process. The consequence of this exploration of time, added to the exploration of space, is that totally new, unpredictable interactions are possible (once again, even if the system is perfectly deterministic). Among these interactions, there can perfectly well exist a situation in which, for an appropriate observer, order has been created. Order here is measured by the fact that an interaction of a new type has been created between part of the system and the observer. In biology this is typically what happens when a *function* is created. Indeed it happens regularly in living systems, in which generally the spatial shape, the architecture of the molecules—that aspect of their form that is responsible for part of their function—is of course the result of an *increase* in entropy, not a decrease! (Remember the pieces of hexagonal tiling, driven by entropy to stick to the flat surface of the cell membrane.) We must also note that in biological processes, the management of time is significant in interactions. A great many biological features and structures, such as the order of the genes and the existence of intergenic regions along the chromosome, have specific characteristics that enable them to introduce the delays needed to make encounters between the structures themselves, or with all kinds of signals, occur at the right time.

Finally, to round off this difficult concept of entropy and its complex relationship with order, we need to bear in mind the implications of scale. When we consider the entropy of a system made up of numerous components, we need to take into account not just the distribution of those components in space and according to their different states (their energy levels), but also the different *scales* on which they operate: there is always a certain entropic contribution in the construction of a hierarchy, for instance. Defining relevant scales is still a relatively unexplored theoretical problem in physics (are the laws of physics the same on all scales?). To a certain extent, we need to know how a system is organized in order to calculate its entropy. This leads to a complete reversal of meaning: entropy has to be calculated assuming an order, rather than calculating order from entropy. And if we want to find out information about a complicated system, it may be important to take account of the structure of its hierarchy.

The Meaning of Information according to Shannon

During the Second World War, the transmission of coded information to military groups was of vital importance. The study of electricity became electronics, and work on the importance of feedback, begun in the late nineteenth century, especially by James Clerk Maxwell, had led to the notions of *control* and *cybernetics*, popularized by Norbert Wiener.[18] Increasingly complex communication networks were developed, and it was natural to be curious about the way the messages were passed along them. We should bear this origin of the mathematical formalization of information in mind—it was motivated by the existence of the telephone and telegraph, and was used to study the transmission of sequential messages along telegraph lines, or by any other means of electromagnetic transmission affected by random interference producing a certain level of "noise." First R. V. Hartley in the mid-1920s, and twenty years later the physicist Claude Shannon, an engineer at the famous Bell Laboratories, analyzed what happened to simple linear messages, mostly composed of sequences of binary or decimal numbers, when they passed along a high- or low-capacity communication channel, both with and without errors.

In the early 1940s it was thought that increasing the number of messages flowing through a communication channel would inevitably increase transmission errors, because the channel would effectively become overcrowded. Claude Shannon set to work to study this specific problem, going back to some of his own earlier work. He had previously been interested in cryptography, the deciphering of secret codes, and had proved mathematically that the new code invented by the American army to protect radio-telephone messages sent first by President Roosevelt and then by President Truman to Winston Churchill would be impossible to decode without prior knowledge of the key. This was a revolutionary improvement not known to the Germans (and in fact the British had discovered how to decipher their messages), and it led to his interest in the role of noise in the transmission of messages.

Shannon was well aware of the limits of his theory. He knew that he was setting out a *theory of communication,* not a theory of information. Shannon's theory is very rich and still plays an essential role in perfecting the high-throughput communication systems with multiple nodes that the public knows as the "information superhighway." The main result of his work, which was surprising at the time, was that provided the channel has sufficient capacity, the number of messages has no effect on the quality of their transmission. This capacity can in fact be calculated easily, if the noise characteristics of the transmission channel are known. Shannon also demonstrated that the messages transmitted, whether they are texts, words, or music, have their own characteristics, which cannot be reduced by compressing them beyond a certain value. He called this value their *information,* and proposed a method to evaluate it.

Because it is a theory of communication and not of information, if we look more closely we realize that Shannon's theory treats the concept of information in a very particular way: what interests him is what is transmitted, which does not need to be understandable. There is one situation in which this view is relevant to life: in the process of DNA replication, DNA polymerase copies nucleotides, one at a time, transmitting the information of their sequential order to a new DNA strand. There is no understanding involved in this. Information here is merely being passed on. Shannon shows that in order for communicated information to be used, two observers and a transmission channel are required. If we want this information to have meaning, the observers must have some means of understanding each other: a common language or the key to a secret code. What is transmitted is not in itself information; it is merely a signal (generally pulses of electricity). The sender and the receiver can speak of sending and receiving information, but this is the result of the *interpretation* of the signal, not of the signal itself. Two readers who speak different languages can read the same message and interpret it entirely differently. French visitors to the United States or Britain are sometimes mystified by signs in shop windows reading "Sale," as this is French for "dirty," and if the sign were to read "Main Sale" they would wonder even more why the shop would want to advertise a "dirty hand." English-speaking visitors to Indonesia are shocked by signs reading "cat oven," until it is explained that this means a car-painting workshop. We can see how important the role of semantics is here, and can understand that the difficult notion of order is related to it.

So we must consider the process of communication in the light of this technical interpretation of information. To communicate a message without any error, everything that can be compressed is compressed (that is to say, everything that can easily be reconstructed at the receiving end of the transmission channel), and all the rest is kept intact. Shannon had a simple way of evaluating the different characteristics of a message. He considered the situation in which a *collection* of messages is sent, and they are analyzed on reception. Certain portions will be easy to reconstruct, while others will not. Knowing the parts that are easy to reconstruct does not gain us much, while knowing the other parts of the message has a high value, and he called this their information. Having a collection of messages allows him to *calculate* the information using the formula $\Sigma\ p_i log_2 p_i$. Here, Shannon is merely following one of the traditions at the origin of probability calculation, the "frequentist" or statistical tradition, which judges the probability of a throw of the dice by calculating the result of a collection of individual events. We should note that this particular tradition is far from being universally accepted. It is based on the law of large numbers, a law which is a probabilistic theorem, and which therefore assumes that the concept of probability is defined, but it has used this concept to define the meaning of a frequency. It is thus a vicious circle, since the concept of probability, via the use of frequencies, is used to define itself . . . It is not hard to see why everything connected with this domain is unstable, and has been interpreted in totally different ways. Once again, as with order or complexity, we find ourselves in a minefield.

In this analysis, it is of course only the discrete form of the message that is taken into account, not at all its meaning. As Shannon points out:

> The fundamental problem of communication is that of reproducing at one point either exactly or approximately a message selected at another point. Frequently the messages have *meaning;* that is they refer to or are correlated according to some system with certain physical or conceptual entities. These semantic aspects of communication are irrelevant to the engineering problem. The significant aspect is that the actual message is one *selected from a set* of possible messages. The system must be designed to operate for each possible selection, not just the one which will actually be chosen since this is unknown at the time of design.[19]

Provided they can be recognized by the receiver as being of the same kind (that is, having the formal characteristics of typical messages, for example one message in English and another message randomly generated, but preserving the same letter frequency, average word length, and syllable formation), two families of equally unlikely messages would be regarded as carrying the same information.

The formal theory that corresponds to this derives from a subset of the general theory of "ensembles," which Shannon defines: "an ensemble of functions is a set of functions together with a probability measure whereby we may determine the probability of a function in the set having certain properties." It takes into account only a collection of similar messages and their *local content,* estimating the probability of the content at each point in the message. A distinction is made between the messages, their reception, and the transmission channel. The messages received should ideally be as close as possible to the messages sent, and the probability of finding a given portion of the message unchanged on reception is measured. Knowing the zones that are particularly sensitive to transmission errors is especially important. In this theory, the *less* certain we are of the probability of a symbol's being at a given position in the message, the *greater* the local information is considered to be. So knowing the value of a symbol at a given position is worth more when that position can take a greater number of different values. This probability operates only at the level of the sequence of symbols in the message; it is not concerned with their meaning. It is measured after receiving several messages of the same kind, which makes it possible to measure the probability in question.

> If one is confronted with a very elementary situation where he has to choose one of two alternative messages, then it is arbitrarily said that the information, associated with this situation, is unity. Note that it is misleading (although often convenient) to say that one or the other message conveys unit information. The concept of information applies not to the individual messages (as the concept of meaning would), but rather to the situation as a whole, the unit information indicating that in this situation one has an amount of freedom of choice, in selecting a message, which it is convenient to regard as a standard or unit amount.

This is the definition of information according to Shannon and Warren Weaver, who added "some recent contributions" to Shannon's work when the 1949 book was published. They went on to state:

> Information is, we must steadily remember, a measure of one's freedom of choice in selecting a message. The greater this freedom of choice, and hence the greater the information, the greater is the uncertainty that the message actually selected is some particular one. Thus greater freedom of choice, greater uncertainty, greater information go hand in hand.[20]

We can see at once from this that for Shannon and Weaver, information means exactly the opposite of its commonsense meaning. Whereas in this sense, information is naturally that of a structured message, according to Shannon information is present only in the part of the message that is normally not recognizably structured. This contradiction between the naturalness of common sense and Shannon's definition of information is certainly responsible for a great deal of misunderstanding, and especially because it would seem normal to want to define the information of *one* message (and, as we have seen, Shannon was aware of this), not that of a *collection* of related messages. However, we can try to understand what Shannon meant: if in a message we are *sure* that a certain character will appear in a given position, knowing what symbol is in that position contributes *little* information. On the other hand, if one position in the message is not clear, knowing its value adds a significant amount of information. Thus, in a way that is opposite to the commonsense meaning, the information associated with a collection of random messages would be very high, because it is never specified in advance.

Wiener's Entropy and Shannon's Entropy

Well before the invention of the transistor, significant progress in electronics, with diodes, triodes, amplifiers, and so on, made quite complex automatic control systems possible. The first electronics engineers rapidly found how to use feedback to create self-regulating response systems. The idea is simple. In a system that has an input, and an output that responds to the input, a small part of the output signal is extracted and fed back into the system; then, amplified if necessary, it is subtracted from the input so as to decrease or even cancel out the effect of the input on the output. An increasing input signal is thus prevented from causing an infinitely increasing output signal, because the effective input will decrease as a function of the output. Systems of this kind are called *homeostatic* systems, because they stabilize around a given state. This is the principle on which thermostats work, and it is also at the root of the vast majority of control systems in biology.

The usually quoted example of a mechanical homeostat is not one that readers will necessarily be familiar with, but it was once a common sight, on top of traction engines. This is the Watt governor, a rotating device in steam-driven machinery, which maintains the engine speed as constant as possible. It is driven by the engine's flywheel, and as this rotates it turns weights (known as "masses" or "fly-balls") that swing outward and upward under the centrifugal force. These are connected to the valve that allows steam through to the engine (the throttle). When the engine speeds up, the masses spin faster and the arms move outward, closing the throttle and cutting back the supply of steam to the engine. When the engine slows down, the weight of the masses counteracts the centrifugal force, so they move downward and inward, opening the throttle again and increasing the supply of steam to the engine. The engine speed is thus kept more or less constant.

Systems of this type have all sorts of interesting properties. They enable a system to be maintained around a particular state, and under certain conditions they produce oscillations. This is an essential basic principle of electronics, which makes considerable use of oscillation. Generally, they produce a response with a level that does not vary much from a certain threshold value. The study of systems in which feedback operates (such as servomechanisms) is called *cybernetics,* a name proposed by Norbert Wiener, who studied them in detail. Drawing an analogy between servomechanisms and the central nervous system, Wiener (and other influential scientists, including John von Neumann) recognized that the brain capacities of humans and animals involved not only communication processes, but also *decision* and *control* devices. Wiener noted the fundamental role of feedback loops in the case of control systems, and also underlined the fact that information measures the probability of the result of a choice between several alternatives, playing a primary role in the case of decisionmaking systems. He commented that "the notion of quantity of information relates naturally to a classical notion of statistical mechanics, that of entropy. Just as the quantity of information of a system is a measure of its degree of organization, the entropy of a system is a measure of its degree of disorganization."[21]

He therefore evaluated information as the opposite of the definition normally given of entropy. For him, information is negative entropy (or negentropy, as it is sometimes called). There was no theoretical justification for this approach, just intuition. It represented a natural way of understanding what information is. But as things turned out, the way Shannon treated information was to lead to a very different mathematical formulation. Remarkably, Shannon's formulation is exactly the opposite of Wiener or von Neumann's interpretation. The confusion between the approaches of Boltzmann, Shannon, and von Neumann (and a few more authors whose contributions were along the same lines), when added to the impression that what is homogeneous is disordered, created a starting point for the analogy between disorder, entropy, and information. The core of the problem is thus twofold. First, there are two diametrically opposed interpretations of what information is. Second, and this is perhaps the main reason for the success of this analogy, the entropy of a perfect gas can be put into equations in a form that is almost identical with that given to Shannon's information. This similarity between Shannon's formula, expressing the information contained in a family of transmitted alphabetic texts as a logarithmic law ($\Sigma\ p_i log_2 p_i$), and Boltzmann's equation, which calculates the entropy of a set of theoretical moving objects bouncing smoothly off one another ($S = logW$), has been misused. The way Jacques Lacan used equations in his *discours* at around the same time is a reminder of the authority they can acquire: if equations are identical, this was thought to imply, by analogy, that the underlying phenomena are identical.[22]

Myron Tribus relates that von Neumann, to whom Shannon had turned to help him find a name for his function defining information, proposed prophetically: "You should call it entropy for two reasons. In the first place your uncertainty function has been used in statistical mechanics under that name, so it already has a name. In the second place, and more important, no one knows what entropy really is, so in a debate you will always have the advantage," thus opening a Pandora's box of intellectual confusion.[23]

It is perhaps not a bad idea to go back to the source, and see what Shannon himself had to say in 1956, when he was very worried by the popular success of his theory:

> Information theory has, in the last few years, become something of a scientific bandwagon. Starting as a technical tool for the communication engineer, it has received an extraordinary amount of publicity in the popular as well as the scientific press. In part, this has been due to connections with such fashionable fields as computing machines, cybernetics, and automation; and in part, to the novelty of its subject matter. As a consequence, it has perhaps been ballooned to an importance beyond its actual accomplishments. Our fellow scientists in many different fields, attracted by the fanfare and by the new avenues opened to scientific analysis, are using these ideas in their own problems. Applications are being made to biology, psychology, linguistics, fundamental physics, economics, theory of organization, and many others. In short, information

> theory is currently partaking a somewhat heady draught of general popularity. And although this wave of popularity is certainly pleasant and exciting for those of us working in the field, it carries at the same time an element of danger. While we feel that information theory is indeed a valuable tool in providing fundamental insights into the nature of communication problems and will continue to grow in importance, it is certainly no panacea for the communication engineer, or, a fortiori, for anyone else. Seldom do more than a few of nature's secrets give way at one time. It will be all too easy for our somewhat artificial prosperity to collapse overnight when it is realized that the use of a few exciting words like *information, entropy, redundancy,* do not solve all of our problems.[24]

Maxwell's Demon

Another image has been added to this general confusion, reinforcing the hasty assimilation between order, entropy, and information. Is it possible to visualize a given macroscopic phenomenon locally, and how would this allow us to modify the way we represent it? If we were to imagine a way of reversing the evolution of a perfect gas toward a homogeneous state, it might be as Maxwell did at the end of the nineteenth century, by imagining a tiny conscious being (which became known as Maxwell's demon) who judged the speed of the gas atoms, and opened the hole between the two gas compartments to allow fast-moving atoms to go in one direction and slow-moving atoms to go in the other, or closed it to prevent them from going in the "wrong" direction. This being would make it possible to return from a homogeneous state to a heterogeneous one, in which one compartment contained fast-moving atoms and the other slow ones. If we imagine that the demon does not use any energy to do this, he would make it possible to contravene the Second Law of Thermodynamics and thus invent a form of perpetual motion. We know that this cannot be done, so we have to discover a reason why it should be impossible. What the demon uses in order to do his job is *information* about the nature of the atoms of gas. This would appear to justify the analogy between information and entropy, because evolution toward homogeneity is governed by an increase in entropy. But it implies that evolution toward homogeneity is the normal form of the degradation of information (in Wiener's and von Neumann's interpretation). It also implies that there is something irreversible about the demon's calculations in judging the speed of the atoms of gas, prompting the question of the relationship between information and the possibility of making a calculation.

> Strigelius is expanding on a favorite theme, a parallel between the poetic imagination and the kinetics of gases. He claims that elementary ideas dance around in the poet's brain, bumping into and bouncing off one another just as randomly as the atoms in Maxwell's container, and that the mind plays the role of Maxwell's little demon, simply opening or closing the trap-door on the ideas that happen to present themselves. So genius itself amounts to no more than staying alert and making a choice. Genius is only a critical vigilance. And inspiration is at most a matter of temperature, which increases the activity inside the container.[25]

This passage from *Les Hommes de bonne volonté (Men of Good Will),* a novel written by Jules Romains in 1932, shows the impact Maxwell's demon had on the thought of a whole era. Before we look at other, more strictly scientific reincarnations of this thought experiment, we should go back to its origin, as the physicist Charles Bennett did. This is how Maxwell describes the problem when he imagines his demon sorting the atoms of gas:

> One of the best established facts in thermodynamics is that it is impossible in a system enclosed in an envelope which permits neither change of volume nor passage of heat, and in which both the temperature and the pressure are everywhere the same, to produce any inequality of temperature or pressure without the expenditure of work. This is the second law of thermodynamics, and it is undoubtedly true as long as we can deal with bodies only in mass, and have no power of perceiving or handling the separate molecules of which they are made up. But if we conceive a being, whose faculties are so sharpened that he can follow every molecule in its course, such a being, whose attributes are still as essentially finite as our own, would be able to do what is at present impossible to us. For we have seen that the molecules in a vessel full of air at uniform temperature are moving with velocities by no means uniform, though the mean velocity of any great number of them, ar-

bitrarily selected, is almost exactly uniform. Now let us suppose that such a vessel is divided into two portions, A and B, by a division in which there is a small hole, and that a being, who can see the individual molecules, opens and closes this hole, so as to allow only the swifter molecules to pass from A to B, and only the slower ones to pass from B to A. He will thus, without expenditure of work, raise the temperature of B and lower that of A, in contradiction to the second law of thermodynamics.

This is only one of the instances in which conclusions we have drawn from our experience of bodies consisting of an immense number of molecules may be found not to be applicable to the more delicate observations and experiments which we may suppose made by one who can perceive and handle the individual molecules which we deal with only in large masses.

In dealing with masses of matter, while we do not perceive the individual molecules, we are compelled to adopt what I have described as the statistical method of calculation, and to abandon the strict dynamical method, in which we follow every motion by the calculus.

It would be interesting to enquire how far those ideas about the nature and methods of science which have been derived from examples of scientific investigation in which the dynamical method is followed are applicable to our actual knowledge of concrete things, which, as we have seen, is of an essentially statistical nature, because no one has yet discovered any practical method of tracing the path of a molecule, or of identifying it at different times.

I do not think, however, that the perfect identity which we observe between different portions of the same kind of matter can be explained on the statistical principle of the stability of averages of large numbers of quantities each of which may differ from the mean. For if, of the molecules of some substance such as hydrogen, some were of sensibly greater mass than others, we have the means of producing a separation between molecules of different masses, and in this way we should be able to produce two kinds of hydrogen, one of which would be somewhat denser than the other. As this cannot be done, we must admit that the equality which we assert to exist between the molecules of hydrogen ap-

> plies to each individual molecule, and not merely to the average of groups of millions of molecules.[26]

In this seminal text we find not only the source of the misinterpretations that have proliferated as a result of a superficial reading, but also the basic elements of Maxwell's thought on order, the nature of information, entropy, and irreversibility. We should note in particular that Maxwell himself points out the inapplicability of statistical thermodynamic principles to a situation such as that inside the cell: "conclusions we have drawn from our experience of bodies consisting of an immense number of molecules may be found not to be applicable to the more delicate observations and experiments which we may suppose made by one who can perceive and handle the individual molecules."[27]

Maxwell's text is perhaps the earliest appearance of the fundamental role of consciousness in the analysis of a natural phenomenon (and remember that it predates the first representations of the world in quantum terms, in which an observer is usually considered to have an active role in the world, even when doing no more than observing). What he performed was a *Gedankenexperiment,* one of the thought experiments now so familiar in theoretical physics, in which we can deduce the behavior of the world simply by thinking and observing ("imagine what would happen if . . ." "and as that does not happen it must be that . . ."). He is imagining a microscopic being who, like us, is able to analyze, monitor, and decide (processes that are difficult to characterize in purely physical terms). This being would measure (the speed of the molecules), choose an objective, and act (by opening and closing the hole between the two compartments).

We know that heterogeneity cannot create itself spontaneously (this would permit the creation of usable energy out of nothing—and we should note in passing that Maxwell does speak of "work"), so we must discover what it is that the demon cannot do, and give reasons for this. The *measuring* can be done without any need for consciousness; all that is needed is an appropriate connection between the gas molecules and a sensor. Is this where the impossible, which would allow perpetual motion, comes in? Next, the demon translates measurement into action: is this possible? Finally, the decision to carry out the action requires knowledge, pertinent information about the phenomenon, followed by a calculation: is this possible without consuming energy? A whole series of physicists have tried to answer these different questions, usually intuitively, without any formalization. Curiously, their conclusions have usually been taken at face value, without any analysis of the reasoning that has led to them, and they have invaded first the scientific literature, and then philosophical and literary writings.

It is beyond the scope of this chapter to review all the different paths that have been followed, but I would like to stress that they seem to be strongly influenced by an ideology of order and disorder, which reflects the way thought has evolved from an acceptance of divine order, to an acceptance of the absence of any organizing principle, via a pessimistic vision in which disorder reigns. It is important to bear this in mind, when we return to considering life, and to trying to understand genome texts, using all the conceptual means available.

In a series of articles that were to play a leading role in the reassessment of the nature of information by the scientific community, the physicist Charles Bennett commented that the first thing to be challenged was the *action* chosen by the demon. He recalls that just before the First World War, von Smoluchowski saw that the demon could not use a trapdoor, because the door would have to be so small that it would be sensitive to Brownian movement. More recently Wojciech Zurek, of the DOE's Los Alamos Laboratory, reconsidered von Smoluchowski's remark, and calculated precisely the position of equilibrium of a microscopic trapdoor subject to Brownian movement in the conditions explored by Boltzmann, and showed that it could not be used to sort individual atoms or molecules, because it would have to be *on the same scale* as they were, and would thus be subject to the same thermal fluctuations.[28]

Even if this challenge were resolved, there remains one essential problem, that of how measurement and decision are connected within the action. This appears to be an act of intelligence: how is it related to energy? The Hungarian physicist Leo Szilard, who was very well known for his work in theoretical physics, but also among biologists because of his contribution to the discussion on the nature of genes and their expression within the cell, published an essay on this problem in 1929, whose very explicit title is often cited in the Anglo-American tradition: "On the decrease of entropy in a thermodynamic system by the intervention of intelligent beings."[29] He thought he had found the answer by showing that the intelligent act of measurement would necessarily be accompanied by an increase in entropy, which would balance exactly the decrease in entropy created by sorting the molecules. He attributed a precise contribution of entropy to the *production* of information. This suggestion had enormous implications for biology. The idea that in order to create information it was necessary to fight the spontaneous creation of entropy, at a significant energy cost, seemed to indicate a need for an external creative principle, a source for this corresponding energy (not potential energy or mechanical energy, but the reversal of an unavoidable increase in entropy). Faced with the marvels of living organisms, biologists had to ask where the corresponding information, transmitted from generation to generation in the genome, could come from. Was it due, as Schrödinger thought, to some kind of production of negative entropy?

Since Szilard's essay, and after much discussion by all sorts of physicists, we are told—without proof of course—that a *bi*nary uni*t* or *bit* of information is worth at least *klog2* (where, as in all physical systems in which a unit of energy is involved, *k* is Boltzmann's constant, used to measure the quantum unit of entropy, that is, the unit of energy per unit of absolute temperature), and that the corresponding energy consumption at temperature T is *kTlog 2,* or in the order of 2.9×10^{-21} joules (about the kinetic energy of one molecule of air at room temperature). As a comparison, to raise the temperature of one milliliter of water by one degree requires 4.3×10^{-3} joules, which is 10^{15} times more. The microprocessors in today's microcomputers carry out basic operations at a rate of more than a billion operations per second, in several million transistors packed onto a minuscule surface. This would still be a long way from causing problems with heating up merely as a result of the calculations themselves, but could ultimately do so if this reasoning were correct, and if Moore's law (that the calculating power of microchips doubles every 18 months) continues to hold.

Does Computation Consume Energy?

Besides this incorrect idea, Bennett notes the influence of another fundamentally important model in the physics of the time, that of black body radiation, and suggests that it added to the difficulty in creating a coherent theory of information. In fact things might have remained at this intuitive, imprecise stage, if another brilliant physicist, Léon Brillouin, had not reemphasized the intelligent act of measurement carried out by the demon, in a book that attracted a great deal of attention, *Science and Information Theory.*[30] But Brillouin's attempt to provide an explanation, which is entirely dependent on the way he imagined the situation, would also have to account for the production of all life's inventions (and indeed, behind all these suggestions we have a little living demon, which would be required in every cell, to perform the appropriate actions needed to produce and manage information). We can begin to see why the reflection of physicists on a very simple problem had such an impact on the understanding (or misunderstanding) of biology.

Brillouin visualized the demon acting indirectly, using a device which enabled him to measure the speed of the gas molecules. To do this there must be some physical phenomenon to measure, such as the exchange of a photon (the demon would have to light up the molecule in order to see it). Is this possible? The demon is enclosed in a container with no exchange with the outside, which Brillouin identified with the *black body,* the theoretical body used as a paradigm in the study of radiation. He pointed out that inside a black body it is impossible to see, and noted that the reason why Maxwell did not take this into account, and thus could not evaluate the information his demon would have to handle, was simply that black body theory dated only from 1905. Brillouin went even further, suggesting that in order to light up the scene, the demon would have to use a light source whose temperature was necessarily different from that inside the container. This would mean that he could get an unambiguous result only if the energy of the photon differed by an amount of the order of *kT* from the mean energy of the gas. He then stated a general law, as if it were a law of nature: "a bit of information can never be obtained for less than klog2 in terms of negentropy." This was a second intuitive justification for the widely quoted but mistaken formula that to create a bit of information requires *kTlog2* of energy. This interpretation, and this use of the value of an elementary unit of information, were then sanctioned by John von Neumann, who had a considerable influence both on the development of computer calculation, and also on later discussions on the functioning of the human brain. He stated very explicitly that a computer performing a basic logical operation, for instance deciding to place a 0 or a 1 at each stage of its calculation, consumed *kTlog2* of energy each time.

The scientific community continued to accept this until Rolf Landauer and Charles Bennett, both of whom worked at the famous IBM research laboratories at Yorktown Heights in New York State, refuted it between 1961 and the mid-1970s. It was certainly not by chance that the question of the production of information was first tackled at IBM. Rolf Landauer was intrigued by the amount of energy used merely to perform a calculation: does carrying out an elementary logical operation such as addition require a minimum quantity of energy, and if so, how much? Information cannot exist without a medium, whether it is a neuron, a sheet of paper, a magnetic disc, or a compact disc read by a laser beam. Once we start to handle real objects, a certain amount of energy must be involved. Is a minimum amount of energy used to *manipulate* information, and if so, at what point is the energy expended?

At the time, this was a purely academic question, because computers could still perform only a very few operations per second. The first computers of the 1940s, which used vacuum tubes, had consumed an enormous amount of energy, around 1 joule per elementary binary operation. Ten years later they were already using one hundred thousand times less. Then the discovery of the transistor effect, and its application to computers in 1955, allowed the electronic systems that performed logical operations to be miniaturized, and it became clear that the local density of calculations in a given medium would increase rapidly. From the mid-1970s onward it became a much more pressing question, with the extraordinary integration of electronic components (and Landauer was one of the first pioneers of this), which enabled more and more basic logical operations to be performed at increasing speed and in an ever-smaller space. Calculations can now be performed at more than one billion operations per second, which would use the equivalent of 1,000 watts at the 1950s rate of consumption for each chain of elementary logical operations, while millions of operations are performed in parallel! At the time, extrapolating the tendency for the energy used per basic logical operation to decrease, it was thought that the theoretical minimum (around 2.9×10^{-21} joules at room temperature, as we saw) would be reached by 2010 to 2020. If the number of logical operations performed by one chip increased, would it heat up merely because it was calculating?

To find out, it was first necessary to break down the problem of the nature of information into a number of subproblems. As with Maxwell's demon, at least three distinct aspects need to be taken into account: the measurement process, the channels of communication, and the computation itself. Naturally, Landauer began by looking again at the demon problem, trying to prove rigorously what Szilard, Brillouin, and von Neumann had thought. To do this he broke computation down into elementary operations, and soon realized that if in some cases these operations did use an amount of energy of the order of kT, it was far from being always true. In fact, thanks to subsequent work by Charles Bennett, it was possible to show that only *irreversible* operations, those which *erase* information acquired on a given state of the computer, produce in their environment a quantity of entropy which can be considered to be the equivalent of the information that has just been eliminated. Provided all its history is preserved, the production of information itself is reversible, and does not consume energy. In principle, the communication process itself can be made reversible, so that it is not necessary to consume energy to transmit each bit of information. Finally Landauer and Bennett showed that the act of measuring does not consume energy either. That leaves only one stage at which entropy (with the corresponding energy, at the temperature T of the computer) is connected with information, and that is when the computer memory is erased.[31] The strategy for reversible computing is clear: never throw anything away! We may wonder whether this is not one important reason for the large amount of "junk" DNA present in genomes (not such junk anyway, as we saw). The expression of the genetic program, which is analogous to the record of the computation, produces "junk" as a by-product of a useful program. If it is never erased, the genome will soon be full of old, now useless junk data that it is keeping just to save itself the free energy involved in getting rid of it. When the genetic program is reused, some of the junk will have to be erased eventually, but in the meantime, it is just ready and waiting to adopt a real role if the opportunity arises, as a timing mechanism or perhaps as something else.

Of course this is true only in theory, in other words in conditions in which operations are performed slowly enough to allow reversible transitions between one state and another state. Computation is carried out mostly through the use of bistable physical systems, so the aim was to produce physical objects of this kind that could minimize energy consumption during processing. CMOS, or complementary metal-oxide semiconductor technology, which Landauer strongly supported, benefited from this theoretical contribution, and microprocessor design took the conclusions of Landauer and Bennett's work into account, optimizing the consumption of energy during calculation. In fact, the energy used per elementary unit of calculation has decreased exponentially over the years, and it can be predicted that between 2010 and 2020 it will become possible to seriously consider making microprocessors that can make completely reversible calculations, and find ways to dissipate the energy used to erase the corresponding memory at a distance from the calculating components, to cool them easily. It is also likely that for large-scale computational facilities, the trend will be toward building computers that run at the temperature of liquid nitrogen, gaining more than a factor of three on the theoretical limit.

The significance of Landauer and Bennett's work, which was totally unambiguous, is not properly recognized. This failure is due to the same error of judgment, based on a combination of superficial intuition and formalization in mathematical terms, that has had such a strong influence on thought since the early 1950s. It has hindered the proper development of constructive discussion in many domains, but particularly in biology, especially where the links between genetics and information are concerned. Their work freed biology from its straitjacket, showing that there was no paradox, and that the properties of life, from simple helices to the most fascinating functions and relationships, could be created without fighting entropy, and without any need for outside help. But it is only now, at the beginning of the twenty-first century, when we can decode the texts of whole genomes, that their line of research is coming back into the spotlight, enabling us to uncover many unexpected features of the richness of life.

Yet quite remarkably, at the turn of this century there are still scientists who believe that producing information costs energy, in order to compensate for entropy loss! The formula ($\Sigma\, p_i log_2 p_i$), which calculates the basic unit of information from the computation of the probability of finding letters in a message, is a very simple one that every intellectual who wants to understand the world even a little should know. But the fact that it derives from an equation having the same *form* as the logarithmic formula for the value of energy used in computing the various states of a perfect gas does not mean that the two concepts *are* the same. An orange is a sphere, and the Earth is a sphere, but the Earth is not an orange. If I have emphasized the point, it is because mathematical formulas have always been abused in the nonsense that surrounds the use of science for political ends, and because this mistaken association between a unit of energy and the creation of a unit of information is repeated in so many textbooks, as well as in popular and philosophical works. It is well known that the political use of science is more concerned with power than with knowledge. A reminder of the historical and conceptual context of its invention seemed appropriate.

Algorithms and Turing Machines

Landauer and Bennett studied the question of irreversibility in terms of physics, and transferred it into the domain of logic. What can we say about the inverse relationship? What connection is there between computation and logic, or, more precisely, the practical use of logic? Starting in the early 1930s, work exploring the limits of computation and logic laid the groundwork on which computing was based, both conceptually and as a technical reality. Several thinkers of the time played a major role in the genesis of the necessary concepts, among them the Austrian Kurt Gödel, the American Alonzo Church, and the Englishman Alan Turing. Between his birth in 1912 and his suicide in 1954, Turing became interested in a great many areas of knowledge, but especially in the foundations of mathematics. Reflecting on the nature of intelligence, he noted its capacity to produce and interpret *symbols*. He was especially interested in the fact that arithmetic and logic are essentially the same, and in the practical possibility of performing arithmetical (or logical) computations, while trying to understand if there were any theoretical limits to the actual *performance* of such computations. In both arithmetic and logic, what counts is the chaining together of symbols, and time—as a succession of events—is implicitly involved in this. The heart of logic is implication, and the heart of arithmetic is obtaining the result of a calculation. Turing's stroke of genius was to find a way of performing these operations simply, using a basic machine.

His studies immediately followed and were parallel with those of the mathematicians Gödel and Church, who were trying to answer the question posed by David Hilbert at the end of the nineteenth century: can all the theorems of arithmetic be known on the basis of the axioms and definitions on which it is founded? In other words, is the theory of numbers just a tautology? The answer, very unexpected and probably contrary to Hilbert's intuition, was that it is not so at all. Knowing the axioms and definitions of arithmetic is far from enough to enable us to know in advance what they may lead to. Stated another way, a theorem does not exist until it has been proven, and working out a proof can have important consequences, because sometimes the conclusion reached is that it is impossible to decide whether the theorem is valid or not. First Gödel and then Church showed that, starting with a set of axioms and definitions, it is possible to form perfectly justifiable arithmetical propositions that can be shown to be valid by using further axioms, but not by using just the axioms and definitions from which they were established. In a sense, a set of axioms and definitions from number theory is capable of creating new propositions that *cannot be reduced* to the initial set of axioms.

The heart of the problem is the difficulty of knowing what connects the abstract, conceptual nature of the way we represent phenomena, and their material reality. It is all very well to talk about arithmetic or information, but *concretely,* how is information manipulated? After he had attended a Cambridge lecture on Gödel's work in 1935, Alan Turing attempted to devise a machine (which was imaginary but could have been physically built) to represent the theoretical way in which all the theorems of arithmetic, and the corresponding calculus, could be deduced mechanically by a succession of basic moves. By 1936 he had established the formal foundations of what has since been known as a *Turing machine,* in which he identified *states* and *types* of objects, familiar categories in physical processes. He used them to develop his thesis, that the world of the mind could be compared to the infinitely rich (a richness that was entirely unsuspected before the incredible discoveries of the 1930s) but well-organized combinatorial possibilities of the states of a machine. What is remarkable about this early work is that the strings of symbols required to make the machine function are exactly what we produce today when we write computer programs—they are lines of "code."

Since the time of Babylon, mathematicians have been particularly interested in the question of calculation. They have sought, and found, numerous ways of making it easier. But above all they have tried to reduce the solution of all conceptual problems to the solution of a sequence of computations, in other words a series of manipulations of symbols. This approach is characterized by the fact that it establishes an explicit, concrete link between the abstract nature of symbols and their physical medium. It refuses to consider a disembodied world of the mind in which strings of propositions would exist entirely independently of reality. The aim of calculation is to gain some mastery—be it infinitely tenuous—over the inevitability of things.

The relationships between theoretical models of computing machines, and the physical reality of the machines that implement the calculations, are still much discussed today. On the one hand we have the inescapable laws of physics, and on the other the Promethean temptation to take control. I must stress here—and we must always remember this when we consider genomes, because we must never forget the real constraints of our material nature—that Turing's way of looking at the problem of computation was deliberately very *concrete,* even if at first it was a thought experiment. His approach distinguishes between symbolic processes, which control the interactions between objects, and the physicochemical nature of the underlying processes. Provided that the machine can actually exist as a material reality, its physical nature is not important, so long as it can establish the necessary relationships between the strings of symbols. This duality of the symbolic and the physical nature of things is a characteristic feature of living organisms: they are compatible with physics, but they cannot be deduced *a priori* from its laws. This type of originality is the reason why attempts to predict the existence of living organisms from the laws of physics alone will always be bound to fail. Physics represents the inevitable and universal constraints on things, whereas life will always try to take control. This explains many misunderstandings between geneticists, for example, and a few physicists who have thought they understood what life is, and have proposed physical theories that cannot help being extraordinarily reductionist and simplistic (naturally, without the least success).

Knowing the newly discovered limits of logic and computation, Turing wanted to understand their implications in concrete terms. This was why his studies were able to serve as the basis for the creation of computers, and the entire world of computer science associated with them. This approach neither leads to a dead end nor limits the questions that can be tackled; on the contrary, it opens the way to extremely powerful lines of reasoning. In particular, the *time* taken to produce a machine and to perform its computation must be incorporated into this conception of concrete reality. What meaning would a procedure have, if it could not be performed in a time compatible with the lifetime of the universe we know? Following Landauer and Turing, I want to persuade you that regarding human limitations as essential is not at all an admission of defeat; on the contrary, it leads us to devise lines of reasoning that have an enormous amount to teach us about the world and the nature of life. Even better, it allows us to get a grip on reality, and especially in the case of computing, it can prompt ideas about how to build material systems that possess some of the properties we recognize in living organisms. It is important that the actions that are contemplated—all of them—can actually be *performed.*

We should note, then, that the most concrete questions can lead to profound mathematical problems. Only in the 1930s, after 2,250 years of trying, was it possible to prove that an angle cannot be divided into three equal parts using only a pair of compasses and a ruler. In differential geometry, René Thom's catastrophe theory can be seen as an attempt to formalize the flux of natural things into forms and symbols. Similarly, Leibniz visualized building a machine that would be able not only to calculate, like Pascal's machine, but also to resolve the most complex formal problems. It was not just a question of understanding the mechanics—although this was particularly fashionable in the eighteenth century—but above all of how to break down the actions performed during the computation into basic actions that could then be simulated mechanically. With time, it became clear that all computation requires the *iteration* of basic operations, and that their performance has to be controlled by a *test* that indicates whether the conditions required for the computation to continue are fulfilled. It was only in the nineteenth century, after more than a thousand years, that the crucial importance of iteration was universally recognized (particularly in the generation of whole numbers starting from 1). Its relationship with the nature of arithmetical objects and with logic enabled mathematicians to reason consistently about a procedure that had long been familiar to them—reasoning by recurrence. The Turing machine was intended to bring all these steps together. Its importance stems from the fact that it established a concrete, direct link between the principles of number theory and of logic, and their operation. Turing proved that any computation can be broken down into a sequence of iterations controlled by a program, using operations so simple and elementary that a machine can always carry them out.

There is a central concept at the heart of this approach—the *algorithm*. As is often the case, the reality this word represents is fairly vague. Esoteric as it sounds (it is derived from the name of a Persian mathematician of the ninth century, al-Khowarizmi), it simply refers to the operation of a set of rules that enable a computation to be carried out from input data, to produce output data. Within the algorithm, we can distinguish a start, followed by a core, made up of a series of computational procedures, together with tests that allow the computation to be oriented toward one procedure or another. Generally speaking, we are interested in algorithms which halt, in other words which can use an appropriate test to decide that the computation is completed.

A doctor's prescription is a simple algorithm. It states the conditions in which the medication is to be used (test: "if there is pain"); the operation to be carried out ("take this medicine"); and a time limit ("for n days"; test: "count n and subtract *1* unit per day"; end: "when $n = 0$, stop treatment"). Similarly, well before the invention of the algorithm, Eratosthenes, in 225 B.C., had invented a simple procedure (a typical algorithm, which can be carried out in practice) to find out all the prime numbers less than a given number n. The procedure is to write out all the whole numbers from 2 up to n, and then cross out one by one the numbers that are multiples of the next number not yet crossed out (all the even numbers after 2; then all the multiples of 3 after 3; then, as 4 has been crossed out, all the multiples of 5 after 5;

and so on). When *n* is reached, all the numbers that have not been crossed out are prime.

In both cases, we can see that there is a question to start with, an iterative program to answer it, and a series of tests that allow us to decide whether or not to go on with each step of the program, and when it is completed (a halt test). At this last stage, the program must decide to reply yes or no to each of the questions originally posed. We should note that an algorithm is composed of extremely simple steps. It is only the way they are combined that makes it appear complicated. As it runs, an algorithm can call other algorithms. It can even call itself, in which case it is said to be *recursive.* An essential characteristic of an algorithm is that it is *generic*—the nature of the algorithm is independent of the nature of the input data. It represents a series of instructions with an input and an output, and which operate on the input to generate the output, without taking into account the nature of the input, provided that it is of a specific *type* and that it is within a *range of values* which has also been specified. A doctor's prescription uses objects of the type "medication," and gives instructions for their use to "patients" according to a protocol involving objects of the type "time," measured in days, hours, and so on. To be operational, an algorithm must produce actions, so it must be associated with a machine, and this is what Turing wanted to represent, in as simple a model as possible.

A Turing machine performs simple logical operations (yes/no, preceding/following), doing no more than reading and locally modifying a medium, such as punched paper tape or magnetic tape, bearing a series of symbols, which in the simplest case are the figures 0 and 1 (or any other representation of presence and absence). What the machine reads, then, is an alphabetic text written in an alphabet of two letters (any message written in a more complex alphabet can be reduced to this using a translation table, a coding rule, just as the letters of the alphabet and all other symbols are represented in the binary code of 1's and 0's in a computer). The machine performs the following operations, all easily carried out (the operations performed by computers are the same conceptually speaking, although they are carried out in a more condensed and therefore more complicated fashion):

changing a symbol in a finite number of places after reading the symbols found there (note that changing more than one symbol at a time can be reduced to a finite number of successive basic changes)

changing from the point being read to other points, at a given maximum distance away in the message
changing the state of the machine

All this can be specified by a finite set of "quintuples," each of which has one of three possible forms:

$$p\alpha\beta Lq \quad \text{or } p\alpha\beta Rq \quad \text{or } p\alpha\beta Nq$$

A quintuple indicates that the machine is in configuration *p* where it reads the symbol α, and replaces it by β, before changing to the configuration *q*, moving to the left *(L)*, to the right *(R)*, or not moving *(N)*.

Turing demonstrated that all machines of this type can be replaced by a Universal Machine, the simplest of all (although it would have been very slow in practice), which can carry out all the calculations of arithmetic. As the calculations and logic used in computers are those of arithmetic and first-order logic, any machine of this kind, and any computation of this kind, can be represented as a Turing machine with an appropriate program. Whenever the problems of computation by computer need to be represented and understood theoretically, this is done by referring to a Turing machine, so naturally the theoretical problem of the consumption of energy in performing a calculation can be restated in terms of a Universal Turing Machine carrying out the same calculation.

What problems can a Turing machine resolve? Remarkably, the analysis by the theoreticians who have looked into this question allows us to state that "any function that can be calculated by a human being can be calculated by a Turing machine." But we should note that this certainly does not imply that the human brain works like a Turing machine (although of course it does not forbid it either)! These studies also led mathematicians to rank imaginable functions according to how suitable they were for resolution by computation. Since the days of the ancient Greeks we have been familiar with two different types of reasoning to solve problems. One way, rarely investigated by the Greeks but developed in the work of al-Khowarizmi, and completely rediscovered in our time, is the *algorithmic method.* The other way is the conventional method we learn at school, summed up under the generic name of the hypothetico-deductive method. This works by combining a set of *axioms* and *definitions,* using logical inference, in order to produce *theorems,* by deduction. For a long time it seemed that the hypothetico-deductive method did lead to the truth. But at the end of the nineteenth century, mathematicians realized that very often, what was thought to be a proof actually relied on a *belief:* alongside the axioms, proofs took for granted certain apparently self-evident properties (such as figures in geometry) that had not been described, let alone taken into account in the axioms. It soon became necessary to define the axioms of the different branches of mathematics more and more precisely, and this task occupied mathematicians for the first quarter of the twentieth century.

It gradually became clear that the main defect in hypothetico-deductive reasoning stems from the fact that it unconsciously adds *meaning* to the words used, and this was the reason for the hidden axioms unintentionally used in proofs. To eliminate this constraint, it was thought that proofs should rely on the use of symbols, but this meant that rules for using this new formalism had to be established. The only simple way to achieve this was to use strings and iterations typical of those used in algorithmic reasoning, which led to a natural merging of the hypothetico-deductive and algorithmic approaches.

An assessment of the internal consistency of the algorithmic system, which is based on a few simple principles of arithmetic and first-order logic, then became a high priority. It was in this context that in 1931 Gödel's first incompleteness theorem threw into disarray a domain which had seemed beyond any kind of uncertainty or dispute, that of arithmetic. We cannot summarize Gödel's reasoning here—Douglas Hofstadter does this in his remarkable *Gödel, Escher, Bach: An Eternal Golden Braid,* a book of more than six hundred pages[32]—but we can simply say that using a string of symbols, and with just the axioms and definitions of arithmetic and first-order logic, Gödel produced a proposition that meant, in plain language, "the truth of what I am saying is not provable." This is the same kind of truth as in the painting by the Belgian artist Magritte, which shows a pipe and, written underneath it, the words "This is not a pipe." We cannot know that it is not a pipe unless we have some contextual proof (outside the picture) that the image formed on the retina does not correspond to a pipe (the retina is flat, so even with a real pipe the eye sees only a flat image of it). What Gödel says is that we have to go outside the set of axioms and definitions used to construct his sentence, to know its truth value. Gödel's paradox thus showed that the program proposed by David Hilbert for the twentieth century, intended to prove the internal consistency first of arithmetic and then of the rest of mathematics, was impossible to achieve.

Gödel's proof relies on the logical exploration of the truth of sentences of the type "This proposition is false" and the apparent paradoxes they express (these are called Epimenides paradoxes). The definitions of arithmetic are unambiguous concepts (*effective* concepts), and it is only by convention that they are given a particular name in one language or another. We could say that they are denoted by a *code.* So why not use whole numbers—integers—for this code? Then, any proposition in arithmetic could be written not in the form of a proposition in English, French, or Polish, but as a whole number. But as we are dealing with an integer, it is possible to transform it according to the laws of arithmetic. This produces a new number, which can be decoded into comprehensible language, because we know the numerical code of the concepts used. This is what Gödel did, and it allowed him to establish propositions whose truth could not be reduced to the axioms that had enabled them to be created.

The Cell as Turing Machine

What does Gödel's proposition mean in real terms? Did it mean that the idea of mathematical certainty had to be abandoned? Turing and his contemporaries were able to explain this apparent failing of mathematics methodically, and in fact to show that the algorithmic approach was extremely powerful and above all *constructive.* Far from leading to failure, it introduced a truly infinite, unlimited opening up of the organization of a process I am not afraid to call creation, because it produces genuinely new objects, which cannot be reduced to those that gave birth to them. Although this phenomenon has often been commented on, it is still very poorly understood today, if only because it is a recent revolution, which changes the way we see reality just as much as logic and the first postulates of science did when they made their appearance two and a half thousand years ago. The aim of this book is to try to improve our understanding of the world, and to show how much this new way of seeing can contribute to our understanding of a phenomenon that seems far removed from the algorithmic method, the phenomenon that led to the organization, development, and evolution of life. I hope I can show that this is much less paradoxical than it seems.

Gödel's proof is possible only because of the use of an explicit but arbitrary correspondence that specifies the basic concepts used, and reveals the meaning of a proposition that has been produced by transforming the initial concepts according to the rules. This transformation assumes a passage through a machine that pays no attention to the meaning, and that the meaning is established only when the message is introduced into a richer universe (which adds connotations to denotations, for instance). The analogy with the genetic code and the cell is particularly clear here: the correspondence between the triplets (codons) of the genetic code and the amino acids is an arbitrary, symbolic one. It is universal, and uses only the properties of each of the amino acids (themselves universal) as its basic domain of meaning, without these properties' having anything at all in common with the codons of the nucleic acids that code for them. And when a chain of amino acids is synthesized, producing a protein, this chain has a meaning only if a biological function can be associated with it, depending on the cell in which the protein is found.

My thesis in this book is that one of the secrets of life lies in the fact that the cell acts as a Turing machine. The way it is constructed means it can use a program, an analog of an alphabetic program, to perform actions as basic as those of a Turing machine, so it must at least have the main properties of this machine at the lowest level. It is precisely because the cell functions using just local, basic operations (of the type connect/disconnect, or presence/absence) that life is possible without there being any *external* causality. This means that we do not need to imagine that life was constructed in any complicated fashion. It is the result of the succession of a very large number of simple events, which became organized essentially because this *worked*. The only systems (organisms) that have survived are those which were able to bring together relationships that were locally extremely simple and probable, and to combine them in the structured way we know today. *Selection by existence* (which is merely a principle of stability) is an infinitely powerful way of discovering precisely what is stable enough over time to be able to survive in a given environment. One property of the stability principle is systematic evolution toward ever-increasing control over the unavoidable physics of the world. And the object of biology is to discover the principles of this evolution toward increasing stability.

Algorithmic Complexity

It is time for a review. We have investigated some of the ideas in which our modern conception of information originated. In particular we have seen that it is an imprecise concept, which carries with it all sorts of implicit assumptions. In the late 1950s, the work of the mathematician and philosopher

John Myhill on this subject helped make this situation better understood, and showed that words can carry a variety of meanings.[33] The nominalism of the Middle Ages, which Umberto Eco illustrated so well in *The Name of the Rose,* has many ramifications. *Stat rosa pristina nomine, nomina nuda tenemus.* The rose is in its name, not in the thing itself.

It is rare for a word to express concepts in a simple way. Myhill identified what he called different levels of "character" (perhaps to avoid using the term *concept,* too closely linked to the connotations derived from the philosophy of Immanuel Kant). A character is *effective* if it can be communicated directly from one person to another without the need for a long discussion to establish complete agreement on what it means. A typical example of effective character would be an expression implying certainty, such as "as sure as night follows day," or the French expression that translates as "as sure as two and two make four." Even though this is not a natural idiom in English, it is immediately transparent when translated into any language and culture, and this unspoken certainty lies behind the joking expression used to indicate an exaggerated interpretation of a situation—we say that someone has "put two and two together and made five." (We should note however that the arbitrariness of the link between the signifier, the word or symbol—which can just as well be 4, four, quatre, τέσσερα, cztery, or quattro—and the signified, the concept of "fourness," is not always obvious to everyone. In a game of lotto, not many players realize that because there is only an arbitrary connection between the name and the thing, they can just as well play 31,32,33,34,35,36 as any other combination of numbers, and still have *exactly* the same chance of winning the jackpot, provided the game is not rigged). A character is *constructive* if its meaning is made more precise over time, using procedures that can be explicitly communicated. This is true of the concept of the algorithm, and of the related concept of the Turing machine. Finally, a character is *prospective* if, although it has a definite meaning for some speakers, there is no means of communicating it (nor ever will be) that guarantees that the exact meaning will be clear to another speaker, still less to all speakers. The word *character* is itself prospective, as we see if we try to pin down its meaning via such words as *concept, idea, notion,* or even *letter* or *sign.* Reading Kant's philosophy is one way of grasping the meaning, but this is only one step on a journey that is better begun sooner rather than later . . . Of course that does not mean that there is nothing to be said about this character, far from it!

For me, the concept of information is a typical prospective concept. It has a basic, rather confused meaning, which I refer to when I say that I use the word *information* in its everyday sense. It acquired a more specific meaning when Shannon grasped one of its components in the context of the communication of information through a noisy channel. But the more we use it, the more precisely defined its meaning becomes, and this will continue in the future.

Some scientists did soon realize that the theory of communication developed by Shannon was quite insufficient to define what information is. Especially because of public discussion of von Neumann's work, it became clear that it would be necessary to relate information to the ability actually to perform calculations. Then, in parallel but independently, in an entirely different domain of science, that of number theory, mathematicians in the United States and the Soviet Union took up the challenge left by the arithmeticians who had been investigating Hilbert's question. We are about to see two new concepts that add to the meanings of information: Kolmogorov and Chaitin's *algorithmic complexity,* and later Bennett's *logical depth.* Of course these are only a prelude to new meanings, which I think will arise out of conceptual advances in biology, and particularly in genomics.

We saw that for Shannon, the meaning of an item of information was unimportant from the point of view of communication. In a sense, once information overload occurs in one place, his point of view is the only sensible one. It is certainly reasonable in the context of the World Wide Web, and the Internet, the network that transmits its content indiscriminately. Here, the information is enormously rich, with one source often contradicting another. But there would certainly be some value in knowing how to sort out information from misinformation in this confusion. It would be useful to have other definitions of information, which could take account of these different aspects. It is the same with genomes. When the molecular machinery responsible for replication, DNA polymerase, duplicates DNA, it takes no account at all of the fact that the DNA plays the role of a program. At this stage, the DNA's meaning has little importance. Yet we know that when the program is expressed, an organism lives or dies. In a different approach, we would like to be able to compare two different molecules of DNA and describe their general properties, so that they can be stored reliably in computer memory for instance. If we can take the meaning of the sequences into account when we describe them, all the better. So we can see that when we want to describe a text such as a genome text, several different definitions of information must coexist.

We need to consider information again, from this perspective. Rather than a collection of messages, as was necessary for Shannon's conception of information, let us take as our example one particular message, the message symbolized by the sequence of bases in a genome. We will try to see how the information can be characterized, taking into account what we have just seen about Turing machines, and using some of the wealth of concepts from number theory (in fact this theory's relevance ought to have been noted very early, since machines that handle information use only integers, binary numbers even).

There have been several detailed reviews of what the exploration of the information content of a message covers in formal terms. The mathematicians Gregory Chaitin and Jean-Paul Delahaye, for instance, and, very early on, the biologist Hubert Yockey and the physicist Henry Quastler all participated in this debate, their contributions complementing and extending that of John Myhill.[34] Briefly, the idea of information is much richer than a single Theory of Information which could be reduced to the theory developed by Shannon and his successors; instead there are many theories dealing with the concept. In particular, it is essential to analyze what the information of a text written in any given alphabet might be. Writing any text can be seen as using an appropriate coding rule to transpose a more basic text, written in a binary language (a series of zeros and ones). This brings us back to the elementary rules of logic (yes/no) and those of arithmetic. Of course this is true only as a first approximation. Introducing a code in order to transpose a text into another one can in itself have important consequences, suggesting that in terms of information, the consequences of what reasoning about binary units has revealed should be applied at other levels of understanding too. However what is already true for bits of information will also be true for more complicated patterns, so for the time being I would like to remain at this most fundamental level.

How can a binary sequence be described? The Soviet Union's school of electronics, with Andronov, and then the school of mathematics under Kolmogorov, as well as Solomonoff and Chaitin in the United States, devised a nonprobabilistic approach to the corresponding information. Even though it is obviously a very primitive metaphor, for our purposes a string of characters expresses the nature of a genetic program fairly well. When we deciphered the sequence of the 4,214,630 base pairs of the genome of the bacterium *Bacillus subtilis,* we were certainly initially confronted with a text written in the alphabet A, T, G, and C, and beginning with

ATCTTTTTCGGCTTTTTTTA . . .

This text contained information, but how could we try to evaluate it?

Shannon's approach had the advantage of making information measurable (if one had a collection of related texts), although this was an abstract quantification. How could we find another way of measuring it, specific to this text? One way of considering a text, a chain of typographic characters, is to try to see how it can be reconstructed, in the most economical way possible, using an algorithm (a program). For example, how can it be reduced in length, or compressed so as to take up as little computer memory space as possible, without modifying any of its properties (at least not in terms of precision, even if it may be modified in terms of time)—or, looked at the other way round, how can it be reconstructed after this treatment, without losing its information content? The algorithm required, if such an algorithm exists, must perform an operation to compress the data in the text into a shorter text. This is not just a thought experiment, but a very common operation, for instance in order to send images, which otherwise occupy a great deal of memory. The advantage of the operation is that images and sound or, generally speaking, any other physical signal can be compressed into a series of binary symbols (1 or 0, black or white, yes or no). This is what we call "digitizing": transforming a continuous signal into a series of discrete binary signals (perhaps we should really say "binarizing," because the decimal system is not normally used). The problem posed is thus how to handle a series of binary signals like this:

011000111110001010101000110101000111101011111100 . . .

Compressing a sequence like this is exactly the kind of operation a Turing machine can carry out, if an appropriate algorithm has been defined. Given a sequence of characters, we can define the shortest formal program (in terms of universal Turing computation—that is, using algorithms that can be run in any machine that operates on integers) capable of compressing an original sequence *S* and of reconstructing it once compressed. Kolmogorov, and later Chaitin, proposed that the information in the sequence *S* should be defined as the minimal length of the shortest universal program capable of representing S in compressed form.

The length of this shortest program is called the *algorithmic complexity* of the sequence. I said earlier that we would see a precise, specific use of the word *complexity,* and here it is. This is very different from the way it is used in the media! A sequence is (algorithmically) complex if a long program is needed to describe it, and algorithmically simple if not. The greatest possible algorithmic complexity of a sequence is therefore equal to the length of the sequence itself. But we should stress that this does not mean that if we know a sequence, we can know its algorithmic complexity. There is in fact normally no way of actually calculating this. The concept of algorithmic complexity is therefore not effective, but it does immediately show us that there are other ways of considering information besides Shannon's communication-based approach. What the concept of algorithmic complexity contributes is only prospective. It suggests we should look for ways (heuristics) that allow us to compress a given sequence as much as possible. What Kolmogorov and Chaitin gave us is thus a strategy, a *research program.* For those like me who are interested in genomes, this would involve understanding the genome text through formal approaches aimed at reducing the algorithmic complexity of the deciphered text. This has an important consequence in terms of understanding what a genome is. Our aim is not simply, as some have thought, to compress the sequence just to reduce the amount of computer memory it takes up (this does not matter, and compression would not gain much space anyway). It is to understand *how the sequence has been generated in the course of evolution.* A genome is not a random piece of DNA, but the result of evolution through duplication, recombination, mutation, and so on, and all these processes could be described in terms of algorithms. Understanding them would make an extremely important contribution to our knowledge of genomes and their meaning.

What is the relationship between the algorithmic complexity (or Kolmogorov and Chaitin's information) of a sequence, and Shannon's information? To establish the link, we need to get away from the fact that we were considering only one sequence, since Shannon's information has a meaning only for a collection of sequences. If we take a family of sequences (which allows us to calculate the corresponding Shannon's information), it can be shown that the algorithmic complexity of any one of these sequences will normally be roughly equal to their information calculated in Shannon's paradigm. In the restricted case in which it is possible to have a family of related sequences, this theorem gives us a rough idea of their algorithmic complexity and allows a basic statement to be made about the nature of a new sequence: if it is not obviously produced by a simple procedure (for example, the repetition of basic motifs), we still cannot state with certainty that it has high complexity. We can talk confidently about a sequence's having low complexity (even if we usually do not have any way of telling whether the complexity we are referring to is the *lowest*) but not of high complexity. This can be only a conjecture, which simply indicates that at the moment we do not yet have any way of discovering whether it really does have high complexity.

So we can consider the sequence of a genome, the genome text, and ask what its algorithmic complexity is. Reasoning in terms of Turing machines enables us to show that a result corresponding to the algorithmic complexity of a sequence does exist (allowing for factors that are independent of the algorithm itself, but dependent on the programming language and machine used). But this logical reasoning (not given here) does not give us any way of knowing which is *the* shortest program or the best program for a given sequence. Once we have compressed a sequence, we know that it can at least be compressed to that length, but except in special cases this does not tell us whether or not a further level of compression will be discovered in the future. We can say only that the algorithmic complexity of the sequence is no more than that given by the algorithm known to compress it. We need to remember this when we consider the information content of genome programs. We will have to find out not only how they have been created in the course of time (this is the aim of the science of evolution), but also how they ensure both faithful reproduction, and the expression of the genome program in the form of specific actions that enable the cell, or the organism, to function.

In fact, to talk about algorithmic complexity, we have to consider the algorithms that describe the sequence, which means we are no longer looking at it locally, but as an organized whole. Considering the algorithmic complexity in this case allows us to give a first measure of the orientation of genome evolution. If we consider the processes that enable DNA to be created *de novo* it seems obvious that it can be done only from existing DNA. This raises an interesting problem in evolutionary biology, because genomes have to evolve by creating new DNA, either through replication errors, without changing the length of the text, or by adding new fragments to the text. The simplest hypothesis is that this happens either at specific points, where DNA polymerase makes a mistake by "stuttering," or alternatively by recombination, insertion, or duplication of long or short sequences. In both cases the algorithmic complexity of the genome text increases (and as it is done by an algorithmic process this is calculable). But we should note that it is the complexity that increases, not the order (in any case we would have to define what that is). This increase in complexity is produced in a direction that is entirely compatible with the Second Law of Thermodynamics, because any duplication, whether local or global, increases the number of possible states of the system, and therefore its entropy. Similarly, we can say that since the genome sequence is recognized by the cellular machinery, which transcribes it and translates the products of transcription into proteins (this is what gives the sequence its biological meaning), there must be a certain number of recognizable features (like the words and phrases of a language, in an alphabetic text) that act as signals. Organizing these signals in relation to each other would in fact constitute an algorithm describing the sequence. The goal of the biologist is therefore the same as the goal of the arithmetician seeking to measure the complexity of a sequence!

On the Best Use of Complexity

The word *complexity* is used with reference to many new areas of knowledge, and these need to be explored one by one. I would like to move beyond the restricted sense we have been considering, that of algorithmic complexity. But first I must eliminate a widespread use of the word, in which *complexity* seems to refer to a property of totality, the Whole. This holistic usage is common in biology, because it is the last refuge of those who want to deny that it is possible to find any explanation for life. Referring to complexity is a convenient way of suggesting that *analysis* is impossible (they label it reductionism), and of promoting new methods of investigation (vague and undefined, but relying heavily on emotional approaches). The main drawback of this usage is the fact that it relies on a truism (a living cell is more complex than a crystal of salt) and that it forbids any kind of comparison, by proposing one qualitative leap, unexplained and unexplainable, between salt and cell. Complexity, in this case, is vitalism by another name.

In a similar but more subtle way, to invoke complexity is to claim specific properties for this qualitative leap, which is said to occur only under certain conditions, and to explain only phenomena that are intrinsically complicated. This is touted as an entirely new way of looking at things—and Thomas Kuhn is often invoked here[35]—a new *paradigm*, requiring new tools, which are themselves complex (and thus complicated). We might call this the elitist use of the word *complexity*. Yet there is indeed something to be said, something new, about life and especially about the genetic program. Of course it is true that it is not as simple as the organization of a salt crystal. Algorithmic complexity has shown us one aspect of this, and the advantage of this concept, even if it is very restrictive, is that it enables us to deal with biological reality head on. It allows us to avoid the trap of the unsayable, the incomprehensible, and throws a genuinely new light on what the analytical method has to offer. It also gives us a new lead, a new way to compare the relative complexity of different genomes. In this situation, if we talk about the complexity of a genome, we are not saying that we must always remain in ignorance; we are exploring its boundaries. When we do this we realize that, as with order, evaluating complexity requires an interaction between the phenomenon being studied (the genome text) and an observer who analyses it. A sequence has a certain complexity in Kolmogorov's sense (although we may not know how much), but different points of view lead to different estimates of its upper limit. We need to specify what kind of observer/phenomenon interaction the word *complexity* refers to, so that it will always be preferable not to use it by itself, but to qualify it according to the interaction in question. Algorithmic complexity is a special type of interaction, in which the genome text is regarded as a numerical sequence, in terms of number theory, Turing machines, and algorithmics.

It took many years for the concept of algorithmic complexity to spread and to enrich our idea of information. In 1989 Wojciech Zurek looked again at the thermodynamic analogies between information (in Shannon's sense) and entropy, in terms of algorithmic complexity. He managed to show that it was possible to compress information reversibly, without relying on a probabilistic approach (which risks dragging us into a vicious circle), and that what Shannon's information measured was essentially the receiver's ignorance. But as Jean-Paul Delahaye stressed, all this by no means exhausts what we can say about information. In this he followed Jean-Paul Benzécri, who, without any formal analysis, gives numerous examples of aspects of information that are not accounted for in existing theories.[36] In particular, the *value* of an item of information is an essential parameter—for instance, what is the value of information that we know exists, but is inaccessible in the time available? (The famous needle in the haystack: we know it is there, but where?)

Depth of Information: Bringing Time into the Picture

Despite its obvious interest, we soon realize that a definition of information as a kind of algorithmic complexity is far from satisfactory if we want to account for the value of the information in a sequence (in particular the text of a genome), and that something still remains to be said if we want to define it in more depth. In fact, the objective of producing algorithms is usually to generate results (which in their raw form are binary series) that have a certain interest, a certain practical or theoretical *value.* As we always try to make them as concise as possible, there are many cases in which a very short program can generate a sequence (or a drawing, as any figure that can be physically produced can be digitalized) that seems far richer than its mere algorithmic complexity—the length of the program—would enable us to predict. This is true of the beautiful fractal figures such as Koch's snowflake or Mandelbrot's set. The programs that generate them can be written in a highly condensed form, no longer than programs describing repetitive sequences (of a sufficient recurring length), which are clearly very "information poor." This unexpected richness is true of all recursively constructed programs, in other words programs in which a routine calls itself as the program runs.

This is the program that traces the Koch snowflake, starting with an equilateral triangle: Divide each side into three equal segments, form an equilateral triangle on the central segment, and erase its base. Repeat the procedure on each segment. Stop when the segments are no longer than three contiguous pixels on the computer screen (or the printer).

This simple program describes a very complicated figure, with an outline that is continuous but cannot be derived, and whose length increases very quickly toward the infinite with the number of iterations of the algorithm. A particular feature of the program is that the figure it describes remains the same whatever the scale, and although the length of its perimeter is infinite, the figure itself does not fill the surface of the plane. The local shape and the length at step *n* of the flake can be predicted very quickly. But it is much more difficult to tell whether a given pixel on the screen will be black or white, at a large enough enlargement (zoom), without a very long calculation. With a

Mandelbrot set, things are even more complex, because the figures on the boundaries, which form all sorts of interlacing spirals, change very quickly according to the scale considered (although when the magnification is increased, the Mandelbrot set itself reappears at regular intervals). Nevertheless, the programs that create these beautiful figures are algorithmically simple.

When we consider a sequence, then, we want to know more than just its algorithmic complexity (by finding an algorithm that describes it). We would also like an idea of the accessibility, and thus the usefulness, of the information it contains. If we compare the decimal places of π and the sequence 01010101 . . . , we immediately notice that although the calculations required to generate either can be described in a short program, there is a significant difference in the corresponding information content (information in its everyday sense), because although it is easy to predict the *n*th digit of 01010101 . . . (or any other more complex but repeated sequence), it is not the same with π. The information value of knowing a given digit of π seems much richer than the information value of knowing the *n*th digit of 01010101. Until the end of the 1990s we could not imagine knowing the value of a digit of π, if it was far enough along the sequence. The situation has changed recently, at least when these digits represent π in binary notation. A new algorithm means that for a given *n*, even if it is large (though obviously not so large that it is impossible to write it explicitly), then the binary units around it can be calculated, without having to work out all those in between. This still cannot be done in decimal notation, but it is not impossible that one day such an algorithm will be found. This shows that, just as we know that every sequence has algorithmic complexity, but may not know how great that complexity is, we cannot be certain of the time required to generate any particular sequence; we can only estimate it. Given a particular sequence, we will want to look for algorithms that will generate it, but we must always keep in mind the parallel importance of evaluating the program's run time. This is why, when we consider genome sequences, it is essential to have an idea of the way they are actually recognized and used in the cell, because if we do, we can envisage constructing algorithms that could show us how a cell behaves in order to understand the genome text. It is the genetic code that makes the connection between the abstract, symbolic character of the genome text and its real, concrete character when it is expressed.

In 1988 Bennett gave a formal definition of this aspect of information, which he called its *logical depth.*[37] The logical depth of a sequence *S* is the minimum time required to generate *S* using a universal Turing machine. It is not the Pythagorean aspect of numbers that appears in this context. On the contrary, logical depth is linked to the Aristotelian distinction between potentiality and reality. It indicates that we should never speak of what is potential in the same way we do of what is real. *It may be meaningless to speak of potential, because it may be impossible to realize that potential explicitly* in the time available. It is only with mechanically generated facts, which are chained together in the same way as a clock marks time, that the potential and the real can be assimilated in this unsophisticated fashion. This mechanical view, typified by the Great Watchmaker who ruled the world in the eighteenth century, is responsible for the long-standing confusion between what is determined and what is predictable. If it were relevant, the consequence of this mechanical view of life, accepted by many biologists who do not really know what lies at the heart of their science, would be a sad and terrible impoverishment of our identity. But we have only to recall the way the genetic program is constructed to see that it simply does not make sense.

Decidable or Undecidable?

Perhaps some readers are rather put off by the apparent complication of what has just been described, or by the mathematical flavor. So I would like to go back over what we have just seen, before we underline how it is connected with the intimate functioning of the cell.

First of all, implementing an algorithm, which is often recursive, is a common practice, and one that we are all supposed to have learned at school, in long division. (I hope that the use of calculators has not become so ubiquitous that these skills have been lost; they were part of the education of generations of schoolchildren, and are in themselves infinitely useful in training the human mind—to design calculators, among other things!) All the essential steps of a recursive algorithm are represented in this procedure. There is a question (in a particular instance, "Is 4,357 divisible by 21?," but in abstract terms "Is the dividend divisible by the divisor?") and a program that allows us to reach an answer. The program uses a routine that runs again and again in a succession of formally identical operations. We see how many times the divisor goes into the first figures of the dividend, we write this at the top, we multiply, we subtract, and we combine the result with the next figures of the dividend. Then we start the series of operations again. Each time the routine starts again, there is a test, a question to which we can answer only yes or no, and as long as the answer is still yes, the routine continues. When the answer is no, the routine comes to an end. If the remainder is zero, the dividend is divisible by the divisor. If the remainder is not zero, then it is not divisible. This is exactly the kind of process I mean, and we should note how universal it is.

As soon as we were able to bring together questions from number theory and logic in algorithmic procedures, it became fundamentally important to know if the final decision was *always* possible. This corresponds to the question David Hilbert posed as a challenge to the hypothetico-deductive method still in favor up until then. Hilbert thought that if mathematics, and especially number theory, was to have any meaning, there had to be an objective criterion by which to decide whether a theorem, written using the axioms and definitions of that theory, was valid or not. Gödel showed that this question could not always be answered, and that for a given set of axioms there would be undecidable questions, but that they could be decided as true or false given a broader set of axioms. Could this be seen in the case of the algorithmic method too? Credit goes to Church, Turing, and a few others for establishing the formal connection between Gödel's propositions and algorithmics, and for proving that here, too, there are undecidable questions, and that this is an inherent property in the nature of questions that may occur to the human mind (remember what we saw earlier, and that Turing stated that any function which can be calculated by a human being can be calculated by a machine). Whenever we have an algorithm that allows us to answer an infinite number of classes of questions, the property $R(a)$, or the relationship $R(a,b)$ or $R(a_1, a_2, \ldots a_n)$, whose truth we want to know, is said to be decidable if there is an algorithm that can reply yes or no to it. The algorithm thus plays the role of a decision procedure.

Of course the algorithm must exist *before* we can apply it to one or other of the questions of the class being considered. This implies that the description of the algorithm must be finite. So an algorithm is a procedure which can be described in *finite* terms, but which makes it possible to answer (or to demonstrate the undecidability of) as many questions as we want, an infinite number of questions, via the description of a finite number of operations. An algorithm is *generic,* and thus creates a link between the finite and the infinite. The formal analysis of this link, which gives algorithms all their power, shows that this also implies that an infinitely large number of questions will remain undecidable, because, in a manner of speaking, the algorithm cannot step out of itself to take the decision, just as the pipe cannot step out of Magritte's painting and become a real pipe. This inability, which is real and shows our lack of control over reality, in fact leads to the systematic production of problems and questions, and new kinds of solutions, about all material phenomena that can be described in algorithmic terms.

When it was first proved, Gödel's undecidability theorem seemed extremely complicated and difficult to understand. But once it was realized that the theorem was based on a particular branch of mathematics, the branch associated with the manipulation of algorithms, things became much simpler, and a great many proofs appeared, parallel to Gödel's, whose conclusions were equivalent. One of these proofs concerns the problem of how an algorithm decides to stop (known as the halting problem), an essential step if it is not to go into an infinite loop. Although it is not possible to go into detail here, the connection between the halting problem and the finite character of genome texts, if they are considered as algorithms, suggests that their formal properties are worth studying in detail, as a source of mathematical conjectures. By the very fact that they exist, they prove that it is possible for an algorithm to have a critical structure, a *critical depth,* which is related to their capacity to reproduce themselves in a given environment, while at the same time producing the machine that runs them.

Data, Machines, and Programs in the Context of the Cell

In theory, a Turing machine can carry out all arithmetical and logical calculations if it is allowed to run without a time limit. It is defined essentially by the fact that it makes a formal separation between the machine itself, the data that specify the conditions under which it functions, and the program it will execute. In actual fact the machine treats data and program as equivalent: both correspond to various states of the tape that the machine reads and modifies as it functions. In a normal proof using a Turing machine, no distinction is made between program and data. But it is more useful to visualize a machine with two tapes, one with the data and the other with the program, as this enables us to consider the case of machines with the same program but different data. As we wish to explore the Turing machine metaphor as it applies to living organisms, which are defined by their inside and outside, separated by an envelope responsible for communication between the two, it is natural to separate what can be considered as data (especially environmental conditions) from the program. The machine itself (the structure that runs the program) must also be specified.

If this distinction is well defined, any logical or arithmetical operation can be programmed and performed. The most remarkable properties of number theory appear only if these constraints are respected. But where is the cell, in all this abstract reasoning? To try to see clearly, it is essential not to be misled by the apparent complexity of living things, but instead to allow ourselves to be guided by a hypothesis (this is the only way to carry out a scientific investigation, because it never appeals to the unknown) and to consider only the simple as the source of even the most complex properties. This "simple" is easy to consider; it just means seeing to what extent it makes sense to talk of a genetic program when we consider the DNA sequence that makes up the chromosomes of a living organism. This is not a textbook on information theory or formal arithmetic; we are dealing with genomes and their meaning. To speak of the "genome text" is obviously a metaphor, and of course it is very restrictive. But do we have the right to do this? If so, it goes without saying that the properties we recognize in logic and arithmetic—everything we have looked at in detail—apply to living organisms as well (but of course they have other properties too). In the case of living things, can we actually identify what would play the role of the Turing machine, and in particular, is it possible to separate data, machine, and program, at least in abstract terms?

Four processes define life, organized into three main functions. These are compartmentalization, which defines the relationships between the inside and the outside; metabolism, which defines the flow and transformation of matter within the cell and its relationship with the outside; and the two processes that contribute to the function of information: memory transmitted from generation to generation, and manipulation or gene expression, the transfer from memory to physicochemical processes. If these functions can actually be physically separated, then it is possible to define the three levels needed for the functioning of a Turing machine with two tapes. To see this, we need to see whether experimental data show a real, physical separation between the program, in other words the genome text, and the rest of the cell.

The experiments on which the techniques of molecular biotechnology are based, and which are used to construct genetically modified organisms, give a first indication that this is actually true. Since the mid-1970s it has been possible to isolate a fragment of the genetic program (in physical terms, a DNA fragment) and insert it into the program of the chosen host (directly into a chromosome, or in the form of an artificial chromosome, in a replication unit such as a plasmid or a virus). The descendants of that host then behave as if they had been reprogrammed to take into account the program fragment which has been added (or which has inactivated a fragment originally present in the host). The formal separation is *effective,* because although scientists initially perform these operations on paper, they subsequently put them into practice physically, in the organism to be modified. This practice is now so taken for granted that the young researchers who do it every day are no longer aware of the mystery, the truly Promethean aspect of what they are doing. Perhaps the only people who can really know the emotion felt on first creating a living organism with the properties they have chosen are those who actually experienced for themselves the hesitant early days of these techniques, and saw their first blue colonies growing on a Petri dish. Blue, because one of the genes inserted during the process of modifying a cell by *in vitro* genetic recombination produced a product that made a molecule in the Petri dish turn a deep blue, thus distinguishing the modified cells from the pale parent cells, and showing that the work had succeeded.

Technically, this is done by characterizing the DNA sequence (the four-letter alphabetic text) and following a procedure similar to word processing, using a "cut and paste" technique to insert or delete letters or pieces of text as required. This text is a symbolic description of the molecule, which the scientist can then work on, physically and specifically, by *really* cutting and *really* pasting chemical motifs, at particular places in the text. The chromosome can thus be locally modified by the deletion or insertion of known DNA fragments, isolated from other organisms or synthesized from chemicals. It is the new chromosome that then determines the behavior of the offspring of the modified host, in the usual stable, hereditary fashion. This kind of modification is routinely carried out in molecular genetics laboratories all over the world. Because of the universality of the genetic code rule, it is possible to have different genes expressed in *heterologous* systems, to make bacteria synthesize human proteins, for instance. In this situation, the separation between the machine (a certain type of cell) and the program (the modified DNA) can actually be seen to be a reality.

One could point out that this concerns only *fragments* of a program and is a long way from demonstrating a real parallel with a Turing machine. However, a few more points show that in nature this separation is made real and exploited in very many ways. One of the most common examples of this is seen with viruses: these are essentially parasite programs, which force their host's genetic machinery to reproduce them by inserting the program into a specialized structure designed to get the viral nucleic acid (its genome, symbolized by its genetic program) into the heart of the host cell. These programs are far more complex than the isolated genes integrated into genetically modified organisms. But there are many more examples, and when the contents of the first genomes to be known in their entirety were analyzed, a very large part (often more than 10 percent) of a bacterial genome was found to consist of semiautonomous genetic elements that are transmitted laterally from one organism to another, and then preserved from generation to generation. We saw some examples of this horizontal genetic transfer with *E. coli* and *B. subtilis*. Another example is the case of viruses that sometimes integrate themselves into the genome of their host, and are then transmitted down the generations, until critical circumstances in their external environment suddenly make them "wake up," kill their host, and escape into the environment to infect other hosts. It often happens that they take a piece of the host chromosome with them in the process, and this may then be transferred into another genome.

But the result of horizontal genetic transfers does not represent an entire functional genome. We might still doubt that it was possible for the machine and the program to be genuinely separated, as would be necessary to establish the validity of the Turing machine as a metaphor for living systems, if the functional separation of a whole genome from the cell that expresses it had not actually been proven. This has now been done in the cloning experiments that caused such a stir in 1997, and are still the subject of much controversy. Cloning a mammal involves taking its genetic program by isolating a cell nucleus and placing it within the enucleated cytoplasm of another cell, using this as a machine to express the program until a fully grown organism is produced. Cloning has now been done over and over again. The first success used the enucleated egg of a sheep, producing the ewe Dolly, who grew up normally and became the head of a new lineage of sheep. Mice have also been cloned, using a fertilized enucleated egg from a white mouse strain, and replacing the nucleus with that of the "cumulus cells" from a black mouse, resulting in the birth of Cumulina, a black mouse born to a white surrogate mother. Of course, in these experiments what has been substituted is a little more than just the program, because a nucleus contains not just the chromosomal DNA, but also a great many proteins that enable it to be expressed. We thus have the program and a significant subset of the data (which Turing regarded as playing the same role as the program), and are therefore in a concrete situation very close to that described by Turing when he invented his famous machine and its programs. The problems encountered with many cloned individuals arise from the fact that although there is usually no change in the program (DNA) transferred into an enucleated egg, the data (made up of other components of the nucleus, proteins in particular) are much more variable, and certainly depend on where and when the nucleus was extracted.

We now have to try to understand the duality between a concrete, material system and the abstract, symbolic systems we have considered in the light of number theory and of the possibility of generating the irreducible. The logical necessity of an effective complementarity between matter and its representation in symbols was discussed at length by von Neumann in the late 1950s.[38] A central question concerned the reality of machines capable of creating ever more complex machines, which has often been called "emergent evolution." Von Neumann notes in particular that normally, matter and symbols are separate categories. He takes as an example the symbolic character of the functioning of living organisms, neurons, and nerve impulses. Neurons produce series of electrical impulses in the form of action potentials, but these impulses do not belong to the same physical or conceptual category as the neurons. Similarly, computers generate and use bits of information, but these bits do not belong to the same category as computers; microprocessors produce numbers, which do not belong to the same category as microprocessors, and so on. For this reason, a machine designed to produce a symbolic output does not normally produce machines like itself, and machines that produce silicon chips do not generate symbols (text for instance). This was a simple observation about real machines and the use of natural language, and not a statement in terms of Plato's ideals. In fact, with the development of computers, machines now exist which produce both text and material components, and we may well wonder where this will lead.

Von Neumann also noticed that these properties appear only at a certain level of complexity (not defined in this context, but implying some sort of organizational hierarchy). This is perhaps one of the origins of the misuse of this word, to refer to the unknowable, a property that some think is inescapably associated with life. The mere existence of living organisms suggests that this borderline needs to be explored further (especially because organisms exist whose genetic program is incredibly short, in the order of 500,000 base pairs). How can we define the borderline between what kind of organization of matter makes it capable of reproducing machines that build themselves, using very simple elements from their environment, and what does not?

Von Neumann and others explored this borderline theoretically. Some have even tried this in practice. Daniel Mange, at the Swiss Federal Institute of Technology at Lausanne (EPFL), draws inspiration from biology to explore this area by using real silicon chips. He has analyzed the properties of universal Turing calculation, and suggested that the construction of a machine would be possible only if the program itself contained a description of the machine, not in physicochemical terms but in symbolic terms, in the form of a map for instance. The replication of the program and the machine would then be logically possible if the program were a kind of universal builder, which duplicated both its own description and that of the machine. But as in the case of the Turing machine, it is important to avoid the logical paradoxes of self-reference, so the machine and its symbolic representation (a kind of program within the program) have to be separated. The very strong implication of this reasoning is that, associated with the program itself, there must be a representation of its environment (that is, a representation both of the cell and of what surrounds it). And this is something truly extraordinary—it does indeed appear that, at least in the case of bacteria whose genome is known, there is a map of the cell, and of the organism as a whole, in the chromosome. A geometrical program is thus superimposed on the genetic program.

The Map of the Organism in the Genome Text

The first genome of a multicellular organism to be known in its entirety was that of *Caenorhabditis elegans,* the model Sydney Brenner chose to study at the end of the 1960s. Before this, nothing was known about its genetics, but it proved to be of particular interest in understanding the functioning of certain animal genes. There are many parasitic species among nematodes, but *C. elegans* is a *free* nematode. This tiny worm, about a millimeter long, feeds on bacteria and lives at moderate temperatures, so it can be cultured in the laboratory in Petri dishes and studied with a simple binocular microscope. At certain stages in its development, the entire worm can be frozen in liquid nitrogen, and after thawing it will continue to grow.

Several characteristics make this species particularly well adapted to genetic study. It is a self-fertilizing hermaphrodite that produces large numbers of offspring, and in optimal conditions its reproductive cycle is two and a half days; but there are also males, so crossing is possible. Because it is transparent, it has been possible to describe all its cells (1,029, including 350 neurons), together with their lineages during development from the egg to the adult. The nematode has a relatively small genome, at a little over 100 megabases. Scientists have had access to a complete map of the chromosomes of the reference animal for a long time, as well as a genetic map showing the various mutants isolated over twenty-five years. All the laboratories that work on this organism effectively start with the same reference animal, which is homozygotic for all genes, so that each one has offspring whose genomes are strictly identical with the parent's. This means that if a mutation is induced, leading to a new phenotype, it is possible to study exactly which genetic modifications are responsible. Given all these considerations, Sir John Sulston's laboratory at the Sanger Centre and Robert Waterston's laboratory in St. Louis (Missouri) embarked on sequencing the entire *C. elegans* genome in 1989, finishing by the end of 1998 (not counting the unclonable or highly repetitive sections).

Scientists observed that the distribution of the genes in the chromosomes does not appear to be random, so that related functions are coded by genes that are close to each other. This situation is reminiscent of the operon structures of bacteria, but different in the sense that eukaryotes do not have single transcription units made up of several genes. This gives further support to the idea that the functional analysis of genomes should be tackled using a "neighborhood" approach. Furthermore, many genes are similar to counterparts found in mammals, and they can be used to explore the features that enable exon-intron borders to be identified, a notoriously difficult task. Finally, as the number of cells reproduced is exactly the same in one individual as in another, it is possible to follow how each individual cell develops from the egg to the adult stage, and thus to show that some of them die. This phenomenon of cell suicide is called *apoptosis.*

We have stressed the fact that its very construction as an algorithmic program enables a genome to systematically produce irreducibly new functions. When we extend this reflection to more complicated organisms, here, too, the processes that lead to the morphogenesis of differentiated multicellular organisms are algorithmic ones. The problem of the spontaneous formation of reproducible forms is one of the most fascinating in embryology. It is most extreme in insects like butterflies, which undergo total metamorphosis, where the egg, the caterpillar, the chrysalis, and the adult imago seem to have very little in common. I remember conversations with the village elders of Lay in Burkina Faso, then Upper Volta, and their incredulous laughter when I mentioned in passing—I collected butterflies at the time—that caterpillars turn into butterflies. They had made the connection between tadpoles and frogs (where the metamorphosis is progressive) but not for butterflies (they ate the caterpillars that fed on the shea trees, but did not breed them).

People have been trying to explain these forms for a long time. Because many reproducible structures exist in physics (branching structures, cells, circles, spheres, and so on), many thinkers looked to certain physical or mathematical principles to explain the genesis of forms in biology. According to these ideas, life has simply rediscovered the general principles that govern physics, and, as Plato suggested, life merely reflects the existence of an ideal world of Forms. This horribly reductionist, Platonist attitude prevailed for a long time. It is still sometimes popular among those who know nothing of biology, because they fail to understand two vital things: first, that the functions which construct, or which ensure control, have an essentially symbolic role; second, that the important form that is preserved in organisms is not the final shape, but the form of *the algorithm that constructs it.* And yet as Aristotle reports in his *Physics,* Empedocles had already proposed that it was the accidental combination of preexisting forms that produced the forms of animals:

> Solitary limbs wandered seeking for union. But, as divinity was mingled still further with divinity, these things joined together as each might chance, and many other things besides them continually arose. Clumsy creatures with countless hands. Many creatures with faces and breasts looking in different directions were born; some, offspring of oxen with faces of people, while others, again, arose as offspring of people with the heads of oxen, and creatures in whom the nature of women and men was mingled, furnished with sterile parts.[39]

Life certainly uses the principles of physics—and we have seen the possible role of the hexagonally tiled plane—but just as a basic vocabulary, a set of elementary processes, organized into a program, not as the main construction principle of life.

It is the fruit fly, *Drosophila,* that probably remains the best example of what genetics can contribute to an explanation of the genesis of forms in insects and, by extension, in vertebrates. Thousands of mutations are known in this organism, and a great many of them affect its development plan. But following on from work initiated by Edward Lewis at Caltech in the 1950s, it was surely the discovery of mutations as spectacular as those strange creatures described by Empedocles, including the *bithorax* mutation, which produces flies with two pairs of wings, or *antennapedia,* characterized by adults with legs growing where the antennae should be, that broadened the scope for an algorithmic explanation of morphogenesis in animals.

These mutations drew attention to forms—such as the leg or thorax or wing—that are not particularly "natural" in terms of physical or mathematical morphogenesis; but above all they showed that modifying *a single gene* could enable a complete, correctly formed organ to be built, but in a different part of the body. This is a long way from the extreme poverty (for biology!) of the "catastrophes" of differential geometry (which in any case can account for only a very few forms), and even further from the "dissipative structures" invoked to explain the genesis of hexagonal cells in Bénard convection, or the oscillations—repeated ad nauseam—of Belousov and Zhabotinski's chemical experiments.[40] We can only be astonished that, confronted by the marvelous variety and sheer *gratuity* of insect forms, scientists have not more often been inspired to explore the mode and timing of their production by starting from reality itself, rather than by hiding it under a veneer of simplistic, reductionist ideas. Such are the vagaries of human thought.

After twenty or thirty years of painstaking work, at the beginning of the 1980s genetic studies of *Drosophila* established that the initial transition from the egg to the formation of a segmented embryo was controlled by the hierarchical organization of the function of a series of genes. Studying mutants with one or more affected genes enabled scientists to plot the development of embryonic cells and segments, providing an initial guide to the likely interactions between certain genes and their products. First it was discovered that it is the organization of the egg itself that plays the major role in determining the sequence of changes in shape undergone by the embryo. And although this organization may appear mysterious and unlikely, the fact that there is a plan of the cell in the chromosome simply tells us that because the egg is a cell like any other, it is organized by the position of the genes in the chromosomes. In particular, a number of essential messenger RNAs are positioned in specific compartments in the egg, where they play the role of factors which trigger the changes that follow fertilization by a spermatozoon, and the events that characterize embryogenesis.

We now have a precise map of the site of a great many genes' activity, and of they way they interact to produce the segmentation of the embryo and the formation of the two main axes on which the insect's body plan is organized, the anteroposterior and dorsoventral axes. Homogenous anteroposterior bands are formed as the result of cooperation between products which can diffuse from one cell to another, and others which remain confined to the cell that synthesized them. These syntheses also have a time limit, and it is the whole scenario of successions of products with catalytic effects and control products that implements the spatiotemporal plan of the organism, in just the same way as the routines of an algorithm would. The processes are all extremely simple in themselves, but the way they are strung together is complex, because it is compartmentalized in space and time. Although the diversity of the control elements is limited, their *combinatorial* possibilities are extremely rich. As always in biology, at this early stage in the development of the embryo there is an infinite variety of combinatorial possibilities. Timing is also important of course. We have seen that genes are so long that their transcription often takes an appreciable time.

To return to the map of the fruit fly, the formation of alternating bands (such as stripes, so often found in nature) is simply explained by a coupling between the synthesis of diffusible products and the positive control of these syntheses. Here, remarkably, we meet Alan Turing again. He pioneered the study of autocatalytic control and reaction-diffusion mechanisms, which can result in the creation of stripes. We know that the layout of the metamers (the ancestors of segments) in the *Drosophila* embryo is generated by a series of interactions between specific regions of DNA (transcription promoter regions) and specific proteins. It all begins with the coupling of three concentration gradients of transcription factors (which implies a spatial organization of these products in the egg itself, in the same way as the organization of the chromosome governs spatial organization within the cell). At the animal's extremities, the environment is very asymmetrical, because there is the animal's body on one side and the surrounding atmosphere on the other. This asymmetry in turn controls the expression of specific genes. Genes with esoteric names like *bicoid* or *knirps* play an essential role at this stage, and some of these are particularly long (as we have seen, this indicates that *time* plays an important role in their functioning).

I will not describe the successive stages of the shaping of the embryo any further (this would need a book to itself), except to say that at the turn of the new century, as the almost complete sequence of the *Drosophila* genome was published by Craig Venter's company Celera in association with an academic consortium, most of these stages were known—enough to account for the way the embryo is organized in segments, and how it differentiates into the elements that will become the head, the thorax, and the abdomen in the adult fly. What is remarkable is that the genes concerned are organized in a hierarchical control cascade, with master genes that allow the expression of other genes downstream from them, according to a strict logic and chronological order. During this development, certain cells are programmed to disappear, leaving room for other cells which are differently differentiated, and which could not otherwise have developed. It is thus important to note that development includes a significant element of *absence,* as distinct from presence, so that a "negative" form plays a role in development that is just as important as that of a positive presence. Once again, the timing and modes of morphogenesis are organized by appropriate routines. The next level of the hierarchy brings into play other pairs of genes with equally strange names like *hairy, runt, hedgehog,* or *even-skipped*. Then comes the moment when the *homeogenes* are expressed—those genes responsible for the specific development of segments or organs: the genes *antennapedia* and *bithorax* are typical examples. These are themselves under the control of genes such as *brahma* or *polycomb,* which act via a series of logic gates that program the development of the insect's shape in just the same way as an algorithm.

We now know that it is the way sequentially expressed genes are combined that presides over morphogenesis. The production and positioning of the corresponding forms are governed by a series of events that unfold in the same way as the strings of procedures in algorithmic programs. Remarkably, in insects such as *Drosophila,* the genes that control cell differentiation, and especially the homeogenes, are positioned on the chromosomes in a *directional* arrangement from the tail to the head, in which each segment of the insect is represented by a set of genes located *at the same place in the genome text as in the adult fly.* It is exactly as if something were preformed, but it is not the preformation of the whole organism, as Charles Bonnet thought when he discovered parthenogenesis in aphids (where the female is full of tiny aphids) and the role of the yolk sac in chickens' eggs:

> Who could deny that the Almighty was able to enclose, in the first Germ of each organized being, the set of Germs corresponding to all the revolutions our Planet was called on to undergo? Do the microscope and the scalpel not show us the generations nested inside one another? The bud hidden away deep beneath the Bark, the tiny future Tree enclosed in this bud; the Butterfly, in the Caterpillar, the Chicken, in the egg, the egg in the ovary? We know of Species which undergo many metamorphoses, which make each Individual take on such varied forms that they seem to be so many different Species. It seems that our World once existed in the form of a Worm or a Caterpillar: now it has the form of a Chrysalis: the final revolution will see it take the form of the Butterfly.[41]

What preexists is not the organism itself, but the *preformation of a development algorithm.* There is indeed what Alain Prochiantz amusingly called a *drosophiloculus* inside the *Drosophila,* but what heredity passes on is not the form, but its construction program. The *successive expression* of control genes, activated or suppressed one after another, enables morphogenesis to take place (while respecting and making use of the constraints of physics, of course, such as the rules of overall symmetry). And in fact experiments show that merely altering the succession of the identified control genes alters the morphogenetic pattern: it is possible to make antennae grow where the legs should be, or to produce two pairs of wings, making the *Drosophila* look like a Hymenoptera (incidentally this also has a lot to teach us about how forms have evolved during the evolution of species).

But what about vertebrates, whose segmented structure is less visible than in insects? First, it seems that the formation of the embryo requires the creation of an anteroposterior axis similar to the head-tail axis in the fly, and also of a dorsoventral axis. Similarly, we know that the segmented structure (metamers) exists during embryogenesis. And we have all seen the segmented flesh of fish. Besides, we only have to look at our own ribs and vertebrae to see that many features of the body are the same from one segment to the next.

Since the late 1980s, it has been discovered that the mouse, too, has homeogenes, counterparts of the control genes in *Drosophila.* These *Hox* complexes are so close to those of *Drosophila* that not only are the amino acid sequences of the proteins they code for very similar, but the genes themselves and the promoters recognized by their products are also clearly related! This discovery has been marked by one extraordinary surprise after another. The organization is so hierarchical that modifying a single gene, *Lim-1,* produces animals without a head. In the South African toad *Xenopus laevis,* widely used as a model for vertebrate morphogenesis, it has also been shown that adding a small quantity of a protein appropriately named *Cerberus* to the abdomen of the tadpole made a one-eyed head grow there. One of the most spectacular discoveries has been that in both *Drosophila* and the mouse, a homologous homeogene (in a sense, the same homeogene) controls the genesis of the eye, whether it is the insect eye with multiple facets or the mammalian eye with its retina and lens. A common origin has thus been found for this organ, which is so different in insects and vertebrates, but nevertheless has the same function, vision, which was thought to be the result of a mysterious convergent evolution. Similarly, the various appendages used as limbs are controlled by homologous homeogenes. And with the anatomical oddities noted by Geoffroy Saint-Hilaire, such as the fact that the main nerve (the spinal cord) is dorsal in vertebrates but ventral in crustaceans such as the lobster, apparently representing a reversal of the overall layout of the organism around its center when comparing crustaceans with vertebrates, this reversal turns out to be reflected in the order of the homeogenes concerned along the chromosomes!

However, there is a significant difference between mammals and insects. In mammals, instead of a single linear arrangement corresponding to the layout of the insect, there are *four* linear arrangements, arranged exactly as in the fly, and also corresponding to the animal's development from the tail to the head. It seems as if the *muriculus* (or the *homunculus*) is made up of a four-dimensional arrangement, in the same way as the notes in four-part harmony are written on a musical score. If an insect is played by a single instrument, we need a string quartet to play a vertebrate. This discovery accounts for mammals' greater complexity compared to insects: the construction algorithm is produced by the combination of four homologous procedures working simultaneously. It also explains how the segmented character so visible in insects (mostly at the larval stage, of course) is much weaker in mammals. We can also definitely see signs of *evolution by duplication* of the genetic program, which suddenly makes new properties appear—the effects of duplication are not only quantitative, they also create new relationships *de facto*. Incidentally, this also predicts the existence of a missing link—which may have produced organisms too unstable for their descendents to be still with us. We can thus expect to discover, between insects and mammals (perhaps among the most primitive vertebrates, or the ancestors of vertebrates), a family of organisms whose *animalculus* is formed by the *duplication* of the plan found in insects.

We are gradually discovering, from the cell to the entire organism, a systematic structuring responsible for the distribution of the various constituents, in time and space. Contrary to what is only too often thought, this is "written" not just in the individual genes, but in the way they are organized in relation to one another, to make up the text of the program that defines the production and positioning of their products and the structures that organize them. What makes the organism in its entirety is not a form derived from those mysterious principles evoked by vitalism; it comes from the program that shapes it. This explains how there can be a constant readjustment between the changing reality of the environment, and a memory that is preserved and reproduced identically. There is no archetype for the form, but instead, it is what *produces* the form that is preserved, and, in producing that form while taking account of the environment, it adapts it anew each time. This is how life constantly accommodates both permanence and change. Life is ready for the future because it does not need to know everything the future will bring. It needs only to find in it some specific characteristics—provided they do not involve a radical change—that will enable it to carry on.

5

WHAT IS LIFE?

Physics, Chemistry, and Genetics

Trying to explain life goes to the very heart of scientific knowledge and links the different domains of science: mathematics, physics, chemistry, and biology. Several physicists who played a major role in the development of what was to become molecular biology quote Erwin Schrödinger's book *What Is Life?* as being responsible for their decision to study life itself.[1] It introduced an entirely new way of thinking about biology, because it brought together the two sciences of physics and genetics into a conceptual synthesis. But to see the whole picture, we have to stand back a little.

Generally speaking, it is fairly easy to build up a picture of the physical world, and to explain it in terms of a combination of simple principles (at least if we do not look at matter too closely), because physics is concerned with reproducible *objects* that cannot normally be distinguished from each other as individual entities. In physics, the explanation of a phenomenon often stems from the way its future is determined—its predictability on the basis of the past and the present. We identify the "things" that make up the world, usually without defining too precisely (at least for the time being) the scale on which we choose to describe them. We study their nature and their behavior in terms of space and time, and we predict their changes from one state to another. We recognize that they interact, but there are not many kinds of interactions, and they can be represented quite simply in terms of mathematics: these are the major laws of physics, such as the laws of electrodynamics or of gravity.

Observers of the night sky have long known that the movement of the planets is so regular that their course can be predicted exactly. This regularity suggests a clockwork image of the universe, which naturally led the human mind to describe the cyclical movement of the planets and the stars in purely mechanical terms. Building on the work of Newton and Leibniz, Laplace and his successors took this approach to its logical conclusion when they said that that merely knowing the initial conditions of any physical system would allow its future to be predicted with total accuracy. It is true we can make such excellent models of the physical world (at least—for Man is the measure of all things—models of material objects on our scale) that we can put satellites into orbit with finely tuned precision. But we have become blind to the beauty implicit in their excellence.

Chemistry is more complex than physics, and begins when atoms combine. Generally speaking, chemistry does not look further into the atom than the outer electron orbit. It is simple at fairly low and fairly high temperatures, but at intermediate temperatures (roughly speaking, the temperature at which water is liquid) many atoms, particularly those in the first two rows of the periodic table, combine into molecules. They form chemical bonds, which sometimes allow the atoms to form quite substantial clusters. These can then behave as individual entities—objects, once again—and it is not normally necessary to understand their internal atomic composition in order to understand their behavior. Chemistry is mainly concerned with their interconversions, as well as with their synthesis and their degradation. It is the object itself, the chemical product, that is deemed worthy of interest. Although atoms come into the description, the catalogue of relationships between them is very limited. Except in very new areas of chemistry (which are often inspired by biology in fact, as with combinatorial chemistry, or with "biomimetic" or "bioinspired" chemistry), there is nothing very complicated in the nature of the chemical bonds between the atoms of a molecule, and knowing them is a sufficient basis on which to predict its behavior and describe its properties.

In chemistry, two individual examples of the same object are usually indistinguishable when they are observed under similar conditions. However, there is one particular characteristic, quite rare in physics but almost universal in chemistry, which clearly illustrates the importance of the relationships between the parts that make up the object in question. There are structures that are identical in every respect but their symmetry, and the link between chemistry and biology was formed after a distinct bias was observed in the symmetries of chemical products produced by living things. It was Louis Pasteur, a crystallographer, who made the astonishing discovery of an original characteristic that went hand in hand with life: molecules produced by living things crystallize into an asymmetrical form, which does not have an opposite counterpart. (On a different scale, a snail's shell—with very rare exceptions—always twists in the same direction.) If they have been chemically produced, these asymmetrical shapes generally have a counterpart in a different shape, similar to the first in the same way as a mirror image is to the original or the right hand to the left. Chemical molecules created artificially produce a mixture of crystals in the two forms, symmetrical with each other. In the same way, those mineral crystals that are asymmetrical exist in both forms, in equal proportions. "Asymmetry is life itself," said Pasteur, as he was led to the conclusion that there is a specific process in the chemistry of life that distinguishes it from ordinary chemistry, and that living beings are endowed with intrinsic, essential properties.[2] He summed this up as *Omne vivum e vivo*—all life comes from life—paraphrasing William Harvey's famous aphorism, and thus introduced a vitalism that persists today. His philosophical convictions prevented him from believing that life could arise spontaneously from matter, either mineral or organic. Life could be born only from life. And because it was well known that broth left out in the air would soon be teeming with life, the obvious implication was that live germs were already present, waiting to multiply in the broth. Pasteur thus stated: "My contention is that I have proved conclusively that in all experiments which are thought to have demonstrated spontaneous generation of the lowest levels of life (that being the level at which this debate is carried on nowadays), the observer has been the victim of illusions or errors, which he has either failed to see or has been unable to avoid."[3]

The fermentation that takes place when wine and beer are produced had first been explained by Berzelius and Liebig, before 1840, as a merely catalytic phenomenon, which did not justify any distinction between organic and inorganic matter. But here was Pasteur, proposing to do away with the idea of spontaneous generation, on the basis of his observation that living things produce molecules that tend to be asymmetrical. The tartaric acid crystals produced when grapes ferment all take the form of a single optical isomer, producing asymmetrical crystals, all oriented in the same direction, whereas crystals obtained through a chemical process make up a mixture of two optical isomers, with equal numbers of both right-handed and left-handed forms. This is known as a racemic mixture, from the Latin *racemus,* meaning a bunch of grapes, to reflect the role that grapes played in the discovery. (In Dorothy

L. Sayers' murder mystery *The Documents in the Case,* the victim is poisoned with chemically synthesized muscarine, the murderer intending to pass it off as mushroom poisoning. He is caught precisely because the poison is shown, by polarization, to be a racemic mixture.)[4]

Specific problems thus arise when chemical objects are drawn from the realm of biochemistry—and when we consider living objects themselves. Although it is still easy to define biological objects (or, more precisely, differentiate them), when considered individually their definition remains very imprecise. It is impossible to distinguish between two atoms of the same object, in the same state, but it *is* possible to distinguish between two individuals of the same species. A species is a population of individuals, a class of objects each with its own identity. This is true even for microbes like bacteria: a look at the way they swim will show that two individual bacteria, which look the same and are genetically identical, can very often be distinguished by their behavior. It is also true of cells, the "atoms" or units of life.

It was over 170 years ago that Rudolf Virchow, Matthias Schleiden, and Theodor Schwann realized the significance of what was probably the most important observation to date for our understanding of the nature of living organisms. Developed into a concept, this observation, that all organisms are made up of elementary partitioned units, the cells, brought life back into the realm of physics. *"Omnis cellula e cellula,"* said Virchow in 1855, himself echoing Harvey—every cell is born from a cell. These basic units vary considerably (in shape, size, or chemical composition), but they all possess the properties necessary to sustain life, and they all share the same general organization. In multicellular organisms such as plants or animals, cells combine with others of the same type to form *tissues,* which are grouped together in *organs,* each specializing in some general task essential to the functioning or reproduction of the organism. In animals, for instance, four main types of tissue are recognized—the muscular, nervous, epithelial, and conjunctive tissues. In turn, the tissues combine to make complex structures, organs such as the heart, the intestine, the lungs, or the brain. Finally the organs are grouped together into *systems* (organized sets of biological functions), such as the nervous system, the circulatory system, or the respiratory system. An animal's body is thus organized in a *segmented hierarchy,* rather different from the *branching hierarchy* of plants. It is very important to understand how all these cell types, sufficiently different from each other to make the various organs or systems, can be built from one single cell, and therefore from one single genome, but although this is an interesting question, it is secondary to the main question of what life *is.*

The first trap for the unwary biochemist is to assume that simply cataloguing a cell's molecules will make it possible to reconstruct the way it functions. This illusion is fundamental to the difficulty many nonbiologists have in understanding biology. Thinking that biochemistry tells us all there is to know about biology, they look at the collection of objects put together by the biochemists and reject its reductionism, often citing the truism that "the whole is greater than the sum of its parts." This ridiculously simplistic holistic vision is obviously meaningless for anyone who really wants to know what life is. Analysis is not the same as reduction. The biochemist Albert Szent-Györgyi pointed this out when he described the path he had followed in seeking to understand how energy is managed inside the cell. In his desire to understand life, his career had taken him from the largest dimensions to the smallest. From animals he went on to cells, from cells to bacteria, from bacteria to molecules, from molecules to electrons. But the irony of the story is precisely that molecules and electrons do not have life. Somewhere along his path, life had slipped though his fingers, and he emphasized that he had to retrace his steps on this journey into the heart of matter, and climb back up the staircase that he had taken such pains to climb down.

In *An Introduction to the Study of Experimental Medicine,* the Frenchman Claude Bernard explicitly declared his ambition of describing life in analytical terms.[5] In 1927, the year that the Cold Spring Harbor Laboratory was established in the United States, his ideas had been the inspiration for the creation of a research institute in Paris, the Institute for Physico-Chemical Biology. Here a group of physicists, chemists, and biologists, led by Jean Perrin, well known for his early work on the identification of atoms, studied the intrinsic characteristics of molecules found in biology and compared them with the usual molecules of chemistry. Biological molecules were initially recognized by their adhesive properties, and were named colloids, from the Greek κόλλα *(kolla),* meaning "glue," whereas chemical molecules behaved like minerals and were thus known as crystalloids. The scientists' aim was that declared by Bernard in his *Leçons de pathologie expérimentale:* "In physics and chemistry, experimental analysis leads to the discovery of the basic mineral elements which make up composite bodies. In the same way, in order to discover the complex phenomena of life, we must delve deep into the organism, analyze the organs and the tissues, and reach down to the basic organic elements."[6]

Progress in microscopy during the nineteenth century had made it possible to identify structures within the nucleus that seemed to be preserved from one generation to another and to be specific to each of the species studied. These were called chromosomes, from the Greek χρῶμα *(chroma),* "color," and σῶμα *(soma),* "body," because they could be made visible by using a colored stain. At Basel, Switzerland, in 1867, Friedrich Miescher isolated a substance he called "nucleine" from the sperm of salmon and pus from soldiers' wounds. Given its specific location within the nucleus of the cells, it seemed likely to have some role in heredity, which was known to be linked to the nucleus. However the second half of the nineteenth century also saw the birth of an approach based on the observable, macroscopic characteristics of living organisms and their hereditary transmission—*genetics*—which developed alongside the chemical approach to life. Genetics was a formal, *phenomenological* approach, in which what is observed is described in a conceptually organized fashion, but without seeking any underlying explanation.

Genetics measured the inheritance of hereditary characteristics, or "characters," as the geneticists called them, without investigating the specific (molecular) reasons for their inheritance. The aim of genetics is to understand the laws that formally govern heredity. It functions at an abstract, conceptual level—the objects corresponding to the characters, such as the color of flowers or the shape of peas, in Gregor Mendel's work, are not considered—and is thus essentially a science of relationships between objects rather than of objects themselves. But if objects count for less than the relationships between them, we are dealing with an unusual kind of *form.* It is not only a question of geometry, of the external shape or form of the organism, or the parts that make it up. This new form that genetics considers is no longer merely an architecture; it is no longer concerned directly with the surface of things, which limits them and defines their individuality, but more with an internal set of relationships, resembling a graph or a structure made up of nodes, meshes, and apexes.

What is remarkable about genetics is that, although it was initially a purely phenomenological science, it soon became explanatory, because as we now know, the description produced by genetics revealed in intimate detail the relationships that make up the rules of heredity. With hindsight, this explains how, at the beginning of the twentieth century, Thomas Hunt Morgan's work on *Drosophila* revealed a set of original relationships between the units of heredity, the *genes.* Everything pointed to the likelihood that these abstract entities, responsible for characteristic traits such as the color of a fly's eye or a flower's corolla, were linked to each other in the same way as the words of a text: in an oriented line. This was the first element of a model of heredity that was to develop into the now-familiar alphabetical image.

We have begun to build up a mental picture of the laws of life by following the path that led to their discovery. For a long time, the definition of life was essentially a question of reproduction (or rather, the production of offspring that resemble their parents) and of the association between the organism's response to and modification of its environment (what was called its "vegetative soul") and its motor activity (its "animal soul"). Plants had only the first of these, while animals had both (and sometimes more, as with man, an animal endowed with reason). This conceptual approach to life was based on an animist vision of a world made up of the four elements (earth, air, fire, and water) rather than on atoms as we know them. The first real steps toward a definition of life began when the cell was recognized as the basic unit of life, and then the chromosomes as the site of heredity. Chemistry on one side and genetics on the other sought to explain the phenomenon of life quite independently of each other, but physics came to play a fundamental role in the recognition of the processes that make life, because it has an intermediate position between the two. Its objective is to understand not only the basic objects making up the matter of the universe, but also the way in which they are created, and their links with the general questions associated with the existence of space, time, and movement. Studied in the light of the spatio-temporal relations they have between themselves and with the intrinsic properties of life, the objects of biochemistry could at last be understood for what they really are.

The Four Basic Processes of Life

What is life, then? Schrödinger suggested that the answer to this question, inspired by his fellow physicist Max Delbrück, would be found by identifying a "memory" process, which would provide a link between heredity, with its abstract mechanisms, and the physical character of a real, concrete object. He concentrated on trying to imagine the material nature of this object, simply based on the intuitive idea one might have of what heredity is. He visualized it in the form of an object that would be very regular (or *periodic*) on a large scale, but irregular on a smaller scale: an "aperiodic crystal," which might have a filamentous structure. Ten years later, this remarkable intuition led to the discovery of the structure of DNA, with its double helix of filaments or strands. However, it is certainly not possible to give an account of life based on an understanding of this structure alone. Applying the analytical method to living organisms in the same way as to the objects of physics has led us to recognize the following phenomena:

The processes that make life are *metabolism, compartmentalization, memory,* and *manipulation.* Metabolism and compartmentalization are organized by small molecules (comprising a few tens of atoms, with a carbon skeleton), whereas memory and manipulation are controlled by nucleic acids and proteins, so the scale of their basic components is that of macromolecules made of thousands, millions, even billions of atoms. The concept of the macromolecule (related to the concept of the polymer, and of the corresponding physics of "soft" matter) defines a new era in physicochemical studies. Two spatial scales are thus interlinked in all living processes, which operate on a *mesoscopic* scale, intermediate between our macroscopic world and the microscopic world of atoms. This is the scale that is revealed in the geometrical program superimposed on the genetic program in the genome.

Reconciling all these processes has seemed so difficult that a physicist like Freeman Dyson even proposed that life originated twice![7] It explains why most molecular biologists have simply omitted to take metabolism and compartmentalization into account as questions that all models of life and of its origin must answer, and have considered only the manipulating objects, proteins, and those that carry memory, nucleic acids. Similarly, at the conceptual level, when comparisons have been made between life and Turing machines, the general principles for the construction of a self-replicating machine have nearly always overlooked the need for compartmentalization and metabolism.

It is important to put the objects and processes of life into a new perspective, so we will take a quick look back through the catalogue they constitute. *Metabolism,* as its Greek name implies, is a state of flux. It represents the construction of molecules from smaller parts (anabolism) and the breaking down of larger into smaller ones (catabolism), and is thus responsible both for building the individual components of the living machine and for managing the energy required to make it go. Dead organisms do not metabolize. There is a special situation, *dormancy,* in which an organism is neither dead nor alive, as in the case of fungal or bacterial spores, or of lyophilized bacteria, but we need to witness metabolic processes, when the spore germinates for instance, to decide whether the dormant organism is really a living organism. *Compartmentalization* is necessary for metabolism, as it separates the inside of the organism from the outside and the different parts of the cell from each other. As organisms have developed, some have retained the minimal single-cell plan in which the cytoplasm is separated from the outside environment by a more or less complex envelope, while others have developed multiple membranes and skins, separating the inside from the outside with increasing efficiency. These two general strategies for occupying the surface of the Earth have taken shape in the evolution of two quite different cell architectures, the prokaryotes and the eukaryotes, while the prokaryotic class has itself been split into two general classes: the usual bacteria of the standard environment, the eubacteria; and the archaebacteria, confined to special niches where extreme conditions prevail (high temperature, acidity, or salt).

Metabolism and compartmentalization are certainly not incidental, but are relevant to all features of living organisms. This means that organisms such as viruses, which do not metabolize, cannot be considered to be straightforward living organisms. They must be studied for what they are: pure parasites, a memory that perpetuates itself at the expense of a genuine life, that of the cell they have infected. Of course they are not similar to the usual nonliving matter found on the Earth; they seem to be artifacts created by life, and perhaps derived from early processes by which life was created from an ancestral metabolism.[8]

When people discuss life nowadays they often remember only the second two processes of the four mentioned above: *memory* of the preceding generations, which is transmitted to the following generations; and the general capability of organisms to *manipulate* a wide variety of objects inside and outside the cell. These processes of memory and manipulation organize and rule the flow of information necessary for life to exist. Associated with memory and manipulation are the two fundamental laws of *complementarity,* which accounts for the vertical transmission of memory; and *coding,* which is the link between memory and manipulation.

The information carried in the DNA controls the synthesis of proteins, which have "manipulative" properties in that they are responsible for the interconversions that take place in metabolism, the intricacies of the control processes, and the construction work required for compartmentalization. We will look at some further aspects of manipulation later in the chapter. But the whole DNA text is not directly expressed as proteins; far from it—only certain segments of the text specify or "code for" proteins. So there must be a syntax (or set of rules of grammatical construction) that makes each of these segments correspond to the sequence of amino acids of a specific protein. The correspondence between the formal object of genetics (the Mendelian gene) and the physical object of molecular biology (the gene in the DNA) is well established, but it is not totally without ambiguity.

The further a prospective concept is investigated, the more different aspects of it come to light, so that different interpretations of the word accumulate, and eventually become incompatible with each other. Historians of science like to say that nobody knows what a gene is, and to speak of the "classical," "developmental," "biochemical," "molecular" gene and so on. However, we now need computers to understand what a gene is, and this certainly leaves little room for vagueness. In this book, a gene is mainly the portion of DNA that codes for a protein (in some cases several genes can derive from DNA segments with some parts in common). Strictly speaking, this definition ought to be extended to a few other roles for DNA (see the Glossary at the end of the book), but it is not necessary to go into more depth here. Because a gene is expressed, it must have various *control regions,* so it can no longer be defined by describing the overall appearance of an individual, its phenotype, as it was in the time of Mendel and his immediate successors (a phenotype is, needless to say, always partial). Of course a mutation in a gene, or in its control region, can result in a phenotype that is almost identical with that produced by the nonmutated gene. But rather than pointing to a difficulty in identifying what a gene is, this simply tells us—and this is extremely important, because this mixing up of concepts has led to public confusion—that an altered phenotype, as compared to what is considered to be the normal state, is *not enough* to define a gene: according to the context a gene can specify a variety of phenotypes, and conversely, an altered phenotype can result from alterations to a variety of genes. Alterations in the DNA regions that act as timers or spacers in the organization of the chromosomes, and in their expression, can lead to all kinds of differences in phenotypes, usually subtle ones.

When the Mendelian gene is responsible for the appearance of a character like eye color in the fruit fly, it is often possible (though not always!) to identify the corresponding segment of DNA. Boris Ephrussi, working at France's Institute for Physico-Chemical Biology, found the gene in the fly that codes for the synthesis of an enzymatic protein, one which transforms the appropriate chemicals into red pigment (although at the time, of course, it was not known that the gene was composed of DNA). Things become much more complicated when genes play the role of the conductor of an orchestra, creating the conditions for the harmonious functioning of the cell.

The Genetic Program

At this stage of the formalization of life, we need to reemphasize the physical, concrete, actual separation between the program, with its material basis, the DNA, and the objects that result from its expression, the proteins. It is the *correspondence* between the sequences of nucleotides and those of the amino acids, which changes the operating level of the physical entities from that of the nucleic acids to that of the proteins, that makes it possible to use the metaphor of a program when we talk of the *formal* content of DNA. The physical creation of an individual organism is determined by the consequences of the expression of its genetic program, carried out by those "manipulating" objects the proteins (which are in particular capable of manipulating the very substance of the program, the DNA, by introducing variations while copying it). One important result of this correspondence between the DNA and the proteins is that it introduces time and space into a self-referring representation of the life of the cell (or the organism, if it is multicellular).

In the sequence of the nucleotides and the amino acids that correspond to them, the genetic program specifies absolutely everything required for the construction, functioning, and reproduction of every organism, when placed in an appropriate cellular context. We are just beginning to identify the nature of the signals responsible for controlling how the macromolecules that maintain and express the genetic program are themselves constructed, in space and time. We do not know all the major functions of genes, although for a long time we thought we did. There is still a great deal that is obscure in our understanding of life.

The metaphor of a program seems to be remarkably well suited to the job of describing the principles that guide the construction of the cell. In this metaphor, we do not consider cells and living organisms in general in the same way as mechanical structures like the automata that were so popular in the royal courts of Europe in the eighteenth century. A mechanism has a fundamental property which a system driven by a program does not share: its progress is entirely deterministic, and above all *predictable,* if we write down all the steps in the system and the conditions of its environment. A mechanical system is thus deterministic *and* predictable. If we know the present position of a cog in the movement of a clock, we can tell in advance what its exact position will be at a given time in the future (provided we know that the clock has been wound). In any programmed system that is at all complex, the situation is fundamentally different. For example, a *recursive* program causes itself to be run within itself without there being any internal contradiction, provided that it is properly designed. This way of constructing a program, though rigorously deterministic, can have unpredictable effects, in the sense that they are not known—and thus not real—*until after the program has actually run.* They are both deterministic *and unpredictable.* With programs of this type, this unpredictability is true only in the strictest sense, that is *a priori.* It is clear that *a posteriori,* the running of the program becomes predictable (if this is not a contradiction in terms). The unpredictability is meaningful only for the *first* run. This metaphor, made real by the use of computers, places the study of genomes in a new light, very different from that shed by mechanics, which still dominates the way we think.

It was in this sense that the announcement of the first complete chemical definition of the construction program of two bacteria in 1995, exactly a century after the death of Louis Pasteur, was a genuine revolution. Because the genome text can be studied only with the help of computers, this announcement also turned the spotlight on a new approach to the study of life, experimentation *in silico.*

The main risk with *in silico* methods is a tendency both to mistake the part for the whole, and to fail to consider important questions simply because they are not explicitly captured in the techniques used to represent and analyze each level of organization. The study of life *in vitro* normally concentrates more on objects than on their relationships with each other. The same might be true of studying the genome *in silico,* if we simply enumerated the properties of DNA, such as its composition in terms of the different types of letters, for instance the proportion of G + C base pairs in the double helix, or the frequency with which particular two-letter sequences appear in a single strand, or other facts of this kind. Very often the genome text will appear to be random when in fact it is not. In order to relate the text to the physical reality of the cell, anyone aiming to construct an *in silico* representation of life needs to be intimately aware of the architectural and dynamic constraints that evolution has imposed on the development of the functional genetic program, as reflected in the genome text. This requires a detailed analysis of the way the objects within the cell have been built up over time, as well as their arrangement inside the cell. These constraints form the basis on which the cell recognizes what is "self" and what is not. The various components naturally present in the same cell compartment are the result of selection, so they have inherent characteristics that allow them either to interact or to avoid interaction. These properties give the different components of the cell the style that differentiates them from objects with the same function but in a different cell compartment. A large part of this internal consistency is the combined effect of the individual histories of each member of the population, with all the incidents and events that may include.

This extends to organisms as a whole. But considering only individuals would be hopeless. Fortunately, we quite naturally tend to group living organisms together in *classes*. More than two millennia ago Porphyry grouped them into families, genii, species, and so on in his "tree," creating a hierarchy of attributes, in much the way we still do, since Linnaeus and Vicq d'Azyr. The very fact that we cannot easily find a name for "living things" (we call them "beings," "individuals," "bodies," "organisms") is actually quite revealing. What comes to mind first, not only when we consider organisms but also when we divide them into appropriate subsets, is the notion of organization, of structure, and above all the notion of function, in other words the link between a biological object (from the simplest to the most complicated) and its capacity to carry out an action. In short, a function links a biological object, with its architecture and its dynamics, to an action that can be seen to have a *goal*. Working from our knowledge of the genome text, genome programs try to reach this elusive property of biological objects, their function, shedding an entirely new light on old questions about biology.

Do Living Organisms Know Where They Are Going?

Finalism and Instructive Theories

Rather than focus on the fascinating discussions about the origin and philosophy of biology, I want to place old problems in the new perspective created by genome studies, with the aim of understanding the nature of biological functions, from the genome sequence perspective. Studying the mechanics of the cell may give us the illusion that we have understood life, but its laws and its origins are still surrounded by speculation. The apparent perfection of this mechanism, especially, seems to indicate some sort of will in the construction of living organisms. Furthermore, organisms as a whole behave in a way that is often surprising, apparently erratic, and not necessarily stable over time. This explains why for a long time they remained outside our representation of the physical world. To explain life, many thinkers have invoked "animating principles" that were of a quite different kind from the principles of physics. There is an instinctive appeal in an "instructive" vision of the world, in which some motivating principle (within or outside the organism, depending on one's philosophical preference) animates material things and gives them life, and this explains why this view has become so widespread and influential.

> I am extremely anxious that you should carefully avoid the mistake of supposing that the lustrous eyes were created to enable us to see; or that the tapering shins and thighs were attached to the feet as a base to enable us to walk with long strides; or again that the forearms were jointed to the brawny upper arms and ministering hands provided on either side to enable us to perform the tasks necessary for the support of life. All such explanations are propounded preposterously with topsy-turvy reasoning. In fact, no part of our body was created to the end that we might use it, but what has been created gives rise to its own function. Sight did not exist before the birth of the eyes, nor speech before the creation of the tongue; rather the tongue came into being long before talking, and the ears were created long before a sound was heard. In short, I maintain that all the organs were in being before there was any function for them to fulfil. They cannot, then, have grown for the purpose of being used.[9]

Thus wrote the Roman poet and philosopher Lucretius in his *De Rerum Natura (On the Nature of Things)*. It is a text of vital importance to Western philosophy, and follows in the tradition of the pre-Socratic atomists and their successor Epicurus. Lucretius tells us that "no part of our body was created to the end that we might use it, but what has been created gives rise to its own function," and warns us to "avoid the mistake of supposing that the lustrous eyes were created to enable us to see." This approach—to ask whether the sequence of cause and effect that we presume at first to be true should not in fact be reversed—goes against common sense, and is rare in our tradition of thought. From childhood, we tend to try to explain the world's mysteries by conjuring up an external motivating principle whose mere existence explains everything.

This way of thinking about things, and especially about life, is prevalent in all civilizations. We want there to be a motivating force behind everything. This explains the dominance of causal explanations, in which a strong principle, normally an external one, but sometimes internal, directs all things in space and time. Of course we would find it ridiculous to claim, as Bernardin de Saint Pierre is said to have done, that a melon is naturally marked into segments so that it can be cut into portions for the family dinner. But we are often happy to say that the evolution of species implies a conscious aim of adaptation to a specific objective (wings were developed in order to fly, eyes in order to see, the corolla looks like a bee so that a bee will fertilize the flower, and the rest of the list refuted by Lucretius). In the United States, there is strong support for an even stranger way of "explaining" things, creationism. Creationism takes absolutely literally the biblical description of the origin of the world, of the Earth, of humans, plants, and animals, as described in translations of various ancient texts and brought together into the more or less consistent picture found in the Book of Books (the standard Bible tells the story of Genesis twice). Of course, creationism is not an explanation of things, but simply an act of faith within a particular religion. In fact it is a tautology, which attempts to explain things merely by naming them, just as in Molière's play *The Imaginary Invalid* the effectiveness of a sleeping pill is "explained" by referring to its "sedative virtues."[10] The reason why creationism has some success (a success that depends heavily on the type of culture: it is rarely found in Latin civilizations) is that it gives every single being a *final cause,* a reason for its existence—the name it bears was given by God. Every living organism has thus been created *for* some purpose. For human beings, this helps to justify the rules of society, seen as laid down by God.

The idea of a final cause goes back to Aristotle, who proposed four different types of cause that made natural phenomena happen: the material cause, the efficient cause, the formal cause, and the final cause. Two of these have names that are rather misleading to the modern English-language reader. We have seen that *formal* is not the opposite of *informal,* but means "relating to form," and formal cause thus describes the idea that structure causes function. Similarly, *final* in "final cause" does not mean "last" or "finished." It comes from the Greek word τέλος *(telos),* which does mean "final" but also means "end," in the way this word is used in the expression "the end justifies the means"—in other words, an objective or goal. (The dual meaning of *final* is also present in Latin and in modern French, but has been lost in English.) "Final cause" thus describes the notion that a phenomenon is caused because the end or purpose requires it: the idea that it rains so that plants can grow, or that the melon has segments so that we can cut it up, or that the eye was formed in order that we might see.

This "finalist" explanation of the world refuses to accept that accident, or "contingency" in its original sense (an arbitrary, fortuitous interaction between objects or processes, not the dependence of one on the other), is found everywhere in the causes of evolution. Theories that postulate the existence of an explanatory agent, capable of intent, are called *instructive* theories. This prescient agent, invented to order, is usually outside the system being considered, and directs its evolution in space and time. It is God the Watchmaker, who created the world, with its own purpose, which we suppose must have a meaning, although we often find it incomprehensible. Theories of this sort are fundamentally animist, inasmuch as the divine motivating power they propose is capable of making the world according to its own vision, creating the soul of the world, which is somehow present in all creatures. There are also more profound finalist theories, inspired by Plato's ideal vision of archetypes, which one might call *"attractionist."* (I have to use this rather ugly neologism because *attractive,* the more appropriate counterpart of *instructive,* has other connotations in English.) They are popular with mathematicians, and with those who confuse the metaphor of Plato's ideal world with reality. In such theories, the motivating principle consists of stable forms or archetypes, which draw the system under consideration irresistibly toward them.[11] This is the approach to biology that appeals most to scientists who are not biologists, and who do not dare to invoke directly the crude finalism of the Great Watchmaker. Unfortunately, it is hard to understand and accept that there is no final cause, and that we do not need a final cause to explain life. However, we must understand this, if we want to find a constructive way of explaining the world.

Selective Theories

In the opposite camp to instructive theories we find *selective* theories, of which Lucretius' *De Rerum Natura* gives a good example. He said "all the organs were in being before there was any function for them to fulfil." Extraordinary as it sounds, major functions do exist before a use is found for them.

The world, even the living world, does not know where it is going. The only material systems that can exist (or at least that we can be aware of) are those that are sufficiently stable in space and time. The only systems that can *survive* are those that have in reserve the potential to innovate, even in an unpredictable future. They are certainly *not* the fittest, as Herbert Spencer assumed. Who should say where, when, how, and who is the fittest? Nor are they the only ones that are fit, but just those that are not totally unfit. What we call natural selection is not active; it works not like a sorting process but like a sieve, letting through only what is able to survive. It also has no real meaning for individuals as such, but works at the level of *populations* of individuals (which perhaps explains why we find it so difficult to accept the laws of biology, for each of us is an individual, not a population). Natural selection is an *exploratory* process, continuously producing and making use of a variety of individuals, all different from each other. Those that can survive without contradiction with themselves or their environment get through the sieve.

This exploration is the simple application of a very general principle of physics, a direct consequence of the Second Law of Thermodynamics. Entropy, or the propensity to explore, is fundamental to physics, and we have seen that although it cannot be reversed without using a great deal of energy or very strictly controlled conditions, life has no need to reverse it. However, this exploration cannot define in advance what will turn out to be the most stable or best suited to its environment. Yes, there is selection, but it is certainly not selection of the *fittest*. To be, it is enough to survive. Many characteristics seen in individuals, although they may seem to be specific, are in fact purely incidental, at least until they find themselves in a new and unfamiliar environment, where they discover an activity that is meaningful in that environment—a *function*. The simple existence of a living organism comes before and is more real than its abstract character, what the medieval philosophers called its "essence." This means that limbs may be legs, flippers, or fins, that a creature may fly with its hands, and with skin covered with hairs, or alternatively with its arms, covered in feathers.

More often than not there are several ways of achieving the same function. Only a few of these are explored as living organisms reproduce. The function itself, originally hidden, will suddenly come to light as the result of some crisis. Imagine that I am sitting reading at my desk. The window is open, and the desk is covered with papers. Suddenly the wind gets up. My immediate reaction, if I don't want my papers to get blown around and mixed up, is to use the book as a paperweight. Not only is it something to read; it is in fact also a heavy block, so it can suddenly acquire the function that goes with that shape, that of a paperweight, even if that is not the original reason for its existence, and even if that new function has been invisible until now. No physical or chemical law has been broken or new one invented, but a *goal* has suddenly been added, to go with the structure of the book.

Goal has a special meaning here: it is not defined *a priori*, but *a posteriori*. *A priori* reasoning is "deductive" or "pure" reasoning, from causes to effects. It does not rely on evidence, but starts from a notion or principle and deduces or predicts a result. An ignorant observer, seeing that books have acquired the function of paperweight, might say they had been made for that purpose (we often find books used as paperweights in antique shops). But if we think about it, we can see that it would not be possible to predict the book's function as a paperweight by *a priori* reasoning: after all, if there had been another heavy object on my desk, or a bunch of keys in my pocket, I could have used either of those instead. Thus the function was not *created for* the purpose in the way that the obvious suitability of form (heavy block) to goal (paperweight) would seem to suggest.

However, it is perfectly possible to explain it afterward. *A posteriori* reasoning is "inductive" or "empirical" reasoning, from effects to causes. It starts from observable evidence and "induces"—works out—what must have caused that situation to come about. It is thus precisely because the book's shape and weight are found to be appropriate in this situation that its function as a paperweight can be explained *afterward (a posteriori)*, and not that it has been selected *beforehand* with a view to the chosen purpose *(a priori)*. This is the way evolution works, opportunistically.

An Illustration: The Transparency of the Lens

Rather than make peremptory and abstract statements, let us look at an example. Take the mysterious transparency of the lens of the eye. It seems amazing that a group of cells should be transparent, and we feel that some final cause must surely have guided their formation, so as to allow light to reach a sensitive surface, the retina. But simple observation soon shows that transparency is common in nature—it is not an exceptional characteristic. Jellyfish are transparent, as is the immature zebra fish. (In this case, the apparent final cause is quite different from that of the lens of the eye: a transparent fish does not look like a fish, so it can escape from predators more easily.)

Transparency is produced in nature in many different ways. If you break a fresh egg, the white is transparent. The reason for this is quite simple and easy to picture—the proteins are arranged in a random fashion in relation to each other, an arrangement that gives them the same properties as the silicates in glass. This is very probably adaptive in itself, as otherwise the proteins would be arranged into complex, more or less regular structures, and they might even form crystals, whose very regular organization requires only a low information content but imposes specific architectural properties. We should note, however, that crystals themselves have some very unusual properties, which can also be exploited for an adaptive strategy. There are actually proteins that crystallize in nature, and this allows them to build structures that are large enough to be stable, and to resist being eroded by their environment. Some bacteria that infect insects, such as *Photorhabdus luminescens,* produce toxins with a crystalline structure, which can thus reach the insect's intestine or hemocoel and poison it. This means that often a property, and another property that is *exactly the opposite,* can *both* be adaptive. There is no project, nor even any final cause, in this.

The proteins of the lens of the eye are arranged in the same way as silicates in glass, although unfortunately they are misleadingly called *crystallins* because they were named after the Latin word for the lens of the eye, *crystallinus,* before it was known that they do not in fact have a crystalline structure. Most cells are not transparent, and this can in itself be an adaptive function, as it allows the organism to filter light radiation, especially ultraviolet rays, avoiding their harmful effects on the genome. So how has the lens of the eye become transparent? A recent discovery has given us an answer to this question. It used to be thought that the proteins in the lens were found only in this organ and were the result of a one-off evolution, especially for this purpose. However, the discovery of their amino acid sequence revealed something extraordinary, and totally unexpected: crystallins are normally quite ordinary metabolic enzymes! They can even be different from one animal to another. If we had known nothing about them but their sequence, we would have identified them as enzymes, with some metabolic function, and not as proteins that give the lens its transparency. Even more surprisingly, these proteins have not lost the ability to act as catalytic enzymes, if they are placed with the appropriate substrate. This would have compounded the error, since *in vitro* analysis of their behavior would have confirmed their identification as enzymes. There is an extremely important caveat here.

Knowing the sequence of a protein (and even, sometimes, some of its biochemical properties) can be extremely misleading when we ask the only important question: what is its function?

How, then, do these proteins, with their function of transparency, come to be in the lens of the eye? How have they been chosen by natural selection? There is an exact parallel with the story of the book-as-paperweight. If we look at the whole group of proteins in the lens, we find that they make up a recognized family. These proteins are usually expressed temporarily, in other tissues besides the lens, when there are drastic changes in the environment, and for this reason they are called *stress proteins,* or heat shock proteins. In this family of proteins, expressed as a group, we find some proteins called molecular chaperones (because they always accompany other proteins!). Their role is to protect cells against sudden environmental changes, such as exposure to radiation or increases in temperature, and allow the cell to put its protective strategies into action in good time. The molecular chaperones play the role of scaffolding, so that cells that have been damaged by conditions of stress can be reconstructed. In the lens, rather than being expressed temporarily, these proteins are expressed continuously, or, to use the technical term, *constitutively.* So it appears that the cells of the lens behave as if they have lost control of switching on and off these stress proteins, and express them all the time.

What makes the cells of the lens transparent is therefore the *loss* of a function, the specific control that allows the cell to take account of sudden changes in the physical parameters of the environment. But it gets stranger still. It looks exactly as though there were a very clever final cause in the development of this process. Not only are all the cells of the lens transparent, but the fact that the chaperones are always present means that any proteins that lose their correct architecture and stick together into an opaque clump can be repaired, regaining their correct shape and their transparency. Remember that the lens is permanently subject to radiation, often containing a significant level of ultraviolet radiation—exactly those conditions that would cause its proteins to denature. So the presence of chaperones means that the lens is resistant to this radiation *from the outset!* It thus becomes clear how evolution has selected this arrangement. The fact that the book was a heavy block meant that it had, built in, a property that allowed it to hold papers down. In just the same way, the set of proteins that protects against heat shock (and which has been selected for this reason) also has, already built in, a property that makes it produce a transparent body resistant to ultraviolet radiation, and this without any prior aim or final cause. It is this hidden, pre-existing function that has been selected during the course of evolution.

Thus once circumstances have occurred in which a function comes to light, it becomes established and passed on to future generations through the stabilizing process of selection. The function (transparency, in this case) was not *created for* the purpose for which it is apparently so well suited *(a priori)*, but because it is found to be appropriate in the situation, it is *selected afterward (a posteriori)*. Of course this supposes that the circumstances that prompted the new function recur from time to time, if not regularly. But it should be noted that in the case of living organisms, even if an event occurs only once, and triggers a specific function, that function remains built in for many generations, even if the trigger does not recur immediately, because the transmission of heredity is remarkably stable.

To follow the paperweight allegory through to its conclusion, I must deal with an objection that may be raised. There is no final cause in the object I used, but it is true that this apparently paradoxical solution was discovered because of my *intention* of keeping my desk tidy. One could then argue that the function was actually predetermined by the way my mind works. So I must continue the story, to show how selective stabilization takes effect without there being any goal in view. Imagine that I have to catch a plane, and that my ticket is on the desk. Imagine that I am so engrossed in my book that I don't notice the time, and now if I can't find the ticket at once I will miss the plane. Imagine, too, that I am young, and that I intend to have children. That if I lose time hunting for my ticket, I will have to take a taxi and ask the driver to hurry, to catch up. And that there will be an accident, and I will be killed . . . The heavy object thus turns out to have saved my life, so I can go on to have those children, even though I used it without thinking, just because it was there. There is no need for the least intention; my survival has in itself given a new meaning to this heavy object, while I remain blissfully unaware of what might have happened if I had not used it. My children and their children, simply by existing, will be the tangible evidence of this new function, which, we must not forget, was latent in an object that originally *(a priori)* had nothing to do with it.

Of course the allegory could also be seen as revealing a much more general function, which would explain why everybody would react in the same way when the wind blew. This general function, which created the "paperweight" function, is the consequence of our general reaction to sudden changes ("when the environment changes in a way we cannot control, try to keep it constant as far as possible and as quickly as possible": use a paperweight, close the window, close the door, put the papers in a drawer, and so on, would be appropriate responses). This *homeostatic* function would be linked to the organization of the human brain, and to its adapted reaction to stressful conditions. We can see this behavior as the origin of a general underlying theme of animal behavior, superstition, a ritual behavior pattern which we follow in the hope of influencing some process that is beyond our control, and which we believe to be the result of some chain of causality. If a particular behavior pattern is "successful" once, we will try to repeat it when faced with the same situation again. When we think about causes and effects, we should always bear in mind this type of relationship between the two.

Genetic and Epigenetic Heredity

Selective theories themselves can be divided into two kinds, of which one is still strongly marked by instructivism, as we have seen with Spencer's universally quoted words on adaptation. This kind corresponds to the early preformationist theories, which assumed that the individual exists already formed in the germ cells or in the embryo. The modern equivalent of preformationism is the extreme but widely held view that the individual is entirely (and above all mechanically) specified by his or her genes, when they express themselves in an appropriate environment. There is a one-to-one correspondence between the units of heredity, the genes (in the formal sense, whatever their physical nature), and the traits that distinguish one individual from another. According to this view, the form is already predetermined, and produces identical copies of itself—or at the very least, the form can be directly deduced, without variation, from the organism's blueprint or, more accurately, its recipe. In other words it is produced as the result of a process that brings it into being already completely defined, and without any alteration whatsoever when it is reproduced. Consequently, simply knowing the initial state of the system is considered to be a sufficient basis for immediately predicting its future. In these theories, much closer to the creationist theories than their authors would like to think, genes are the entities with intentionality. This explains why their authors can call them by the curious anthropomorphic qualifiers of "selfish" or "altruistic," for example.

The second kind of selective theory corresponds to *epigenetic* theories, which distinguish between genetic and epigenetic heredity. *Genetic heredity* refers to what is preformed, or is constructed in a way that can be reproduced exactly with each generation, *whatever* environment the organism may find itself in. *Epigenetic heredity* refers to characteristics that will be transmitted from one generation to another *provided* the conditions are right. They are not directly contained in the genes, but only in one particular state of their expression in a given environment. We could draw a parallel with wealth, which is passed on, but not in the genes!

The transmission of visible characteristics is not limited to living things: adding a small cubic crystal of salt to a supersaturated solution of sea salt will produce a mass of crystals with the same geometry. With living things, this transmission poses critical questions, not only because what is transmitted is very complex (there is a huge difference between this and passing on the crystalline structure of cooking salt), but also because living systems actually produce the conditions that govern their own reproduction. It is therefore essential that we understand the nature of what is transmitted in genetic heredity quite separately from what is transmitted in epigenetic heredity.

The question first arose in terms of organisms such as plants and animals, simply for reasons of scale—the visible should be explained by the visible. It is clear from the start that what is preserved is *a kind of form*. To explain this transmission of form, the Swiss naturalist Charles Bonnet hypothesized that the laws of nature are the same at all scales, so that if the scale is changed, the form remains more or less constant. If aphids are dissected in summer, they can be seen to contain tiny aphids inside them. The string of eggs developing inside a chicken is another example that supposes a preservation of form, though not of scale. Did this not show that nature is made up of organisms nested inside each other like Russian dolls? However, William Harvey, famous for his discovery of the circulation of the blood, took a completely different approach. He noticed the difference in shape of the egg and the chicken, and came up with his famous rule *omne vivum ex ovo*—all life comes from an egg. In contrast to the preformation hypothesis, his rule assumed that individuals develop from an egg, involving a significant change of shape, a *metamorphosis*. As this was a developmental process, this implied an *epigenesis* that determined the transformation from the embryo to the adult state. This was in line with what we see in birds and mammals, but also of course with the fascinating metamorphoses of frogs or butterflies and moths.

A conceptual difficulty inherent in the preformationist approach did not become apparent until much later, when Georges Leclerc de Buffon easily demonstrated that imagining a nested set of identical forms rapidly led to an infinite regression, which was obviously impossible. There must be some change of shape associated with the transition from one scale to another.

How can the unchanging nature or form of heredity be reconciled with the change in shape? Buffon's approach preserved this coexistence by imagining it not as simultaneous (which it clearly is not) but as developmental, expressed as time passes. The egg develops progressively toward an adult state, as the result of a spatiotemporal process. This produced the idea of the existence of an "internal template" and a mechanism by which the form is produced from this template. Since Buffon, we have distinguished between a structure that directs, rather like the conductor of an orchestra, and the carrying out of the directions—the performance. In his work on the transmission of observable characters, Mendel was implicitly looking for the rules determined by the internal template. He discovered that the color of a flower or the shape of a pea is preserved as if there were "atoms" or units of form: indivisible particles that could combine with each other in every possible arrangement.

It was not until recently that the nature of the internal template was understood. With the work of Hugo De Vries and others at the beginning of the twentieth century, and especially Thomas Hunt Morgan a little later, Mendel's laws became increasingly unshakable. But they remained formal laws, independent of the substance of the entities whose action they described; they indicated profound relationships existing within the egg, but their physical nature remained a mystery. However, with the discovery of the chromosomes and of nucleine, it gradually became possible to distinguish between the hereditary substance as a physical reality, and the manifestation of hereditary characters. The Dane Wilhelm Johannsen proposed a distinction between the *genotype* (what is stable and transmitted from generation to generation) and the *phenotype* (what is expressed at a given time, in a given environment). The units of heredity, the genes, were contained within the chromosome, in an arrangement and a physicochemical structure that were still to be discovered. Morgan's studies of *Drosophila* showed that genes can be formally defined, and that they are arranged in a linear chain, with one linked group corresponding to each chromosome. The concept of *mutation* was revived by De Vries to account for not only the large, sudden changes that sometimes appear in an individual's offspring, but also small changes. It was compatible with observations which suggested that one and the same gene can exist in several forms, called *alleles,* which are all functional, but vary from one individual to another.

Thus it was that just before the Second World War, formal genetics was reaching its peak. Since the work of Alfred Garrod at the beginning of the century, followed by that of Linus Pauling between the two wars, it was known that different alleles of the same gene correspond to different modes of biochemical activity. Scientists had even attributed these differences to different forms of the same macromolecule, made from a specific chain of amino acids, a protein. But how was this related to heredity? Nothing was understood about the rules of transmission. They were not known until ten years after the war. In 1944 came the discovery of the role of DNA in heredity by Oswald Avery and his colleagues; then the work of Boris Ephrussi in France and George Beadle and Edward Tatum in the United States demonstrating the link between the structure of DNA and the hereditary expression of specific characters; and last the discovery of the structure of the double helix in 1953, which finally made it possible to understand DNA replication. These discoveries ushered in the era of molecular genetics. Only since then have we understood the significance of the distinction between the *genetic program,* the structure that directs, and its implementation, its *expression,* represented by the set of proteins coded for by the DNA and expressed in a given set of conditions.

A central element in this distinction is the fact that there is no one-to-one correspondence between a gene and its expression. In particular, a gene may or may not be expressed, depending on the cell's environment. This is obvious in multicellular organisms such as mammals—a skin cell does not express the same proteins as a brain cell, and when it divides it produces more skin cells, not neurons, despite the fact that both of them must have the same DNA content, and therefore the same genetic program. This same program can thus produce different outcomes, demonstrating that the external environment is an intrinsic part of the way the program is expressed, because it contains the *data* that determine the outcome. A cell can be defined as a machine that puts the genetic program into operation *according to the data provided by its environment.*

We can see how absurd it is to try to separate what is hereditary or *innate* in an individual from what is *acquired* or due to the environment, although unfortunately this is often attempted. Everything is innate, *and* everything is acquired. The most we can say is that what is innate makes it possible to select from the environment what will be perceived and therefore taken into account in the development program. What belongs to genetic heredity is the DNA sequence and the program that goes with it, of which we can see only the part that is expressed in a given phenotype. It may even happen that one part of the program is only rarely expressed, so that it "jumps" a generation. There are other ways in which heredity shows itself, which are not genetic in the sense that they are not a modification of the DNA text, but which can be seen as a heredity of the *expression* of a part of the program. This is what happens in the differentiated cells of an organism when the time comes for them to reproduce: the conditions that allow their program to be expressed are transmitted from generation to generation of these cells, without there being any change to the program itself. This is what makes muscles produce more muscle, and skin produce more skin, normally without any alteration in the genetic heredity. How is this possible?

When we actually look into it, we discover that there are many different possible mechanisms. When a cell divides, it duplicates not only its chromosomes, but all the cellular structures as well. It is clear enough that if one of these structures is duplicated (for instance a protein that might be present in the nucleus), that could alter the way the genetic program is expressed. A simple mechanism would be that the presence of the protein itself was sufficient to trigger its own synthesis. Once present, it would transmit the conditions needed for its own synthesis from generation to generation, without any alteration in the text of the genome.

In selective theories, explanation, in other words the search for links between causes, begins with a phenomenon, then looks for the processes that underlie it. It discovers that these links are the result of the organism's history, and that this history usually involves not just one individual but a population of individuals, all more or less the same. Throughout this history, connections between different functions have been repeatedly established and broken again. And as a result of the form one of these connections or disconnections may take, an entirely new function may appear, which could not have been predicted, but which can be understood and explained perfectly well *a posteriori* (as with the cristallin proteins).

Perhaps one of the reasons why we find it difficult to accept selective theories as the explanation of life is that, as individuals, we cannot bear the painful discovery that individuals are of no importance in the direction the world takes; only populations count. Nobody is irreplaceable; but we find this truth so unpalatable that we produce endless flattering biographies and other signs of the cult of the personality. Besides, it is only in looking back that we are able to tell how "cause" and "effect" are oriented in relation to each other. More often than not, all we can see at the time is that there is a systematic association between two events—it is not clear which is cause and which is effect. Adaptation may give the impression of a certain order, imposed by its own "final cause." But this kind of explanation requires, in addition to observation, a hypothetical motivating principle that contemplates things and dictates how they should be ordered. What selective theories show is that the goal is created by adaptation, and not the other way round—and this is confirmed by history, independently from the phenomenon.

Teleology and Teleonomy

Ernst Mayr suggested that a goal can be defined in at least three domains. *Teleology* is the name given to explanations using the concept of final causes, and is typical of instructive theories. *Teleomatics* is typical of attractionist theories, and corresponds to the physical constraints biological objects are subject to, such as gravity or light, which make the stem of a plant grow upward, while its roots grow down into the earth. Finally, *teleonomy* is illustrated by the goals we see in selective theories (and, as we saw, sometimes retains a teleological flavor). As life has evolved, nature's teleomatics and teleonomy have combined to produce organisms so highly evolved that they can play God. In domesticating animals, and more recently in making clones and genetically modified or transgenic organisms, humans have shown teleological behavior.

One of the reasons why biology became a science only quite recently is that, not only are we living organisms ourselves, so that everything we can say about life concerns ourselves first and foremost, but also that organisms often seem to be so well adapted to the environment in which they live that they seem to have evolved according to a project. But there are several ways of looking at this project. It may simply be driven by a final cause, but it may also be the result of an *intention*. Jacques Monod used the word *teleonomy* slightly differently, to represent an apparent final cause, without intention. Where intention is implied we speak of *teleology*. Distinguishing between these two aspects of finalism, teleonomy and teleology, allows us to ask more profound questions about the nature of life. Rather as we did with instructive and selective theories, we can try to see if there is really any need for a motivating principle, guiding life toward its goal. Jean Piaget made a clear distinction between teleonomy and teleology in his book *Epistémologie des sciences de l'homme:* "We soon realized it was possible to give a causal interpretation of processes acting toward a preset goal and to find 'mechanical equivalents of final causes,' or as we say nowadays, a 'teleonomy' without 'teleology.'"[12] The adjective "mechanical" in this description is significant, as the question of final causes arises when we consider a mechanical vision of the world.

François Jacob put it slightly differently in his book *The Logic of Life,* stressing in his summary of life that the project of the cell is to make another cell.[13] But what is a project? And is it realistic to think of this as a project? More recently, in *Of Flies, Mice, and Men,* the same author spoke of the need to manage the unpredictable.[14] But any idea based on finalism requires that we identify both a goal to be achieved and the means to reach that goal. Because language makes symbolic representation possible, then where humans are concerned we really can believe that there is a genuine, intentional project, which uses appropriate means to reach its goal. However, this is more difficult to justify with animals (although some animals, such as the crow *Corvus moneduloides,* use tools, which they know how to use again). It is harder still with less complex organisms such as microbes. We may well say that the project of a plague bacillus is to infect a human organism, using the digestive system of the flea as a means of access, but this is all very metaphorical. How can the apparent finalism of adaptation be explained?

The Organism and Its Environment

However difficult it may be, to see things clearly we need to get away from the "instructive" way of thinking, pervasive as it is. We must accept that there is no motivating principle guiding or orienting the actions of living organisms, but rather that the orientation is established *a posteriori,* because of some selective action by the environment. The overall survival of an organism can be explained only because its parents survived in an environment not totally alien to the present environment. In this sense, the only organisms that can survive are those whose genetic program has some representation or internal image of their environment, the result of eliminating a whole series of behaviors that were *impossible* in that environment. But there is nothing in this image resembling a project. It is made of bits and pieces, put together out of all the opportunities that have presented themselves, usually as the result of accidental interactions. This means that adaptation often appears arbitrary, even if it seems appropriate to a given function at a given time.

This image a given organism has of its environment, or more precisely that part of its environment it has become *aware* of in adapting to it (simply remaining perfectly unaware of the rest), has been selected down the ages by the harsh, relentless trials its ancestors have gone through. We often feel that the real meaning of our life lies in the idea that it was not just desired but was somehow meant to be. True lovers feel that they were made for each other since the beginning of time. In a way this is true, but it is only part of the story, and the reality is sadly quite different from this beautiful "poetic reasoning," if we can call it that, to distinguish it from "scientific reasoning." If we are alive today, it is because our grandfathers survived the First World War (especially for the countries most heavily involved, France, Germany, Britain, and Russia). Everyone will be able to find a reason for this survival—which was surprising on the face of it, given that so many men were killed in the war—and they will be astonished to hear that the others also have a similar explanation, some lucky chance that spared their grandfather. But it's quite simply that those who died have no descendants to tell their story! If the human race exists today, it is certainly partly thanks to our intelligence, but it is above all because our ancestors survived infestation by large numbers of parasites, as well as cholera, smallpox, or the plague. Even if we have eliminated most of these diseases from our natural environment, we are still ready to fight them off, just as our closest descendants will be if they should reappear. Human intelligence had nothing to do with it. There is no design in this, no ultimate aim, each of us is no more than a (lucky?) accident. Even better, as heredity cannot foresee exactly what the future will bring (diseases evolve as all living things do), our apparent project to survive these attacks is a *generic* project. What has been selected during the course of evolution is not so much a specific system, which would allow us to resist specific diseases, as a general program of resistance (imperfect, of course) that allows us to survive a class of diseases—those caused by pathogenic agents, bacteria and viruses in particular. This is an abstract class, because it is impossible to predict the future of microbes and parasites.

The apparent final cause in the adaptive behavior of organisms is thus the result of the selection of genetic programs that gradually incorporate an image of their environment. The image varies according to the way in which that environment is taken into account. The organism may simply be unaware of the environment it is multiplying in; in other words, that environment is "transparent." The organism may not perceive one characteristic or another, and, as there is no interaction, it is not sensitive to that characteristic, and is therefore adapted to it. Alternatively, the organism may take specific characteristics of the environment into account in its internal image. Or it may construct multiple barriers between itself and the environment, so that it interacts in only a superficial way, without having to cope with all variations.

The first example of "transparency" that comes to mind is that of the many kinds of electromagnetic radiation that, at the temperature of the Earth, do not interact in any significant way with living organisms. They are effectively simply background noise. This is true of long-wave radiation (most radio waves for instance). For many animals, it is true of colors, as they can distinguish only a small part of the visible light spectrum (or they may see parts we cannot see, which is why we see many birds as black, while other birds see them as colored). It is also true of a large number of chemical products that, for many animals, are simply odorless, tasteless, and invisible. But transparency for some is not transparency for others, and this feature is exploited when insecticides are sprayed in buildings, where they are supposed to trouble only the insects . . .

In the second case, the organism is immersed in its environment, and is adapted to it in that it takes its features into account as far as possible. This is true of most microorganisms such as bacteria, which are separated from their environment by a single envelope of varying complexity. In this case, the barrier presented by the envelope monitors the environment and carries out a kind of sorting process, passing the results to the genetic program, which reacts appropriately, perhaps by absorbing a molecule, by guiding the cell to those parts of the environment that seem to be best suited to it, or alternatively by moving away from places it recognizes as harmful. Its behavior as a whole will respond to changes in the environment, so we find that what has been selected over the course of evolution is rapid adaptation to varying conditions.

Let us take a look at what happens to an *E. coli* bacillus, for example. It develops slowly in the external environment. Suddenly it is ingested by a mammal and undergoes a series of brutal changes. Its temperature rises dramatically from a relatively low point, about 20°C, to 37°C. Coming from an environment that was rich in oxygen, it finds the level dropping sharply, and from the neutrality of the outside world it is thrown into the strong acidity of the stomach (pH 1.5), where all kinds of destructive hydrolytic enzymes are at work. Then, in the second duodenum, the acidity changes abruptly to alkalinity, accompanied by strong detergent agents (the bile salts) and degradation enzymes. Then the bacterium arrives in the intestine, where it finds a rich diet, but no oxygen. As it passes along with the digested food it meets other species of bacteria, in increasing concentrations, while reproducing at a fast rate itself. Finally it finds itself out in the open air again, and exposed to the harsh shock of oxygen, while the temperature drops steeply and nutrients virtually disappear. It goes without saying that most highly organized living organisms could not survive a fate like this. The bacterium adapts itself by starting and stopping a large number of different processes as the conditions around it change.

Animals, on the other hand, attempt to avoid direct contact with environmental changes of this kind. They achieve this by constructing multiple protective barriers between their inside and outside environments. This could be seen as an adaptive strategy in which many layers of membranes and skins are produced, up to and including nests for animals and clothes for humans. But we need to be careful here: *adaptive strategy* is certainly the term used, and it is an evocative one, but it is dangerously finalist, as it takes no account of the extraordinarily *incidental* nature of the adaptation. Nature makes use of whatever it can find, so we must always be prepared for the unexpected, in discovering adaptive biological processes. The high point of this evolution has now been reached, with warm-blooded vertebrates that are able to keep the internal temperature of most of their cells virtually constant.

Of course it is still difficult to explain *all* adaptive situations, precisely because they are incidental. But when we look at them one at a time, it does become possible to understand how they have appeared, while avoiding any explicit finalism that would lead us to think that organisms really do have a project. What looks like a striking intentionality in the way organisms behave is in fact the result of a selective process that has allowed organisms to survive only if they are capable of working within not only a changing environment, but one that is unpredictable. The illustrations I gave, first the book-as-paperweight and then the lens of the eye, show that it is opportunism that allows such basic processes as association and dissociation to take effect, and to be the basis for other processes, even those that appear to be the most highly evolved.

There are many examples of this recruitment of preexisting functions to perform entirely new functions. One very surprising example is seen in the way animals are built up from a single egg cell. Let us go back to the early days of life on Earth, when an organism's basic function was to find itself a niche on the Earth's crust. This was initially an easy task, but as time elapsed, life evolved, and of course organisms met other organisms. Initially the simplest strategy to cope with this situation was to invent some way either to cooperate with the others or to get rid of them, and this is presumably how nature's own antibiotics were invented. Several functions are needed for this. First of all one organism must recognize the presence of the other, and this requires a sensor. Once it is recognized, this information must be passed to the program, which must trigger the synthesis of the antibiotic and its secretion into the environment (of course this also requires the synthesis of an appropriate immunization system against the antibiotic; otherwise the organism would be committing suicide). This is carried out by a cascade of chemical reactions mediated by proteins with appropriate structures and properties. So much for the early days of life. In a different domain of biology, scientists investigating the immune system of present-day insects ask: How do they react to infection by bacteria or fungi? They do what they did eons ago: they synthesize antibiotics. For example, *Drosophila* challenges a fungal infection by synthesizing an antifungal agent, aptly named *drosomycin*. Scientists investigated this cascade of events and identified the corresponding genes. But they made an absolutely incredible discovery: this cascade was already known, but in an entirely different field. The same genes, and the same cascade, are involved in a brief episode in the shaping of the *Drosophila* embryo, when its dorsoventral axis is determined! An old function, making an antibiotic, had been recruited for an entirely new, completely unrelated function, constructing an insect's body plan. This is how natural selection proceeds, by opportunism. We should be aware, too, that this story of *Drosophila* has very surprising implications if we extrapolate to humans. It suggests that the way we react to infectious diseases today might in the distant future influence the shape of our descendants! This should be a warning for those naïve people (to put it kindly) who believe in eugenics.

So what have we discovered, as we conclude this overview of what life is? We have seen that after more than a century of thought, questioning, constructing models of living organisms, and now deciphering their genomes, we are now in a position to put forward a (provisional!) definition of life. Furthermore, studying genomes has led to the discovery that there must be a geometrical program, superimposed on the genetic program, linking the architecture of the cell to that of the genome and placing living organisms right at the heart of physics. Life does not require any special physical properties of matter, and neither does it rely on the far-fetched conclusions sometimes drawn from the uncertainty principle of quantum mechanics. The laws we identify as specific to life explain living organisms in terms of original properties that are absolutely determined by the laws of physics, and are necessary *a posteriori*. Yet they cannot be *predicted* from those laws alone.[15] These original properties allow organisms to produce not only offspring identical with themselves but also, occasionally, other offspring which are not identical, but which live, and cannot be *reduced* to their parent organisms.

We have also discovered that cells maintain the memory of a program for building appropriate manipulating objects, and that the set of processes responsible for the uninterrupted production of those cells is capable of continuously monitoring all the properties of the objects it comes across, in such a way that it becomes adapted to the environment in which it lives. But even if we are sometimes tempted to accept its apparent finalism *a posteriori*—its teleonomy—this adaptation is no more than a combination of preexisting properties, simply brought together by natural selection into an association that merely appears coherent at first sight. What shall we say, then, about the relationships that constitute life?

Objects and Relationships between Objects

As living things ourselves, biology naturally seems familiar to us, and we would like to find it easy to understand. But in many ways, rather than being the concrete science we like to think of it as, biology is more like mathematics, and thinking about life demands a great deal of mental work, much of it unfortunately very abstract. Besides, although living organisms are made up of a *finite* number of basic objects, there are still a great many of them. Unlike in physics or chemistry, where the numbers are small, in biology we have to deal with somewhere between a few hundred and several thousand objects, even if we simplify things as much as possible. An initiation into the science of life means much more memorizing than for the other sciences.

What is more, the objects created by living organisms have a series of original properties that make them immediately recognizable, and quite distinct from those of physics. They are artifacts, intricate products that do not seem natural in the usual world of mineral chemistry that accounts for most of the structures found on the surface of the planets. They are best described as having some goal. On our human scale, man-made artifacts are easily identified—there are numerous examples in the world of technology. But among "natural" objects, all those produced by living organisms, even the simplest, can be recognized by their complex and enigmatic structure, when compared with that of even the strangest minerals. There are of course some physical constraints that limit the originality of the structures living things can come up with, and these produce a few coincidences. Branching structures, spheres, and hexagons, as well as the five regular solids, Plato's polyhedra, can be signs of life, but they are also found in purely mineral structures. However, although they can be found anywhere, they are not as important as other shapes found in living things. It is originality that scientists look for when they try to find out whether life has existed, or still exists, on other planets, such as Mars, although these efforts are unfortunately much too limited and basic. In 1996, NASA scientists tried to demonstrate that some microscopic formations found in meteorites, which were very unusual for minerals, had been produced by the activity of living organisms. The French biologist Jacques Monod gives a remarkable illustration of this originality when he describes how fossils are identified and contrasted with typical minerals, at the beginning of his famous book *Chance and Necessity*.[16]

Our knowledge of physiology tells us that a living organism is in the same situation as the Delphic boat: the result of metabolism is that all the body's parts, even bones, are gradually renewed during the course of its life. The boat in the riddle, whose planks have all been renewed, is still the same boat. What is unchanged is its construction plan, expressed in the various relationships that the planks have with one another. We can ask the same question about a living organism, even a human being: At different stages in its life, when all its cells have been renewed, is it still the same being? Where is its identity? During sleep, in dreams, or when the human brain is affected by the early stages of disease, can we really say that we are still dealing with the same person? If we can, it is once again because it is *relationships* that count.

Generally speaking, this is true of the cell, the basic unit of life; it is true of multicellular organisms, too, but it is also true even of the most advanced functions, such as brain functions. Our priority should be studying the relationships that make up life, rather than remaining at the level of the objects themselves; and we should do this by looking into the nature of whatever it is that gives these relationships their permanence. Studying relationships is essentially what Georges Cuvier was doing—and what paleontologists still do—when he took a few bones of a long-extinct animal, or even sometimes a single tooth, and proposed a reconstruction of the entire creature. This importance of relationships is not a trivial property, to be noted in passing, but a hard fact with considerable practical and theoretical implications.

The fundamental importance of relationships, which represent a particular interpretation of form, was noticed more than 2,500 years ago by Empedocles and many of the pre-Socratic philosophers. St. Thomas Aquinas also refers to it when he analyses the philosophical status of the concept of creation: "When motion is taken away, only different relations remain." He even says in his *Summa Theologica* that what is important in the act of creation is the fact that it produces relationships. He has to go through some very complex intellectual maneuvering to remain faithful not only to the spirit but to the letter of the two texts of Genesis, according to which there was a single act of creation at a specific time, so in principle it could not be a continuing process. Briefly, following the reasoning of Aristotle's *Physics*, this would lead him to say that creation—in other words the act of making something new appear out of the Nothingness that is the negation of Being—came after the world, even though this was in direct contradiction with his readings of the holy scriptures:

> But as "action and passion coincide in the substance of motion" and differ only according to different relations . . . it must follow that when motion is taken away, only different relations remain in the Creator and in the creature. But because the mode of signification follows the mode of understanding . . . creation is signified by mode of change; and on this account it is said that to create is to make something from nothing. And yet to make and to be made are more suitable expressions here than to change and to be changed, because to make and to be made imply a relation of

> cause to the effect, and of effect to the cause, and imply change only as a consequence.[17]

We must nevertheless steer clear of the purely "idealist" approach, which would suggest that these relationships can exist in themselves, in a world without objects, separating the body from a transcendent soul. To show how this separation would lead to a paradox, here is an analogy. A computer program is a text, written using a specific syntax and vocabulary. It can be interpreted and then compiled in a machine, so that it can be used: once certain initial conditions are defined, the program can be run. Remarkably, the same program can run in different machines, and even in computers built from different materials, for example one based on silicon chips and one based on gallium arsenide chips. But you will probably agree that this does not give us the right to say that the program represents the soul of the computer! Material objects are always needed to express the program, and it cannot be said to exist without them; indeed problems of portability and "bugs" do show that the program is to a certain extent dependent on the matter. But although these objects are made of different materials, what they have in common is that each can establish internal relationships of the same kind. In living organisms, the program is physically part of the organism itself, even if the fact that it is abstract means that, in formal terms, it is separate from the rest of the organism. It has not been put into the organism by an outside agent (except in specific experiments, such as those carried out by biotechnologists when they create new types of cells, and create genetically modified organisms or clone mammals).

It is certainly not a question of looking at just any relationship, or even at all possible relationships. The relationships that are at work in the processes of life have a certain well-defined form; they do not connect everything with everything else, and they operate on a specific scale, using specific structures. Life is a long way from the test tube, and it is aptly said (though often forgotten) that *in vitro* experiments are "cell-free." To explain life it is essential to recognize which relationships are relevant. This is the very purpose of biological research. We cannot pretend not to know about relationships, and treat a cell as if it were a black box, as if we were unable to study anything but its general behavior. A living organism, or a cell, is *never* a black box. Nor is it a "whole" that we need to know as such, without concerning ourselves with what goes on inside it. If we manage to simulate some behavior pattern, as can be done by building increasingly elaborate robots, this simulation, interesting though it may be, is only phenomenological—descriptive, in other words. It cannot be used as an *explanation* of life. This is another reason for caution when we use analogies to describe life.

Neither is it a question of an infinite regression, looking for components, and then the subcomponents of those components, and so on. Is anyone really simple enough to believe that life can be defined merely on the basis of a catalogue of the objects in a cell? There is often confusion between *analysis* and *reductionism*—the suggestion that explanations are to be found only in a reduction to the most elementary microscopic scale. This confusion, influenced by the negative connotations of the word *reduction,* serves as a convenient refuge for those who want to deny that it is possible to understand what life is. We need to understand exactly what form the spatiotemporal relationships within living organisms take, by analyzing them, not by just considering the "whole" as such. To do this, we need to *identify the scales of time and space* that are relevant to the description of life, and in doing this we are absolutely not reducing life to its component parts. In fact, at the Earth's temperature, at which life operates, the limits of scale are relatively easy to define. Since the forms we are attempting to describe are stable over a limited but not too short period of time, we can discount (except as averages) any statistical phenomena that take place over a significantly shorter time scale. The same is true of the scale of spatial structures. Except perhaps in the specific case of energy management, where photons and electrons are involved, the objects and the relationships that make up life fall into the domain of chemistry, which implies that the smallest scale we need to look at is that of molecules. More often than not, this is realized implicitly rather than conceptualized, but it is the basis of molecular biology. The corresponding approach is an analytical one: it analyses the contents of the cell, not just to list the objects in it, but above all to understand how these objects interact, in space and time. This is not reductionist, as some would have it.

However, we also know that at this very small scale, that of molecules, life is expressed only in the objects that make it up and in the relationships they have with one another, represented by the traces it leaves, in other words the biochemical artifacts it produces. The intrinsic properties of life, the four processes of metabolism, compartmentalization, memory, and manipulation, as well as the fundamental laws of complementarity and the genetic code, do not operate only at the molecular level; they also have to function *together.* Understanding life thus means identifying the whole set of spatiotemporal relationships between the molecules that make up the cell, and the way they are coordinated as the genetic program runs. This is where we find the underlying idea that suggested the metaphor of the Delphic boat.

When we think about the idea of a program, we need to bear in mind that it is a series of instructions, and thus of operations, in a specific order. The full significance of their order cannot be understood until the program runs, because they are not merely formal, but must be expressed, at some time or other, in the sense that they actually make a machine do something—either a computer, or in this case a chemical machine. This is why I want to give an explicit, concrete illustration, starting by looking more closely at the specific process that shapes proteins, the central objects in the expression of life. These represent the physical nature of the Delphic boat.

The Algorithmic Construction of Proteins

The genetic code is responsible for the translation of messenger RNA, itself copied from portions of the genomic DNA, into the chains of amino acids that make up the corresponding protein. Once we have made this observation, it seems natural to ask what would be the justification for the translation process if it were no more than a simple transfer from a text written in a four-letter alphabet to a text written in a twenty-letter alphabet. It would not seem to have much point, except perhaps concision, and it would be hard to justify its existence, if only for reasons of economy. In particular, translation would be a purely syntactic process, a transformation that excluded the corresponding semantics or meaning. But we want to look into the meaning of the genetic program, as an integral part of what life is. The implication of this necessary introduction of semantics is this: it makes no sense to study the genome merely by statistical analysis, describing the probability of the appearance of one letter or another, or this "word" or that (a word in the sense of a sequence of letters). Clearly, if such a description were relevant, it could not account for the stages of transcription and translation, which would be simply redundant—but we *know* that these stages exist. Even if the distribution of "motifs" in DNA sequences or in proteins appears to be random, we need to find out whether it genuinely is random (words are distributed just by chance), or whether there is in fact an underlying regularity but it has been masked by the superimposition of constraints at two levels. Constraints on transcription and on translation operate independently of each other, and cannot be reduced to each other in the way that parts of simultaneous equations cancel out to reduce down to one equation. Because of this, the final distribution may appear to be random, although in reality it is regular, at the level of transcription and again at the level of translation.

This brief description highlights a physical phenomenon that operates as a result of the interaction of molecules (physicochemical objects) that have quite different general chemical properties, depending on whether they are nucleic acid molecules or protein molecules. The role of the coding process is to make the transfer from a chemical world in which, broadly speaking, the objects (in this case segments of DNA) can be regarded as exploring only one dimension of space, to a world in which other objects, proteins, explore it in three dimensions, or even four if we include time, because proteins can change their shape. This distinction is so important that we distinguish several levels of structure in macromolecules. Their *primary* structure is the way the basic motifs are chained together—the sequence of nucleotides in nucleic acids or amino acids in proteins. *Secondary* structure refers to an initial spatial arrangement of some of these sequences into architectural features or conformations. They do not depend on the exact nature of the motifs that make up the primary structure, but they are common enough to be recognized. The helix is very important, but sheets, loops, turns, and barrels are also found. *Tertiary* structure is the way the whole primary structure folds up in space, taking account of the elements of secondary structure that have formed within it. It is not very different from what happens in the construction industry, where we have the blueprint, the basic materials, prefabricated elements, and their final combination into a building.

All these objects, these proteins, must appear in the cell in the right place at the right time. Their production and their behavior thus need to be carefully orchestrated. Strands of messenger RNA are the necessary intermediaries that play the central role in this orchestration. In size, they are between DNA and proteins, and their own folding properties are used in the processes that regulate gene expression—they dictate when, where and how much of the product is synthesized. This is a particularly abstract and complex kind of relationship, and we are still a long way from fully understanding it. However, it is possible to summarize it, if we are prepared to accept some imprecision: whatever determines the expression of life has to move, metaphorically speaking, from the "unfolding" of a concise structure, DNA, toward the "exploration of space" by proteins. One might think that going from a very restricted, one-dimensional space to a much more open, three-dimensional space could lead only to disorder; but the existence of life (and of all forms of exploration, in fact) is proof of the contrary.

The spatial shapes that proteins adopt when the strand formed by their amino acid chains has folded itself up are very precise, and sometimes very complicated. It is because they are the building blocks of proteins that there are so many different amino acids—far more than the nucleic acid bases. Translating mRNA into protein makes the meaning of a gene tangible, by expressing it, materially, as an object with a specific shape. This object is then able to interact in space and time, in a specific way (often in many ways) with other objects, whether they are chemically the same or different. So the movement from nucleotides to amino acids is effectively an exploration of an entirely new world, involving action on the surrounding environment. This is the world of protein function (or, as the philosopher Claude Debru so evocatively put it, their "spirit"). How is this exploration of forms in space achieved?

Anfinsen's Postulate

In 1973, after he had analyzed the way the chain of amino acids in a model protein folded, Christian Anfinsen suggested that this folding is solely and completely determined by the sequence of amino acids in the polypeptide chain. In reality, it was very easy to carry out an experiment to show that an unfolded protein "knows" how to fold up again correctly all by itself. The shape of biological molecules is essentially determined by the fact that they exist in water—and water, despite what we might think from the fact that it is all around us, is a very unusual kind of solvent. It is very highly structured, and most chemical formations within molecules are either hydrophilic or hydrophobic. Very roughly speaking, we can say that amino acids divide into two classes, distinguished by the way they interact with water. In a protein, the hydrophobic amino acids show a strong tendency to take up a position on the inside of the structure formed by the polypeptide chain, while the hydrophilic ones remain on the outside. The cavity in which catalysis takes place is very often entirely free of water.

If we degrade the structure of water, by adding salts or other appropriate molecules or by heating it, the structure of the proteins within it is immediately significantly changed, too, because the forces that make the hydrophilic amino acids remain in contact with the water while the hydrophobic ones avoid it can no longer take effect. We then say that the protein is *denatured.* Observation shows that a denatured protein takes on a more or less random form; the chain folds up in all directions, often producing a roughly spherical structure, known as a "random coil."

Anfinsen chose a model protein, placed it in a solution, and added a denaturing agent to transform the protein into a random coil. Naturally, the protein lost its function, and using various physicochemical tests he demonstrated that its functional conformation was completely destroyed. He then performed the experiment in reverse, removing the chemical elements that had damaged the structure of the water, and found that, quite quickly, the protein regained both its form and its function. From this he drew several important conceptual conclusions: first, that form is definitely correlated with function (or even that it *causes* function, at least in part); and second, that the protein contains all the information it needs to regain its form when it is put into the right solution.

The second part of the first conclusion is debatable, as we have shown with the book-paperweight. The second conclusion is also debatable, but, strangely, it has been accepted without controversy, not just for *one* particular protein, the one Anfinsen chose for his experiment, but for *all* proteins. The effects of authority—Anfinsen is a Nobel Prize winner—are particularly dangerous, as they lead writers to repeat this conclusion entirely uncritically. It is stated in most textbooks on protein biochemistry, and cited in an article in the well-known *New York Review of Books,* on March 6, 1997, by Jared Diamond. He writes: "If the hypothesis [that a protein can fold up into several different spatial conformations] does prove to be true, it will in my opinion be the most astonishing discovery by molecular biologists since Watson's and Crick's discovery of DNA structure in 1953. Until now, we have been assuming that a protein's three-dimensional structure is uniquely determined by its sequence of amino acid building blocks that supposedly specify how the protein will fold up."[18]

Common sense alone should throw doubt on this. The only thing that Anfinsen's experiment proves is that the *particular* protein he studied contains, in its chain of amino acids, all the information it needs to fold up correctly in water. What right do we have to generalize from this fact to *all* proteins? The simplest real-life phenomena suggest that it cannot always be true; for instance, once an egg white is cooked, it is hard to see how it could be "uncooked" and made viscous and transparent again. In old age, the lens of the eye can develop cataracts. Proteins in the lens, which are transparent in their native state, become denatured. The molecular chaperones also tend to denature with age, and are thus no longer able to repair the lens proteins. Once cataracts form, it is clear that they are irreversible. Of course one could argue that denaturation by heat or light has particular consequences. But shouldn't we wonder, all the same? And should we not look more closely at the protein that was initially chosen as a model, if we are suggesting there is a link not just between structure and function, but between structure, function, and the evolution of proteins from ancestral genes, and thus from ancestral proteins, over the course of the history of species?

In order to understand this properly, we need to know more precisely which protein was chosen and what it is like. Anfinsen chose a *small* protein for his experiment, which is already an important constraint. It has only 124 amino acids in its sequence, whereas most have an average of 300 amino acids, although this can vary from a few tens to several thousands. A further constraint, probably even more important because it is subject to the selective pressure that determines the evolution of species, is the fact that Anfinsen's protein is not an intracellular protein; it is an enzyme *secreted* by the pancreas in mammals. It is a pancreatic ribonuclease, and has a very simple catalytic function, that of a hydrolase: it uses water to cut ribonucleic acids (RNA) into small fragments.

The fact that this is a secreted protein represents a particular set of constraints. It is synthesized in the cytoplasm of a cell, like all proteins, so it has to cross the cell membrane, and probably has to change its shape in order to do this. This fact alone should make it ineligible as a paradigm for all proteins. But there are other important facts that make it an even worse model. The outlet from the pancreas is in the second duodenum, just after the stomach, and just where the gall bladder releases bile, a fluid rich in detergents. Under natural conditions, Anfinsen's protein is thus secreted into an environment that is bound to denature it, since detergents have the same property as soap—they make hydrophobic molecules soluble in water. So pancreatic ribonuclease must necessarily have been naturally selected to have just the ability Anfinsen's experiment required—the ability to *renature,* or regain the active form it had lost, once the concentration of bile salts in the intestine drops! This is not an incidental property; it is definitely the result of considerable selective pressure, because only proteins that were able to regain their shape in these hostile conditions can have survived. So it is only to be expected that Anfinsen's enzyme should be able to fold itself again spontaneously after denaturing. The logical consequence is that its amino acid sequence must contain all the information it needs for this refolding, because without this it could not be active.

The obvious conclusion from this is that pancreatic ribonuclease, the protein chosen by Anfinsen to establish his postulate, is a particularly *ill-suited* example from which to derive general laws that would be valid for all proteins. On the basis of this example alone, there is nothing to prove that all the information needed for most proteins to fold up properly is contained in the amino acid sequence of their polypeptide chain. In fact, since 1988 a large number of experiments have shown, initially in a few cases but then in most cases, that protein folding follows a particular sequence (together with synthesis, beginning at one end and finishing at the other) and normally requires the presence of molecular chaperones. These unusual molecules, whose form and function we are only beginning to understand, play the role of scaffolding, or intermediary templates that help the protein to take up its final active form. Looking at the vault of a church or at the Pyramids cannot tell us exactly how they were constructed. In the same way, knowing the conformation of an active protein cannot give us an exact idea of how it was formed *in vivo.*

Protein Construction

There is an epistemological lesson to be learned from these observations: we must beware of generalizing too hastily. But there is also a conceptual lesson. Proteins undergo a construction process; they are the result of the coordinated action of a complicated kind of machine. From beginning to end, their formation follows a sequence of well-defined stages, as if it were a factory process or a process controlled by a computer program. Folding does not happen by trial and error, with the amino acids exploring all possible conformations (there would be infinitely too many of them, in any case).

It now seems clear that there is some sort of *protein folding program* that directs the way in which each one folds up from its primary sequence into its final, tertiary structure. This process is usually guided by auxiliary proteins, the chaperones, which indicate the correct folding pathway; but the place where they are synthesized and the whole set of proteins and cellular structures found nearby all participate in the formation process. These constraints lead to the prediction that in a given organism, we should expect the macromolecules, and especially the proteins, to have a certain number of features in common, a common *style* characteristic of that organism. The way chaperones assist in folding is still imperfectly understood. But since the mid-1990s it has been possible to describe the shape and behavior of some of them, and to throw some light on the way a protein passes from a stage at which it is not yet correctly folded to its correct final structure. Here again it is essential to understand the detailed structure of the cellular machinery, on the molecular scale, in order to understand how genes are expressed, right through from the DNA from which they are made, to the proteins they code for, and the place in the cell where these proteins play their role.

It is clear from these observations and considerations that in many cases the architecture of a functioning protein depends on the way it has been constructed. It is perfectly conceivable that the same sequence of amino acids, sent down different pathways, can result in different spatial structures. Working from the template of the messenger RNA, the ribosome puts the amino acids in sequence to make the protein, and functions both as a guide through which the mRNA gradually passes, presenting each codon to the device that reads it, and as a structural template, feeding out the new protein strand from a particular point on its surface. From the moment it leaves the ribosome, the protein being formed begins to fold up as it interacts with other proteins that guide it toward its functional conformation, preventing the chain of amino acids from exploring the immense number of possible conformations it could adopt if it were left to itself. This procedure works like a chain of calculations, a process organized in the same way as an algorithm. The most important constraints are those which *prevent* the exploration of all sorts of conformations: it is not so much a question of directing the protein toward its final form as one of making sure that it does not stray from the correct pathway. It should be noted that Anfinsen's postulate is in a sense a resurrection of the finalism of instructive theories (the protein "knows" which form it is aiming at), but that in reality their construction is selective (the molecular chaperones act as a sieve). It is the fact that *a posteriori* this or that chain of amino acids has taken on such and such a function, compatible with a given structure, that determines its existence.

Finally we should note that proteins are rarely associated with only one function—they have several (some of which are simply architectural)—and that, like the book-paperweight, they are always ready to adopt another one, simply because they are in the right place at the right time. Folding is also the means by which the protein is guided to the correct cell compartment, without interfering with the many other syntheses all going on at the same time. The amino acid sequence contains not only the structure associated with a specific function (such as catalyzing a metabolic reaction), but also an *address* to a given location within the cell. However, this address may be indirect, arising simply from the fact that the amino acid sequences produce proteins that have a specific tendency to associate with one another. The formation of multiprotein complexes is probably of major importance in the construction of a cell.

One conclusion from this line of reasoning and from the experimental observations that go with it is that, although a sequence of amino acids in a protein may well appear superficially to be more or less random, it is in fact the result of a very stringent selection process, which has only "let through the sieve" a very small number out of all those that were theoretically possible. A quick calculation will make this clear: taking into account all the organisms that exist on Earth, there are probably no more than 10^{15} (1 million billion) different types of proteins. However, even for a very small protein of 100 amino acids, the number of possible sequences is enormous: 20^{100}, or about 10^{130}, which is infinitely more than all the electrons in the universe. This shows that out of all the theoretical possibilities, hardly any have actually been explored, and nor could they have been in any case. The sequences that have actually been produced have been selected for resistance to all the constraints, whether physical, chemical, or biological, to which they are inevitably exposed in the cell compartment in which they function, or in the surrounding environment if that is where they are active.

We began by looking at genome programs and the analysis of genome sequences, and endeavored to see whether we could predict gene functions from the sequence, "bottom-up." Scientists face considerable difficulties when they want to do this: remember that the sequence of the HIV genome has been known since 1983, but despite the fact that tens of thousands of scientists work on the virus, we still do not have a full understanding of its biology. Now we must approach the question in a more concrete way, "top-down," and think about what a gene function really is. To illustrate this, we shall take an example that has rarely been out of the headlines: bovine spongiform encephalopathy, or BSE, otherwise known as mad cow disease.

A Protein that Folds Up in Different Ways: The Prion Protein

Biology never ceases to surprise us. Every time we think we have exhausted the possibilities for a behavior or an architecture, we discover another one that was completely unexpected. This is also true for everyday experiments: the scientist asks a question, expecting there to be only two possible answers, but reality comes back with an unexpected reply—not yes or no, but *something else.* This is as true for very specific biological phenomena as for general ones.

Pathogenic Agents

The first pathogens to be identified were microbes, which breed and multiply at the expense of a host. The earliest organisms of this kind to be identified were mostly unicellular organisms, usually bacteria. Though minute, these are still on a large enough scale to be separated out from a medium by appropriate filtration techniques, and this is the basis for a whole series of methods for purifying water, for example with porous ceramic filters. Then, a little over a century ago, scientists discovered a new class of pathogenic agents—viruses.

In 1892, the Russian botanist Dimitri Ivanovsky made a very interesting discovery. He took leaves from tobacco plants that had a common disease called mosaic, extracted their sap, and rubbed it onto healthy leaves. These leaves soon showed symptoms of the disease. He performed the experiment again, but this time he filtered the sap through ceramic filters with pores so small that they retained all the bacteria known at that time. He rubbed more tobacco leaves with the filtered sap, but once again they were infected. Ivanovsky, who was a prudent scientist (as all scientists should be), did not interpret his results as a discovery. He simply recorded in his laboratory notebook his opinion that this contamination was probably caused by a toxin that was secreted by the bacteria and was not retained by the filter. A few years later, in 1898, the Dutch scientist Martinus Beijerinck surmised that this was indeed an infectious agent *(contagium vivum fluidum);* and in Germany, Friedrich Loeffler and Paul Frosch made the same observation, but in animals, with foot-and-mouth disease. They showed that the disease could be transmitted by a filterable agent, but also that an animal inoculated with a minute quantity of saliva from an infected animal could produce a large amount of infectious substance, and that the size of the agent responsible, which seemed to be able to multiply from almost nothing, was very much smaller than bacteria were thought to be. They formed the hypothesis that the germs in question were intracellular parasites, since they could not multiply except in the presence of a cell. Loeffler and Frosch called these new microorganisms "filterable viruses" (filterable because they were not removed by the normal filtration methods, and *virus* from the Latin meaning "sap," "venom," or "poison"). However, after a few years the adjective "filterable" was dropped, as all viruses had this property.

In 1935, an American scientist, Wendell Stanley, took up the work of Ivanovsky, Loeffler, and Frosch. He ground more than a ton of tobacco leaves infected with mosaic disease, and extracted a few grams of the virus, which crystallized. He observed that however long the virus remained sealed in a test tube, it did not increase in quantity. On the other hand, if he rubbed healthy tobacco leaves with a solution of the virus, they soon developed the symptoms of the disease, and the quantity of the virus collected after the infection was greater than the quantity he had started with. This confirmed Loeffler and Frosch's hypothesis that viruses are parasites that multiply at the expense of cells. In the same year, the invention of the electron microscope brought virology into the limelight. This new microscope's spectacular magnification made it possible to study the structure, shape, and characteristics of viruses, to classify them into various families, and to observe their activity.

As with bacteria, viruses reveal their presence through different symptoms, reflecting the different ways they can work as pathogens. Robert Koch, famous for his discovery of the agent of tuberculosis in 1882, summed up the properties of pathogenic agents as follows:

It must be possible to take a sample of the agent from a diseased organism.

If injected appropriately into a healthy individual, it must produce signs of the disease, which must appear after a specified interval.

It must then be possible to isolate the agent and purify it by dilution.

The quantity of the agent must be increased after having infected a host.

These characteristics are observed in diseases caused by both bacteria and viruses. But viruses were still an enigma, and remained little understood until it was possible to isolate them, and to realize that they were not alive in their own right, but that they replicated at the expense of the living cells of the organism they had infected. Viruses are essentially genetic material (DNA or RNA) surrounded by a protective structure (the coat or *capsid*), which allows them to resist variations in their environment and to penetrate the host's cells. Once inside, as they are genetic material, and recognized as such by the host cell's biosynthetic machinery, they reproduce and multiply (often killing the host cell in the process). After producing a new capsid, they are released into the environment, where they can go on to infect other cells. They are pure genetic parasites, and it is worth noting how unusual this is. Viruses are foreign bodies reduced to a simple *program* and able to use a host organism as a Turing machine. This is why it was not possible to understand their behavior until we understood the way the hereditary material is reproduced at the molecular level.

Viruses vary considerably, and the way they reproduce is often surprising. Given the wide variety of methods discovered, it seemed reasonable to think that these two types of microbes, cells and viruses, represented all possible biological objects capable of acting as transmissible pathogenic agents. However, this is probably not the end of the story, as we are about to see.

An Enigmatic Degeneration of the Human Brain

Many years ago, in 1957, the virologists Vincent Zigas, of the Australian Public Health Service, and Carleton Gajdusek, of the American National Institutes of Health, described a new disease that had long affected the Fore, an ethnic group living in the mountains of eastern Papua New Guinea. Jared Diamond wrote, in the article already quoted:

> In 1964, while studying bird evolution on the tropical Pacific island of New Guinea, I happened to set up camp among a tribe known as the Fore. I soon found my attention drawn away from birds to a human tragedy unfolding around me. Many of the tribespeople, children as well as adults, were limping on crutches, or unable to control their facial muscles, or lying semi-paralyzed in their huts. When I asked what was wrong, their healthy relatives answered with the single word "kuru," as if no more explanation were needed. Kuru, the Fore way of death, is now internationally notorious as a neurological disease, always fatal, and confined to that

> one tribe living in a group of mountain valleys only a few hundred square miles in extent.[19]

This fatal illness was marked by the progressive loss of motor coordination (ataxia) and later by the degeneration of higher brain functions. It was strange because it normally appeared very late in patients, and very probably as a result of ritual cannibalism. It was known that the Fore honored their dead by eating their brains. Gajdusek was a Harvard graduate and had been head of the Laboratory of Central Nervous System Studies at the National Institute of Neurological Disorders and Stroke. He was rather a nonconformist, and was particularly interested in diseases that appeared in unusual circumstances. At this time kuru was thought to be a viral disease, caused by a yet-unidentified pathogen. Failure to identify pathogens is fairly common with viral diseases because the pathogens are so small. Gajdusek quickly realized that kuru must be spread by the Fore's ritual cannibalism, but he did not succeed in formally identifying the pathogen. Since his work the ritual has been abandoned and the disease has disappeared, proving the connection between consuming human flesh and contracting the disease.

However, in recent years, kuru, a disease limited to one place and one cultural phenomenon, ritual cannibalism, has been brought back into the spotlight because of a new disease, *bovine spongiform encephalopathy,* or BSE, often known as *mad cow disease.* Much still remains to be discovered about BSE, which in some ways is similar to kuru; but to understand what is already known we need to make a detour via another disease Gajdusek studied, Creutzfeldt-Jakob disease, or CJD. The fact that it can be transmitted from humans to animals in the laboratory suggests that it also involves a pathogen, according to the criteria defined by Koch. But when these diseases are studied, the very definition of a pathogenic agent must be reconsidered and refined.

Discovered at the beginning of the 1920s by Hans Creuzfeldt and Alfons Jakob, CJD is a neurological syndrome, which they called "spastic pseudosclerosis" or "subacute spongiform encephalopathy." Unlike kuru, Creutzfeldt-Jakob disease is found throughout the world, and its first symptoms are the degeneration of higher brain functions, and dementia, rather than ataxia. CJD generally appears sporadically, affecting about one person in a million at around the age of sixty, but some populations are particularly affected, and places have been identified, in central Europe, in England, and in the United States, where the disease is much more common. This is a very strange disease. On the one hand it seems that it can be contagious, a classic sign of a pathogen (some cases are caused by medical errors such as contaminated intracerebral electrodes, corneal grafts, or injections of growth hormone made from contaminated human pituitary gland extracts), but on the other hand it can also be hereditary (10–15 percent of cases), and it can even appear here and there for no reason (though very rarely). All this suggests that there is a pathogenic agent. In order to understand it we need to find an explanation that can reconcile its sporadic appearances, its inheritability, and its transmission from one individual to another. This may not in fact be as strange as it appears. In order to infect an organism, a pathogen must enter a cell, which implies that it can recognize a cell type and cross a cell envelope. Human polymorphism is known to be such that all cell envelopes have characteristics that label them as belonging to an individual, and which thus differ from one person to another, so they are not equally susceptible to a given pathogen. However, if the number of susceptible people is large enough, the disease can be contagious. It is thus clear that a disease can have both environmental *and* genetic origins.

There are at least two other diseases related to kuru that affect humans. One was described in 1936 in Austria by two neurologists, Joseph Gerstmann and Ernst Sträussler, and a neuropathologist, Isaak Scheinker. This disease first shows in ataxia and other signs of damage to the cerebellum, followed by dementia. In this disease, the damage to the cerebellum is visible in its spongy appearance. The second is "fatal familial insomnia," which always appears in more than one individual in a family, and in which dementia follows the inability to sleep. This was discovered more recently by two scientists at the University of Bologna in Italy and one at Case Western Reserve University in the United States. These two diseases are thought to be hereditary, and they appear in middle age. All of these diseases show more or less the same kind of anatomical signs, whether in the cortex or in the cerebellum. The brain matter degenerates into plaques, forming zones with holes in them, rather like a sponge; hence the name "spongiform" encephalopathies.

From Humans to Animals: The "Kuru-Scrapie Connection"

The diseases I have described are specific to humans, but what about animals? A quick look through descriptions of similar diseases in animals shows a fatal disease of sheep, *scrapie,* known at least since the beginning of the eighteenth century. Sheep with this disease suffer from increasingly severe convulsions (hence its French name *tremblante*) for about a year before they die. These convulsions can lead to frenetic scratching in which the animal scrapes off tufts of wool, giving the disease its English name. It was initially attributed to excessive sexual activity rather than to a pathogen (which just shows how the cultural context of an era can influence thought, and thus research). It was first described in neuroanatomical terms by the French veterinarians C. Besnoit and C. Morel in 1898, but it was not until 1936 that Jean Cuillé and Paul-Louis Chelle, also in France, identified it as a transmissible disease, and they published their findings in 1938.

The disease has been studied in some depth in Britain, and was at first attributed to an "unconventional" virus, because it had not yet been possible to identify it, and its replication was difficult to understand. It was found in flocks of sheep, but it was never the case that every animal in the flock was infected. It seemed to be more common in some breeds of sheep than in others. In some places, Iceland for instance, it could be shown that the fields in which the sheep grazed could spread the disease from one flock to another. However, the fact that flocks in Australia and New Zealand were free of the disease, despite coming from farms in Britain that had either been contaminated or had bred genetically predisposed animals, seemed to show that the disease was transmissible, and that it could in principle be eradicated.

Many other animals besides sheep suffer from similar diseases. Mink farms are sometimes affected by a disease of a similar type that is clearly transmissible, but mink are exclusively carnivorous and very aggressive, which means that they are particularly prone to pass on diseases. It is also found in cats and in herbivores such as goats, deer, and elk. More recently it has appeared in cattle, possibly as a result of the contamination of meat-and-bone meal, an animal feed additive prepared from sheep carcasses. It became very widespread in the United Kingdom in the early to mid-1990s, but the number of new cases is now very low. Known as mad cow disease because the affected animals have serious behavioral disorders, the disease is very worrying, because it seems to have been transmitted orally and may thus have crossed the species barrier.

Unlike kuru, scrapie has not been known to cause any disease in humans, despite its chronic presence in flocks of sheep for at least 200 years. However, the fact that it may have been able to cross from sheep to cattle inevitably raises the worrying question whether it could be transmitted in turn from cattle to humans. In fact the transmission of diseases between species is not normally directly transitive (in the sense that if a disease passes from species A to species B, and from B to C, it will not necessarily pass from A to C). But we know that sometimes a disease that is not directly transmissible to humans can be passed on if an intermediate host is a carrier. This is probably the case with flu, for example, which affects birds such as ducks without really making them sick, but then is transferred to humans, probably after infecting pigs, as this selects variants of the pathogen in a way that allows humans to catch it. It then passes very efficiently, and often dangerously, from one human to another.

This is why BSE is particularly worrying. It was identified in 1986 by Gerald Wells and John Wilesmith at the Central Veterinary Laboratory at Weybridge in England, after it had begun to strike British herds, and was thought to be the result of a recent change (at the end of the 1970s) in the heat treatment of animal food additives. It is thought that the previous very harsh treatment undergone by carcasses was sufficient to eliminate all pathogenic agents, and especially the scrapie agent, but the more recent treatment, though sufficient to destroy bacteria and viruses, was much gentler, and may have left some resistant agent still able to propagate. The aim of putting additives in animal feed is to accelerate weight gain by improving the amino acid content, because the plant matter in normal feed usually has both a low protein content and a set of amino acids which is not properly balanced for the requirements of animal protein biosynthesis. The new treatment was intended to kill bacteria and viruses but not to break proteins down into small peptides or amino acids. If, as is now accepted, scrapie and BSE are caused by a particularly resistant form of protein, a pathogenic agent entirely different from anything previously thought possible, this new agent would be much more resistant to most physicochemical treatments than agents such as bacteria or viruses. If this hypothesis is true, it means that, in itself, a protein cannot be considered innocuous until shown to be so by appropriate experiments. It is also worth noting that the farther away from humankind a pathogen is, the less likely it is to be highly virulent. This shows that we must be particularly cautious about animal-derived products, more so than about products derived from plants or bacteria.

At the beginning of the 1960s, scientists began to establish the first experimental models of these spongiform encephalopathies by infecting mice, using the most direct route, injection into the brain. At about the same time, the American veterinary pathologist William Hadlow made the connection between scrapie and kuru, in an editorial in the medical journal *The Lancet,* which often launches controversies in the medical domain. Tikvah Alper, D. A. Haig, and M. C. Clarke in 1966, and then R. A. Gibbsons and G. D. Hunter in 1967, showed that the scrapie agent was remarkably small, smaller than all known viruses, and that the physical techniques that normally rendered viruses inactive (heat and ultraviolet radiation) had no effect on the nature of the agent responsible for scrapie. This was a new and completely unexpected situation. An explanation for the phenomenon had to be found.

An Unconventional Cause: The Prion Hypothesis

At that stage, the only known causes of disease were either organisms made of cells, or viruses. The prevailing idea was therefore that all these diseases were caused by viruses, of the rather mysterious class known as "slow" viruses, because of the long incubation period before their effects showed. It was thought that these agents must necessarily contain genetic material, either DNA or RNA, which would allow them to reproduce. It ought to be possible to identify them, because all the studies of scrapie, CJD, and kuru had shown that they had been transmitted experimentally to healthy animals by injection (and sometimes even orally), using extracts from the brains of diseased animals. Yet their resistance to ultraviolet or ionizing radiation was incompatible with the presence of nucleic acids, which are very sensitive to radiation when they are of the size required by their role as the genetic program. Faced with this mystery, scientists grouped these agents together under the enigmatic name of *unconventional infectious agents.*

In 1974, Stanley Prusiner, of the University of California at San Francisco, had undertaken to characterize these agents. Everything seemed to indicate that they contained nothing but insoluble aggregates of proteins, possibly chemically modified, and for this reason he called them prions (from "proteinaceous infectious particles"). Of course, as these particles contained no genetic material as that is usually understood, the prion hypothesis took a long time to be accepted (and although Prusiner was honored with a Nobel Prize for his work on this subject, it is still contested at the turn of the millennium, despite the existence of strong if not entirely compelling arguments in favor of it). The fact that extracts from the brains of sheep suffering from scrapie remained infectious after irradiation implied that prions did not contain nucleic acids, and that they were neither viruses nor any of the other agents known at that time. We know, however, that epigenetic processes can take effect over and above genetic processes. We should therefore be cautious in considering genetic phenomena alone as the cause of diseases. Of course they always require an underlying genetic process, using nucleic acid, at some point. Only nucleic acid can be reproduced from a single particle (thanks to the complementarity of the sequences of the two strands of the double helix, during replication) and can result in the translation of the properties of its template into proteins, once enough copies have been produced by a chain reaction. The scientists working in the domain had all sorts of ideas, but it was only after many years' hard work, using highly purified extracts of hamster brain, that Prusiner and his colleagues found themselves obliged to conclude that the infectious substance seemed to be no more than a protein, which stuck together easily to form insoluble clumps, or *aggregates,* which could be responsible for the nerve cell degeneration observed in the victims. Of course we must still be open to the idea that something associated with the protein, and extracted from the environment, contributes to its pathogenic activity.

All Prusiner's laboratory results pointed in the same direction, and especially when he carefully replicated Alper's experiments, they showed that there was no nucleic acid in the agent responsible for the infection. The agent was of a proteinaceous nature—that much was demonstrated by the fact that techniques which inactivate proteins severely reduced the activity of prions. All the attempts to purify it led to the identification of just one protein, called PrP (for prion protein), which seemed likely to be the agent responsible for the disease. But we know that proteins are specified by genes, and yet there is no genetic material in prions. For a long time there was no answer to this paradox, and it would have remained unanswered if the techniques of molecular genetics had not evolved in parallel with Prusiner's work. Even if it had been possible to visualize conceptual models of infection via prions, without more precise information about their biochemistry it remained impossible to make valid comparisons between these models and reality, because the very advanced techniques this required did not exist when research into this area first began.

Because the genetic code is virtually universal, since the beginning of the 1990s it has been possible to reconstitute the order of the bases in a stretch of DNA (a gene) from the protein it specifies (or a fragment of the gene from a fragment of the protein). This is not very simple in practice, because the genetic code is partly redundant, but it is possible to synthesize chemically a "population" or range of short fragments or oligonucleotides that includes the particular gene fragment that codes for the protein fragment being studied. The technique of molecular hybridization makes it possible to "fish out" the gene in question from the nucleus of the cell, using this artificial gene fragment.

This is what Prusiner and his colleagues did, in collaboration with Leroy Hood at Caltech and several other laboratories, including Charles Weissmann's laboratory at Zurich. They isolated the first 15 amino acids of PrP and constructed appropriate probes, to fish out by molecular hybridization any gene that might code for the corresponding peptide, in tissues from both healthy and diseased animals. The simplest hypothesis was that the gene would be found only in affected animals, in association with the prions. But the scientists found to their surprise that although there certainly is a gene that codes for PrP, this gene is normally present in the healthy animal. What is more, it was clear that this gene is expressed in the normal way in healthy individuals (hamsters, mice, humans, for instance) and that the corresponding protein is present in their brain in the normal way, without their being at all diseased. One paradox after another . . . unless there could have been a systematic error in the experiments, and the preparation used had been contaminated by a very small quantity of the pathogen. Although this possibility cannot be conclusively eliminated, the best experimental techniques available at the end of the 1990s seemed to indicate that there was no gross error in the approach. This drove scientists to look for an understandable explanation for the presence of a normal PrP gene in healthy animals.

Faced with this paradox, we need to remember that the nature of heredity may be either genetic or epigenetic. At the same time we have to reconsider the concept of a pathogen, so as to understand that it can have several fundamentally different categories, depending on their mode of transmission. We are used to pathogenic agents that cause disease by exploiting the usual mechanisms of (genetic) heredity—this is what unicellular microbes and viruses do. But it is not hard to imagine that other agents exist, which exploit the mechanisms of epigenetic rather than genetic heredity. Once they have been expressed, they pass on *the conditions of their own pathogenic expression* from generation to generation. How might this happen? Until the prion hypothesis, there was no known pathogen of this kind. But it was this reasoning, based on epigenetic heredity, that Stanley Prusiner and Samuel Cohen turned to in order to explain the pathogenic effect of prions.

A Protein as Two-faced as Janus

When scientists analyzed the protein of prions and found that it was a single normal protein found in the host organism, they were faced with a paradox. As normal individuals are by definition healthy, something very strange must have happened to this protein. Then a discovery was made which confirmed that the prion hypothesis could be considered seriously. In hereditary cases of Creutzfeldt-Jakob disease, PrP is different from the protein found in healthy individuals: its amino acid sequence differs in one or more places from that of the healthy protein. This correlation between the appearance of the disease and a genetic mutation is unlikely to occur by chance. It was reasonable to draw the conclusion that PrP is the *cause* of the disease rather than a consequence of it. A number of hypotheses can be proposed to explain the role of PrP. One of the simplest would be that the normal protein undergoes a chemical modification in the brain after the normal translation process from its messenger RNA (this is something that often happens), and that it is the *modified protein* that leads to the disease, perhaps by making the protein likely to aggregate easily. No experiment has been able to show this convincingly, and besides, in this case we should expect to find genes that correspond to the modification system, and mutants of these genes should interfere with the disease. They have not been found. However, this is not necessarily a fair objection, as CJD is so rare that it is not impossible that mutants of these genes exist, but that we have simply not yet been able to identify them. Or perhaps the protein itself is not modified, but is synthesized in the wrong place or at the wrong time, perhaps as the result of a change in the structure of its mRNA (for instance its leader sequence or its 3′ terminal end).

Of course it is not possible to use the nonexistence of an object or a phenomenon as a conclusive argument, so in the mid-1990s the viral hypothesis could still not be eliminated altogether. On the other hand it was possible to propose a model in which PrP plays the role of pathogen unaided. In this model, proposed by Cohen and Prusiner, the PrP protein is assumed to exist in at least two different spatial conformations. The strand of amino acids that makes it up is thought to be able to fold up into two significantly different shapes (this is a common situation and numerous cases are known), and one of these is thought to cause the disease. It is also thought that when the pathogenic form—the prion type—comes into contact with the normal protein, it is somehow capable of provoking it to *convert into the pathogenic form.* A simple contact between the brain tissue of a normal individual and a prion would thus lead to the gradual conversion of all its normal PrP proteins (which are soluble, and already exist in the nervous system before infection) into prion type PrP proteins (which cause the disease), in a gradually accelerating process leading to the precipitation of plaques, and finally to death. The advantage of this model is that it is simple enough to be tested. First it must be shown that several different architectures exist for this protein. Next—and this is the most difficult part to prove, because it requires the purification of enough PrP protein in each of the forms stipulated by the model—scientists must be able to show that adding a small amount of the protein in its diseased or prion form causes a significant amount of the healthy protein to convert into the diseased form and then aggregate. This model assumes a direct interaction, but of course the phenomenon could be indirect, for example if it relied on an auxiliary protein (which might be the PrP itself), or if the diseased form of the protein influenced its destination in the cell, which then led to a different folding pattern. Demonstrating a process of this sort is obviously more difficult, but in principle it is possible.

If it were proven, this hypothesis would account for the transmissible aspects of the disease satisfactorily, provided an appropriate route for transmission were discovered (of course the agent cannot be directly injected into the brain!). Sporadic cases would be explained by the low, but not nonexistent, probability of spontaneous conversion of the healthy form into the pathogenic form. Hereditary diseases such as CJD would also be easily understood. In this case, the PrP is different from normal PrP, and it is quite reasonable to suppose that it has a higher probability of changing spontaneously into the pathogenic form. It should be possible to show this experimentally too. Finally, the fact that the incorrectly folded protein acts as a catalyst, triggering the conversion of normally folded protein into its own pathogenic form, would explain how it can act as a pathogen. If this were the case, it would mean that spontaneously mutated protein in affected animals would become a source of the disease if this protein is included in the diet of normal animals, which is a worrying possibility. However, it is still not clear how it reaches the brain of the host after being eaten (most experiments have used prions injected into the brain or into the blood).

There are several very convincing experimental arguments in support of the prion model. First, normal, soluble PrP has been found to be very sensitive to proteases, the enzymes that break down protein, whereas the protein extracted from prions is very resistant to the same enzymes. The spatial structure of PrP was established in 1997, and it shows that the protein is made up of a rigid end followed by an extraordinarily flexible part, able to adopt a large number of different conformations, depending on the conditions of its environment. Several experiments with genetically modified or *transgenic* animals—mice with the gene for PrP deleted—show that unlike their normal parents they are resistant to injected prions. They do not develop the disease. But the experiments that are the most convincing at present—conceptually at least, because they demonstrate the epigenetic character of the transformation of certain proteins—are those that use a model microbe, brewer's yeast. Thanks to the sequencing of the yeast genome, a yeast gene has been found that produces a protein very similar to PrP. This has allowed scientists to experiment on the model protein. Its function in yeast is still a mystery, but the fact that it exists in a variety of organisms suggests that it must be significant, at least on an evolutionary time scale. Studies of its distribution inside the cell show that in standard growth conditions it is distributed in a soluble homogeneous form. In contrast, if the yeast protein is placed in contact with PrP extracted from prion, it appears to be insoluble, and to aggregate in various locations in the cell. This behavior then becomes stable across the generations, which is understandable if the aggregated form triggers the conversion of the soluble form into the aggregated form, either during its synthesis or immediately afterward. This is a typical illustration of epigenetic heredity. The conversion was shown to require the presence of an auxiliary protein, probably a molecular chaperone. In yeast, the protein appears to interfere with the normal termination step of the translation process, causing errors by making the messenger slip within the ribosome. If this is true in mammals, it is easy to see how certain kinds of alteration to the translation process could have very serious pathogenic consequences, and would explain why prions cause dangerous diseases.

This observation substantiates the hypothesis that prions are pathogens of a type that is radically different from the usual agents—bacteria, viruses, protozoa, or fungal agents. A particularly important difference is that these new agents have no genome, no hereditary genetic memory as we know it. They are significantly altered forms of a normal protein, able to convert the normal form of that protein into an abnormal, pathogenic form. They are thus *triggering* agents, and not reproducing agents (as is the crystal of salt that triggers crystallization in a supersaturated solution). The chain of events leading to the disease is completely different from the normal situation, because the agent does not multiply, but converts a preexisting healthy reservoir into a pathogenic product. The pathogen triggers the conversion of normal protein into another form, or prevents it from folding up into its healthy conformation, without needing to reproduce in its own right. Koch's rules for the definition of a pathogen still apply, but the meaning of the term is slightly different. The spectacular example of prions highlights once again the fact that we cannot understand what life is without understanding the relationships that form between the objects that make up cells. Of course these relationships can exist at any level in the hierarchical organization of living organisms, and we can also be sure that it is the very existence of these interactions between cells or organs that gives the living world its originality. The genome sequence quest is just the beginning of our exploration of what life is, not the end.

A Provisional Conclusion: The Difficulty of Predicting Gene Functions

I have chosen to illustrate the case of a simple biological object, a protein, because this has enabled me to show how a simple object can be involved in the most profound properties of life. It has also allowed us to see the complexity of the relationship between its structure and function. We have also seen that the function of "being a pathogen," the cause of a disease, can be extraordinarily varied, so much so that it is impossible to know in advance where an attack on a living organism will come from. We have seen that there is no well-defined category that marks the difference between, say, a poison—this is no more than a simple molecule that imitates a normal object needed for survival, takes its place, and kills—and a predator, which is itself alive.

Starting with the relationship between predator and prey, we can trace a continuum, through parasitism (which is pathogenic), commensalism (living side by side, either happily or virtually unaware of each other's existence), and symbiosis (which implies the simultaneous development of two organisms) to a point where the relationship breaks down. Here, the pathogen is no longer alive in its own right, but draws life from life. Genetic material itself displays a similar continuum: it can ignore its context (as with certain viruses and "insertion sequences"), it can be symbiotic (as we are about to see), or it can be predatory (as with most viruses). We may wonder if it makes sense to go further, and to suggest that there exist stable structures with no genetic material (no nucleic acids), but which, like viruses, can make an organism produce copies of themselves.

Gene Migrations

The same series of questions was asked in the mid-1940s, at the time when *cytoplasmic heredity* was discovered. In 1949, at a meeting on "biological units endowed with genetic continuity" organized by France's CNRS (the national scientific research organization), Boris Ephrussi described a mutation in baker's yeast *(Saccharomyces cerevisiae)* that gave rise to "petite" colonies when the yeast was cultivated in Petri dishes. The small size of these colonies reflected the lack of certain enzymes that the yeast requires for cell respiration. The rate at which mutations appeared increased considerably if the cultures were treated with a chemical dye from the acridine family. Nothing out of the ordinary so far: after adding a mutagenic agent, mutations appear, with the loss of a function. At the time (and we must not forget that this was four years before the discovery of the structure of DNA), it was thought that genes were always found in the nucleus of the cell, and never elsewhere (yeast cells, like animal cells and unlike bacteria, have a nucleus). However, Ephrussi noticed, more or less by a lucky chance, that certain cytoplasmic particles, mitochondria, were absent from buds of yeast grown in the presence of acriflavine, and that this phenomenon coincided with a high mutation rate. Of course it was impossible to exclude *a priori* the idea that it was an unusual kind of mutation in the nucleus. Following the work of Paul Portier, which he had outlined in *Les Symbiotes* in 1918 (a book which for a long time served as a reference, and was much discussed, even in English-speaking countries), Ephrussi thought that mitochondria were probably degenerate bacteria, leading him to prefer the hypothesis that these were mutations that directly affected cytoplasmic particles.

Two years later, in 1951, at a meeting at the Cold Spring Harbor Laboratory near New York, Luigi Provasoli and his colleagues described a new experiment on a unicellular alga, *Euglena,* in which they had destroyed the *chloroplasts,* the green corpuscles it used to process carbon dioxide and produce oxygen via photosynthesis. They had done this simply by adding an antibiotic, streptomycin (at the time, nothing was known about how this functioned), and the cells treated with this antibiotic rapidly turned white and died. At the same meeting, other scientists described a protozoon very similar to *Euglena* but which has no chlorophyll, and so can grow only in a rich medium, because it cannot use carbon dioxide. This showed that whatever makes up a cytoplasmic organelle, it is an autonomous entity, distinct from the nucleus, and with its own rules of heredity.

In 1951, nobody suspected that chloroplasts and mitochondria have their own genome, in addition to the almost complete set of genes found in the cell nucleus. The precise nature of cytoplasmic heredity remained shrouded in mystery. Later, the discovery of chromosomes in the mitochondria and chloroplasts accounted for cytoplasmic heredity and made it possible to understand why it follows the maternal line. Because these organelles are found in the cytoplasm, they are present in the egg cell, but the paternal germ cell is almost entirely reduced to a nucleus, with only a minute amount of cytoplasm and normally no organelles. However, because most of the alterations occurring in the genome are harmful, the accumulation of mutations will at some point lead to a phenotype that is no longer viable, unless some mechanism can erase many of the mutations at the same time, in one big step, allowing the system to proceed further. This mechanism might be the result of sexual reproduction or some other means that allows the lateral transfer of genes. The phenomenon is known as "Muller's ratchet," since Muller pointed out its importance in the mid-1960s, and it indicates that cytoplasmic heredity must at some stage become difficult to maintain, unless the corresponding genes have been able to migrate to the nucleus. This complicated phenomenon therefore has to be taken into account when endeavoring to predict a gene function from the genome sequence.

Many other structures are preserved from generation to generation as these organelles are, but the way this happens was still not understood. They are not all associated with DNA. It was not until the mid-1970s that scientists began to accept an idea that had been proposed much earlier, and developed at length by Portier in *Les Symbiotes*. This was that these organelles are the result of the evolution of bacteria living in symbiosis within nucleated cells. The genome of these *endosymbionts* has degenerated until it is left with only the genes that code for a few of the enzymes needed for respiration or using carbon dioxide. But we now know that this degeneration can go much further, to the point where the genome simply disappears, and the genes that are essential for the synthesis of the organelle are placed in the chromosome in the nucleus. This is the case with the *hydrogenosomes* found in some protozoa that live in environments where there is no oxygen, such as the first stomach in ruminants: all of their genes are now in the cell nucleus.

Classifications Based on the Historical Origins of Functions

Again and again, we have had to refer to the central notion of function in order to discuss life's aptitude for specificity. A function represents an ability to engage in one specific action among a great many other possible actions. And it is this specificity of action, an original characteristic of life, that gives the impression that living things organize their activity within a *project*. This gives them their apparent "final cause" and explains why early observers of life were unable to conceive of it otherwise than as animated by a magical objective, a foreknowledge of the future. It is hard not to admire the often remarkable adaptation of living organisms, and the ingenuity of the solutions they have found for the most unlikely situations. This is the fundamental reason behind the instructive explanations that always dominate our first attempts to account for the phenomenon of life. But relationships between organisms, and the lines of descent they assume, mean that any attempt at explanation soon comes down to a search for origins. Genome programs, which ask about the heart of heredity itself, are not exempt from this fundamental questioning. We must assume that for a great many scientists who want to know exactly what genomes are, their interest centers on this question of origins. Speaking for myself, it was the completely unexpected discovery that two enzymes needed for the synthesis of two amino acids, cysteine and tryptophane, which are chemically very different, were actually related to each other, and clearly descended from a common ancestor, that reinforced my conviction of the value of whole-genome sequencing projects, and led me to take part in one myself.

We have seen the importance of hierarchical classifications in understanding biological functions. Knowing the genome text enables us to use the genetic code to translate sequences of nucleic acids into amino acid strings, and thus to predict the primary structure of the proteins coded by the genes. We have seen how difficult it is—difficult in principle, not just technically, so technology will not be able to resolve it—to predict not only the spatial shape of the protein from its primary structure alone, but also of course its function. However, once we know several genomes, or sometimes even just one in its entirety, we can see that the primary sequences of certain proteins are related, in the sense that the sequences of their amino acids are similar to each other. And if these organisms are close enough to each other, it may be possible to extrapolate from one to another, at least as a first hypothesis. We are thus in a position to begin an initial exploration by establishing families of genes whose products can be aligned and compared according to the similarity of their amino acid chains.

However, this is not an easy approach. In a given protein, some amino acids are understood to play a much more important role than others (for example because they are directly involved in the protein's function), while others could be replaced by almost any other one of the twenty amino acids, without the function's being affected. Multiple alignment of related sequences is one of the most useful tools for exploring the evolution of functions. On the basis of these alignments, we can construct family trees that tell us about these sequences' *past.* This is how the chemist Steven Benner was able to artificially reconstitute ancestral proteins and to study their activity. The conclusions of this study indicated that their original activity was probably broader in scope, and less specific, than the activities of their modern descendants, confirming the hypothesis set out above, that evolution often comes about through preexisting activities' becoming more and more specific.

It is particularly useful to try to compare the trees produced as a result of analyzing different proteins from the same organisms, and to study the way they have evolved. But if we try to understand sequences that have diverged more radically, their interrelationships no longer have the hierarchical organization seen in tree diagrams. There are still domains in an alignment that can be similar, and can thus provide information about the obligatory constraints connected with an activity (for instance, in the case of enzymes, amino acids preserved throughout evolution have a key role in folding, in architecture, or in catalytic activity itself). Understanding the origins of present-day sequences will certainly establish a specific range of combinations of ancestral sequences, and we have seen how important this is in all the specific properties of life. Using these, Claudine Devauchelle, Alain Hénaut, Alexandre Grossman, and their colleagues discovered a way to make transition matrices describing each family of homologous proteins, and thus to describe their rate and mode of evolution.[20] It then became clear that although there is, roughly speaking, a kind of molecular clock that tracks the evolution of the proteins of various organisms in a relatively uniform fashion, their evolution can be speeded up or, in some cases, slowed down. Besides, some organisms have an evolutionary position outside the main stream that connects organisms with one another. Then we will want to know where the various protein families found in a given organism come from, and to research their origins. A recent discovery, thanks to the work of Jean Lobry, Eduardo Rocha, and Alain Viari, showed that in fact in many bacteria, the evolution of the genes found on one of the DNA strands was not the same as on the other strand! It is not hard to understand why establishing family trees is full of pitfalls, and this may explain the bitter quarrels between scientists who are obstinately convinced that one tree or another must reflect *the* truth about evolution . . .

It is still too early to analyze the origin of archaebacteria, eubacteria, and eukaryotes in this way, but it is very likely that unexpected results will improve our understanding of the way variations down the generations and lateral genetic transfers have been combined. This will probably tell us a little more about the origin of mankind as well. Thanks to the work of Eric Lander and his colleagues at MIT, it has been possible to study the "neutral" polymorphism of several megabases of the human genome in various populations. One interesting observation, which seems to lend more weight to the idea that humans originated in Africa, is that polymorphism is much greater in Africa than in Europe or Asia (and incidentally this means that it is wrong to put all Africans together in one group just on the basis of skin color: there are several groups, each at least as different from each other as they are from Europeans). The family-tree approach should also enable us to understand the evolution of enzymes whose catalytic activity involves numerous variations on the same theme. Using them as our inspiration, we will be able to direct evolution, in order to carry out specific chemical reactions, for economic ends. This is important both in helping to break down the thousands of new chemicals released into the environment every year, but also in producing new molecules for use as medication, cosmetics, or even as precursors of products needed in the major industries.

From Function to Structure

The apparent paradox of this title is intentional: the idea of a cause-effect relationship between the structure of biological objects and their function is so well established that it motivates the work of hundreds of thousands of scientists around the world. But I have tried to show that the causal relationship between the architecture of biological objects and their function is often arbitrary and accidental. More seriously, cases in which function has actually been deduced from structure are extremely rare. There is one of course, probably the most famous one, that of the double helix of DNA: when Watson and Crick made a model of DNA they realized that in the two strands of the helix, A was always opposite T, and G opposite C, and it did not escape their attention—to use the famous understatement in their 1953 article—that the function of preserving and faithfully transmitting heredity was possible thanks to this structure. But in actual fact, apart from the immense impact it had in orienting our understanding of life, this article is deservedly famous precisely because it drew attention to this unusually simple relationship between structure and function.

But it is practically unique in this respect. Even in the simplest cases of enzymatic catalysis, knowing the exact three-dimensional structure of a protein does not always reveal its function; far from it. More often than not, it is through comparison with objects that are already known, for which we have not only the structure of the enzyme, but that of the enzyme in combination with its substrate, that we can discover what characteristics of its architecture and which amino acids are important in the catalysis. There is a very simple reason for this: as we have seen, adaptation occurs *a posteriori,* and not *a priori,* because there is no final cause. The living being that survives a critical situation did not know in advance what would save it, but, having found the solution, and because it has survived, it passes that solution on to its descendants, thus preserving and multiplying it. Such a solution is always some way of establishing a link or relationship between processes, events, or objects. The link is part of a structure, but it was the function that revealed it, so it is the function that ensures that the structure is retained. Function does not create structure, but discovers it when it is needed.

This is also true of the way we apply our biological knowledge. It is how we go about creating new medication, for example. Imagine that we have to develop a new antibiotic to fight tuberculosis. To begin with, bearing in mind the problem of antibiotic resistance, and the way it is horizontally transmitted between very different organisms, we will not want to look for a molecule that attacks a broad spectrum of microbes, but just one, *M. tuberculosis.* Of course we will also want the molecule to be harmless to the human host. We will begin by looking for a specific function of *M. tuberculosis* (or mycobacteria in general). For example, the mycolic acids, complex molecules on the surface of the bacterium, enable it to protect itself against the immune defenses of the host. This is the function we should change, by preventing the synthesis of these protective molecules. So we must characterize the stages of the synthesis, find out the genes involved, and the specific functions of their products (noting especially those that do not exist in the host). Knowing the structure of these enzymes or the regulatory proteins associated with them—and it is only now that structure comes into the picture, *after* we know the function—we will then try to produce all sorts of molecules that interact with them. We will test their activity, their stability, and their toxicity. Clearly, this will take a very long time, particularly during the various clinical-trial stages. After appropriate experiments have been carried out on animals, clinical trials always have to be undertaken to take account of human characteristics (and the oddities produced by our very polymorphous heredity). The molecule that passes all these tests, because of its inhibitory function, is the one that will be chosen. Once again, it is the function (satisfying all the test criteria) that determines which structure will be selected.

If we want to reap the benefit of our knowledge of genomes by drawing applications from them, we must reverse our usual approach, and study structures only when they are firmly linked with a function. The relationship between structure and function, so often cited, is certainly an interesting one, but it is exactly the opposite of the way it is usually regarded.

Genome Engineering

I have said that because of the way life constructs itself, it is impossible to know what the future will bring. However, we can have some idea of the areas life might explore. After all, the future is not totally unpredictable; it normally changes slowly enough for life's construction algorithms to adapt their output to suit it. And although we will not be able to find out what life will be able to *create*—by definition, creation has no other past than the conditions of its birth—at least we will probably be able to imagine what we can make from living organisms, if we create a new discipline, both theoretical and practical, an artificial evolution, which we could call genome engineering.

The alphabetical metaphor, with the genetic code, which rules the correspondence between the genome text and its expression, suggests that the present state of every living organism is the result of a recursive process that has put together the corresponding information bit by bit, by rewriting it generation after generation. The fact that the program and the machine that runs it have evolved simultaneously allows information to be extracted quickly and efficiently, enabling an individual to be created by the concrete application of an algorithm, through a process that allows only those individuals that are not totally ill adapted to survive. To understand its function, extracting information from a DNA sequence requires much more than just identifying genes using the genetic code rule and a few contextual descriptors, as is usually done. We also need to understand its *history.* Exploring the rules (if there are any rules) that have contributed to the success of those individuals that have survived during this history will enable us to turn them to our own use.

I would like to look at one of these rules, one that underlines the concrete implementation of a *control* process. With the evolution of ever more complex animals, numerous regulatory mechanisms have appeared. Gradually they have combined to ensure the stability of the environment surrounding the cells within the organism, rather than leaving them subject to major environmental changes. These protection mechanisms seem to be necessary because the enzymes that catalyze the cells' chemical reactions are highly sensitive to their physicochemical environment. This sensitivity is an effective selection procedure that explains how the organism as a whole can normally maintain the local environment of its organs and cells within narrow limits. For instance osmotic pressure in the blood or in sap plays this role, as does temperature control (homeothermia) in warm-blooded animals. To arrive at homeostasis, a variable such as temperature must be regulated so as to minimize the difference between the desired value and the observed value. There are many ways of doing this. Often it is done by monitoring the difference between a reference value and the actual value, for instance by using a thermometer to measure temperature, and introducing feedback into the process responsible for heating or cooling, so as to make it change in the direction required. One method is to measure the value of the variable with a *receptor.* The receptor sends a signal to a control center, which then activates an effector, which produces a change in the variable. If the temperature drops, gooseflesh, and then shivering, produce muscular activity, and our temperature rises again. Often the control center compares the temperature produced by the effector with the desired temperature (it needs a reference to do this) and activates the reheating process by negative feedback. Sometimes homeostasis is achieved without any comparison: changes in body temperature produced by shivering are not sent to a control center. This is an open control loop. Finally, there are a few situations in which an increase in a variable leads to an even greater increase, in a positive feedback mechanism. This is what happens when blood clots, for instance: a very small local effect leads to an explosive cascade of reactions that very quickly form a clot to block a hemorrhage. But problems with this system can also have harmful effects, as in thrombosis, and can even be fatal. Here we have three different kinds of control in which the agents have very different functions. It is important to identify them in real situations, and to see how they are linked to the nature of one gene or another.

In particular, the receptor function, coupled with that of an appropriate mediator (which controls by activating or repressing the expression of specific genes), is essential. It is this function that enables the interior of the cell—and especially its genetic program—to be simply and specifically linked to the characteristics of the environment. In a way it is the equivalent of a sense organ for the cell. This kind of sensory function (and receptors are often called "sensors") exists in organisms as simple as bacteria, and is the result of an evolution that has entirely separated the notion that a molecule can be the *substrate* of an enzyme, from the notion of a *signal* (which then acquires a *symbolic* function).

The regulation of gene expression comes into this category. The control of transcription can be implemented via the activation of promoters that are recognized by specific factors, either repressors or activators. Finally, there must be a link between gene expression and the way the activity of preexisting proteins is modulated, depending on the variable characteristics of the environment (this was the first model proposed for β-galactosidase, although as we saw it was the wrong one). If we have catalogues of these functions and the way they are implemented in reality, then it becomes possible to create cells that act in circumstances whose parameters we can control at will. This is the aim of genetic engineering. Using this type of control of gene expression, pharmaceutical companies such as Sanofi have been able to make *E. coli* express the gene for human growth hormone cloned by Willem Roskam and François Rougeon at the Pasteur Institute in Paris. By constructing artificial genes whose expression was increasingly cleverly adapted to the conditions of industrial production, this company was soon able to produce very large quantities of a hormone identical with the human hormone, but free from the risks of contamination by pathogenic agents that are associated with any use of human products, or animal products in general.

Numerous other proteins of interest to medicine have been produced in this way, but certainly fewer than had originally been hoped. The problem is that the most commonly held view of life is a mechanistic one, in which we just need to change the "cogs" in order to make the cell's machinery do what we want. It is nothing of the sort, because a program, an algorithm, has its own properties, which cannot be reduced to mechanics. Our knowledge of genomics is sure to change this eighteenth-century view and enable us to make better use of cells and their products in the future. Knowing how to take account of the position of genes in the chromosome or the style of the organism will definitely enable us to vastly improve what we can achieve. Among the objectives worth exploring, we will certainly have to consider all the human gene products, starting with the components of blood, which are universally useful, but also very dangerous because human pathogens are well adapted to this particular environment.

Expressing genes in heterologous organisms does no more than make use of life as it is. But we can go further than this, and in two different directions: toward the past, imagining what life was like a billion years ago, and toward the future, creating new life. The exploration of present functions shows us what trails to follow.

Acquiring New Functions

Understanding the nature of biological functions makes it possible to understand how they arose. Although, as I have stressed, it is not generally possible to understand *a priori* how new functions appear, their necessary compatibility with the laws of physics and chemistry makes them comprehensible *a posteriori*. A process that makes innovation possible when there is no objective will not stand in its way when the objective is fixed. So it is not unreasonable to want to "domesticate" genomes. This was done long ago with brewer's yeast, and more recently with animals. But it was done in an entirely natural way, simply by selecting from the variety available, without having any power over its creation. If we can—and still should—make use of the immense power of selective pressure, we can also direct it, through the choices we offer it. This is the privilege that language gives us, enabling us to have some power over the future.

Let us look at an example. Bacteria in the *Streptomyces* family are famous because they produce a very large number of antibiotics (as we can tell from the name of streptomycin, which cured tuberculosis for a time), as well as other molecules used in medicine, including immunomodulators such as cyclosporin, which overcomes problems with transplant rejection. These active molecules are all produced by the same family of proteins. These proteins combine several domains that are similar to one another, and which assemble amino acids or lipids according to a rule governed by the way these domains combine, together with small local variations in their amino acid sequence, allowing them to select from the cell environment different specific lipids or amino acids (according to the sequence of the protein domain concerned) and then combine them. The evolution of the corresponding genes is simply understood. It has combined domains with similar structures; these are variants of the same amino acid sequence, which has been modified by mutations. This suggests that over the course of evolution, there has been an interplay between a mutation process within individual gene fragments, producing different mutated domains in a gene's DNA sequence, and recombination of different genes, producing fragment shuffling. Given a few domains and a few mutations, we very soon arrive at an immense range of combinations, and an immense variety of molecules is produced by a very simple principle. From this variety, nature has selected those molecules whose biological effect helps the bacteria to survive in hostile environments. But obviously this evolution could not be oriented. On the other hand, in a laboratory it is clearly possible to make use of the natural process that couples the production of mutants and the selection of those that are active, while at the same time *orienting* it, at least partially, and accelerating it considerably. And if we go about it the right way, we can redirect this artificially adapted and accelerated but natural process, by giving it other substrates to work on besides the natural ones of amino acids and lipids, for instance by adding to the medium various analogs of these natural products, which will be recognized by the catalytic sites of the new enzymes. A large number of laboratories and industrial groups have been working on this since the early 1990s.

There are many other examples of this combinatorial approach, which was not applied systematically until 1990, although the central idea had been known for some time. This idea harnesses the inventive power of the combination of variation and selection (so important to life) to produce new molecules in great variety and very large numbers. Initially, research concentrated on peptide and oligonucleotide synthesis. It is probably fair to consider Mario Geysen, a scientist at Glaxo Wellcome's famous laboratory at the Research Triangle Park in North Carolina (famous because it was here that trimethoprim, the first entirely artificial antibiotic since the invention of sulfamides, was invented), as having initiated this research. In 1994 his group developed a technique of peptide synthesis on a support made from fine needles, so that a great variety of peptides could be produced simultaneously. The real difficulty in this extremely simple and productive approach is being able to *amplify* the result of the selection sufficiently to identify and reproduce it. This is exactly what living organisms do, when *a single* mutant individual is multiplied to the point where it invades the environment and becomes noticeable. More precisely, what is multiplied is not the individual but its genetic program, its chromosome, or in this case a part of the program, the gene that codes for the active protein. If we know how to transpose the chemistry into chromosomes (generally into nucleic acids), then the crucial step of amplification can be achieved. Then the combinatorial potential of fragments of nucleic acids can be explored directly. Several biotechnology companies in the United States are doing this with RNA (especially with introns). It can also be done with peptides, by making segments of DNA code for their sequence; then cells will be able to produce them and, in multiplying themselves, amplify them. Clearly, other ideas will need to be invented, so that completely different classes of molecules can implement the winning triple formula of variation, selection, and amplification.

Future Lives?

To round off this panorama, I would like to look at what will be easy to achieve in the near future. First—and the businesses involved are working on this—it must be possible to produce "clean" domesticated organisms, stripped of a miscellany of genes and sequences that they need to survive only in very varied environments, whereas the commercial laboratory environments they are kept in are necessarily as constant as possible. The quality of a product depends on how easily it can be reproduced, and thus on how easily the conditions needed to make it can be reproduced. More important, the organisms modified in this way will be safer, less potentially harmful, as they would clearly be unable to survive long outside the industrial sites in which they were used. This factor is more important than it may seem, once we start to modify organisms using foreign genes. It is certainly safer to make them as dependent as possible on the living conditions imposed by their designer. We saw with *B. subtilis*—but it is also true for *E. coli,* and is probably very common—that recently domesticated organisms are just ready and waiting to become pathogenic if circumstances should allow it (especially by the horizontal transfer of virulence genes).

While developing such "clean" organisms, it will also be worth producing bacteria with the smallest possible genome, but which are able to grow well in the laboratory or in an industrial fermenter. This objective is desirable not only for theoretical reasons, but to permit their metabolism to be reconstructed from basic elements, in order to produce a variety of molecules (or simply to produce biomass for animal feed, or even human foodstuffs. Yeast extract is already widely used as a condiment, particularly in English-speaking countries).

We probably now understand life well enough not only to attempt to reconstitute a scenario of ancestral life, but also to create other forms of life. To do this, all we need to do is bring together the processes we have seen to be necessary, and especially the existence of the genetic code. It is already theoretically (and practically) possible to make structures analogous to DNA but with different chemistry (we mentioned PNA), or proteins formed of different amino acids from the ones nature's proteins are made from. In fact, certain proteins contain a twenty-first amino acid, selenocysteine. Nobody yet knows what these new organisms will be and what they will be capable of, but they could certainly provide further answers to the questions we have been asking.

One last word before we come to the epilogue, which will only ask more questions without answering them. Making hybrid genomes could also be a very rewarding route to new properties, for example in order to transform otherwise harmful bacteria into vaccines. As Miroslav Radman and his colleagues showed, controlling the genes that correct mismatches enables the species barrier to be ignored, and real hybrid genomes to be produced. If this type of construction were extended to the world of domestic animals, it would probably be possible to produce new animals, better suited to their human environment, and better suited for whatever use, whether as food or as labor, that we want to make of them. We can see a fundamental question emerging here, which I have not yet tackled, but with which I will finish: that of the moral problems associated with manipulating what makes life, the genetic program.

EPILOGUE: MORALITY INTRUDES

One of the most profound motivations (not the only one!) behind the human aptitude for producing ever more knowledge is that applying it may improve the welfare of humanity. Louis Pasteur's philosophy was just that: to produce knowledge first, and then to develop applications of that knowledge, which he considered to be the heritage of humanity. He was thus interested both in a question from the realm of pure knowledge—the origin of life—and in a great many applications of the research he had carried out in order to understand how microscopic life could propagate. Pasteur pursued the most effective ways of applying the theoretical knowledge generated by his work, which ranged from the diseases of beer and wine to human diseases, via those of the silkworm.

The Human Genome Initiative has no meaning in itself; it acquires one only as its objectives become more precise. There is no point in sequencing a genome unless we do something more than merely line up three billion bases to make a text. Why sequence this genome? Apart from the starting point, when the Atomic Bomb Casualty Commission was concerned about the future of the populations exposed to atomic radiation, it was perhaps the work of Renato Dulbecco that gave the first signs of a real justification for the enterprise: understanding the human genome should enable us to find a cure for cancer. Then this was extended, first to genetic diseases, and then to all diseases, since we know that the factors which predispose to disease are always heavily dependent on the genes. The initial motivation was thus diagnostic, with the aim of knowing more about the fate of human populations. The deepest motivation is therapeutic, but unfortunately, unless appropriate new basic knowledge is created, we will be able to fulfill only the first of these, predicting the fate of those unlucky enough to be affected.

In reality, there is a lot more to understand than this. Knowing a genome text means having access to the recipe book that gives instructions for making an organism, but this recipe book is written in a language we do not understand. This attitude implies that we attribute a *value* to certain kinds of knowledge. Over and above knowledge itself, a moral choice has to be made. The knowledge we gain through genome programs illustrates particularly clearly this dichotomy between knowledge and its applications, on the one hand, and, on the other hand, their value, represented by the recognition of something else beyond knowledge itself. Here I will just give a few examples of what it is that makes these programs so important, and suggests that this importance can only grow and be confirmed. No doubt our imagination will not stop short at these examples, and will come up with far more unexpected applications. We will have to ask ourselves constantly what it is that decides whether the choices we make are the right ones, apart from their scientific interest itself. It should not be taken for granted that knowledge for its own sake is a valid moral choice.

In very general terms, we can say that knowing a genome means getting to know the program responsible for the construction, development, and survival of the corresponding organism. Can we use this knowledge, and if so, what for, and why? Understanding enables us to reason, to recognize, to discover what the future holds in store. This is the aim of diagnostics: to discover the limits of an individual destiny. Thanks to the techniques of polymerase chain reaction (PCR), for instance, we can now find out whether or not a given individual's genome carries a particular form of a gene, a certain allele. When those individuals are a means to an end, when they are plants or domestic animals for example, the main problem is essentially one of efficiency. Economic considerations come well before moral questions. In our consumer society, value is mainly seen in commercial terms. But when we are dealing with human beings, we have to stop and think, and understand what it would mean for society if human beings were to become a means to an end. After considering the commercial value, we cannot avoid the issue of moral value.

This is the crucial point in all the questions that have been raised by the idea of patenting living things. It is an old conflict, but it was brought back into the spotlight when the National Institutes of Health (NIH) made a patent application in June 1991. This application made no attempt at all to consider the ethical angle, only the legal aspect of the patents. Is there any connection between patent law and ethical questions? The question is not as straightforward as it looks (one could simply say that there is no connection), and this becomes clear if we ask the question in a military context. Every arms patent registered can be seen in these terms, and we can illustrate the moral question by asking whether it is possible to register a patent on a letter bomb. In the strictest legal terms, the answer would be yes. But even in a strict legal context it is not so easy to give an opinion, unless we separate the ethical question entirely from the question of commercial protection. Is it possible to register a patent to protect an invention that cannot be sold? Patent law (which is not the same everywhere) requires not only invention, but also usability. So we could interpret the law as saying that anything whose use is forbidden by law cannot be patented. This immediately brings us back to the relationship between law and morality. In the case of genome sequences, there is the additional factor that life cannot very well be considered in the same way as machines (and the heart of this book shows that the eighteenth-century mechanistic image of life can no longer be considered valid). More specifically, the real distinction between the program and the machine that reads it leads to paradoxes as soon as they are separated. When we look closely at the specific distinction that brings morality into it, it becomes purely academic to try to distinguish what is human in a gene, which in material terms at least is no different at all from a plant gene or a bacterial gene. In fact, as far as I know, nobody has taken this fact into account in defining what patentable invention means, where living things are concerned. Yet this is what the public senses, in its reaction to this question (a reaction that is full of common sense, though often aimed at the wrong target), and in the worry or the enthusiasm aroused by genetically modified organisms.

More for political reasons to do with prestige and profile than for economic reasons, some institutions have registered large numbers of patents rather than concentrating on the rare few that will bring in royalties. This attitude has obscured not only the ethical questions but also the commercial questions themselves. For one thing, it is not at all obvious that patenting human genes would have any legal meaning (quite apart from the need for a definition of what is meant by "human genes"). For a patent to be granted, there has to be an "inventive step," and it is hard to see where any inventive step is involved in the work of the researcher who merely reveals what nature alone created. For another thing, there is no guarantee that these patents will have any economic value. In the majority of cases, all that is needed to get round the patent restrictions is to clone the same gene from a plant or an animal (provided it can be shown that it really is the same gene). Besides, from the moral point of view, a human gene cannot be regarded as embodying any aspect of human dignity unless genes are considered in a purely mechanistic fashion—and we have seen that this makes no sense. It is only the way a gene is *used* that could be shameful, for instance if it were used for selective diagnosis; but sadly, ironically, it is this use that seems to be regarded as the least problematic. Humanity is not to be found in a collection of genes.

In reality, more than anything else, the debate over the registration of patents by the NIH revealed the competitiveness surrounding control over the techniques and results that might derive from our knowledge of genomes, both at the national level (between the Department of Energy and the NIH) and internationally (between the United States, Europe, and Japan, and later between Celera and the International Public Consortium). The question was, would the data be published, and accessible to the whole world, or would they be kept private, thus allowing the scientific and genetic stage to be completely dominated by a few players? The debate certainly had a considerable impact on certain countries, which were cautious at first, but soon became aware of what was at stake in setting up a genome program.

In reality, this question of patents—on which Europe took a very different line from the United States—marks a new awareness of the commercial potential of biological processes. The situation is quite similar to the early years of the electronics industry, whose commercial potential was only gradually realized. The most important of the first techniques on which genetic engineering is based were not protected by patents. Neither Frederick Sanger nor Allan Maxam and Walter Gilbert patented their DNA sequencing techniques, just as the transistor effect was not covered. The cloning technique that uses the functional complementation of two β-galactosidase fragments has never been patented, and it is used everywhere. Only Stanley Cohen and Herbert Boyer's technique for DNA recombination in a plasmid has been patented, after long discussions, and licenses to use it are not expensive. It was only after Cetus discovered PCR in 1983 that things really changed, especially when the patent was sold to Hoffmann–La Roche in 1991 for $300 million. This time the patent was taken out, and licenses were issued for a certain number of diagnostic applications, for very high fees. This prevented the use of PCR in systematic diagnosis for a considerable time. This situation may change rapidly as the patent reverts to the public domain. Similarly, the fluorescent sequencing techniques used in automatic sequencing machines are patented, and their high cost is reflected in the cost of reactants and of equipment (often as much as $150,000).

The history of patenting policy is full of contradictions, and in reality it is quite difficult to decide whether or not patenting is economically worthwhile. In addition to the profit a few rare patents may bring in, the cost of accumulating unusable patents has to be considered. On the other hand, it is indisputable that treating life as a means to an end has and will continue to have considerable moral consequences. To tell the truth, this attitude is less a consequence of the technical and especially theoretical progress in biology than of the enormous social changes that have taken place, first in the West and then throughout the world, since the beginning of the twentieth century. These changes have brought wealth to an increasingly large part of the population, demographic explosion, and domination by the American model of individual democracy, which invades societies everywhere, sweeping away civilizations thousands of years old, and leading to demented reactions such as the attack on the World Trade Center. However, this view of life as a means to an end necessarily relies on an outdated mechanistic image of behavior, which is incompatible with life. Throughout this book I have tried to show that living organisms can never be considered in the same way as mechanical objects like those found in industry. By their very nature, inherent in their makeup, they are capable of producing the unexpected. This provides us, in fact, with an optimistic view for the future: yes, there still is room for humanism, because a human being cannot be reduced to any other type of living being, and certainly not to any type of mechanical automaton. There is even more room for morality.

I have been speaking only of genetics so far, but of course when we consider what makes an individual we also need to take into account the ephemeral nature of epigenesis—broadly speaking, the interaction of the genome with its environment. We should note once again that the main scientific justification for genome projects is not to explain the *whole* of biology. If we pursue the analogy with a computer program and Turing machines, the structure of DNA does not enable us to account for the *concrete* characteristics of the machine that reads the program (the cells, the organs, and the way they are arranged to form a whole organism). Still less of course does it explain the remarkable properties of some kinds of behavior, such as that of the central nervous system, or relations between the individuals who make up a society. Since its earliest days, biology has made a distinction between different levels of organization in living things. It has always considered what relates to the direct expression of the genetic program (which can virtually be considered to be *preformed,* at least in an abstract sense, because the program reproduces itself identically from one generation to the next) separately from what expresses the program indirectly, what is *epigenetic* (such as an individual's memory, the formation of specific social bonds, or a particular language—but not the aptitude for language, which is universal in humans). Of course these properties, just like those of all the functions expressed by living organisms, depend on the nature and functioning of genes. To master a spoken language, we need to hear it, and thus to have properly formed hearing; we also need to speak it, so we need the appropriate organs. And it is important to identify the genes involved in establishing one function or another. But it goes without saying that a large part of our adaptation to the environment, or of epigenetic phenomena such as the language or the culture of human societies, has no direct link with the genes. There are no genes for speaking English or Chinese. I have said several times that what is special about the development of living organisms through the course of evolution is their progressive discovery of the immense capacity for adaptation that the symbolic function affords. More than this, in the case of the nervous system, it is the capacity to produce the symbolic, on the spur of the moment, and without any transmissible genetic support.

What makes an animal is a great deal more than the simple expression of a few hundred thousand genes. This is demonstrated by the complexity of the brain, with its tens of billions of cells, all connected to each other by tens of thousands of synapses; by the extraordinary adaptability of the immune system, able to recognize and respond to completely unpredicted and unpredictable attacks from the environment; or by the balanced and harmonious relationships established between the different cells and organs that make up an organism like a mammal (and we can see the effects when this goes wrong, in cancer for instance). More precisely, it is through the expression of these 40,000 or so genes that the data generated by the specific constraints of the environment come to define each particular individual, complete with its unique characteristics and the traces in its memory. The role of genes is thus essential, even in these specific phenotypes. But of course there is no one-to-one correspondence between a gene and its expression. Each gene is a link in the chain of a procedure that enables the specific traits of organisms to be established. It cannot determine entirely which *particular* trait has been produced. But it does impose limits on their potential, and the most direct way to understand the ins and outs of these traits is in fact to know the genes involved. This is why genome projects are revolutionary: they give us direct access to the complete genetic programs of the organisms studied. The first stage is to identify the instructions in this program one by one. Later, it will be essential to understand how the instructions are connected when the program runs in a given environment. And above all, it will be crucially important to understand the nature of the geometrical program that is superimposed on the genetic program. This is what makes the machine a Turing machine.

An identity cannot be formed solely on the basis of the history of a species. We also need to take the individual history into account, even if it is not perpetuated. Animals have evolved with a nervous system that has become increasingly rich, enabling new levels of symbolic representation to be defined, most of which last only as long as the individual or the society (not as long as the genome, which exists, unchanged, over many generations). It would seem that if the human central nervous system has some algorithmic properties, and if the symbolism of language allows the *separation* of a *neural program* from the *data* used to express it, then with human beings the same constructive process is applied a second time, on a higher level. This idea is not a pointless dream, and it would have the advantage of doing away with certain enigmas or paradoxes, such as Jean Piaget's *Gestalt:* what is imprinted on the brain, so that the "wholeness" of an object is ready to be made available in an instant, would no longer be the image of that object, or of a sign or a gesture, but the image of its *construction algorithm.* Writing or recognizing the letter *e,* whatever its dimensions, would not indicate the presence of a *Gestalt* of *e,* just of an algorithm that allows it to be created: the conditions under which the algorithm is applied would define its scale (or the means with which it was created, for instance with a pen, a paintbrush, or a piece of chalk), but the algorithm would remain the same. If we follow this reasoning it becomes clear that, as we have seen with the properties of the genetic program, there would be nothing mechanical about the properties of the central nervous system; in fact they would distance it immeasurably from the automatism dear to eighteenth-century thinkers like La Mettrie. This would in itself justify our special human status in relation to other animals, even in relation to anthropomorphic primates. In a different way, the creative principles at work here, as I have described them, also make sense in terms of the continuing interaction between an individual and his or her environment. The ability to produce relationships shows that every individual has the potential to create, even if only for his or her own lifetime. And so, by constantly producing identities that cannot be reduced to those that gave them birth, life shows that creation came after the world. Is this not what the poet Rimbaud sensed, when he said: *"Je est un Autre"*—"I" is an Other? The poetic voice exists in its own right, and what we create cannot be reduced to what we are.

Morality applies to knowledge itself: knowledge does not override it. This might sound like a futile, academic reflection, but it is a question that soon arises for any scientist who works on objects that could be used for military ends, for instance. If an object is impossible to control—think of a disease that is established everywhere and cannot be eradicated—it seems to me that in this case, knowledge is not just theoretically justified, but morally justified, and that it must be shared (although this knowledge certainly makes it possible to do harm, it also enables us to prepare a defense against it). However, when the organism studied is both dangerous and contained, we have to ask: Should we know? It seems to me that the answer is no, that the organism should be destroyed, *without* our knowing the mechanisms by which it causes harm. This was the case with smallpox, as I have already said elsewhere, and unfortunately this was not the choice that was made, which only shows how morally impoverished we are. We may have eradicated the disease in theory, but we have preserved it in a form that is much more contagious, because it cannot be quarantined. The virus now exists in the form of knowledge, accessible to anyone, anywhere in the world, and eventually it will be possible to recreate it. There has been a great deal of talk about the Ebola virus, but smallpox is every bit as contagious, as swift, and as lethal. This paragraph was written before September 11, 2001, and now it looks like a terrible prophecy . . . Alas, mankind is entirely devoid of reason!

Last Word

Here this text comes to an end. In our time, now that the written word has lost the almost sacred status it had for so long, I wonder about its significance. Should I have written it? Or indeed, what place is there for scientific writing? This is a minor work; what benefit can it bring? Choose a major work, such as the 1905 article in which Einstein explicitly sets out the foundations of the theory of relativity, and compare it, with its enormously destructive consequences—do you remember Nagasaki?—with Mozart's *Don Giovanni,* with Mickiewicz's love poems, or with Michelangelo's *Pietà*. Ask yourself which you would destroy, if fate allowed you to save only three of these works. Science is, and remains, anonymous. Anyone might rediscover Einstein's findings, one fine day. Science is anonymous, and this is probably why it is both feared and disdained. What ambition, what mad hope drove me to dedicate my days and nights to thinking through what is set down in black and white in this bundle of pages? Astonishment, no doubt. Astonishment at finding myself one of the billions of human beings taking part in the last doubling of the population of humanity, at a time when we can all look around us at this planet we have devastated. At having witnessed the destruction of all my hopes and all my beliefs, and yet finding myself still standing on this transient Earth. At knowing that a day will come, not for me but for others, when a new form of creation will replace what we have destroyed, only to disappear in its turn into the night of Oblivion. I wanted to leave some signs of this new hope of an impossible future, to help those mysterious Others who will one day give it birth, to help them find their way, or rather to know that the way exists, without me.

A BRIEF GLOSSARY

ACKNOWLEDGMENTS

NOTES

INDEX

A BRIEF GLOSSARY

My aim is to make this book accessible to anyone interested in genomes, whether they have a scientific background or not, but you cannot please all of the people all of the time. Some readers will inevitably find this glossary too basic, but I hope some will find it useful. In the following definitions, words in italics have their own entry.

Allele: A particular variant of a *gene,* found in part of a population of individuals of the same species. Most alleles of a given gene are functionally equivalent (and therefore invisible, except when the gene is sequenced).

Amino acid: The 20 different building blocks used to make the *proteins* found in living organisms. They are represented by all the letters of the alphabet except B, J, O, U, X, and Z. Each amino acid is made up of an amino group ($-NH_3^+$), a carboxyl group ($-COO^-$), and a "side chain" whose chemical composition distinguishes one amino acid from another. These groups can react with each other, forming a covalent bond and releasing a molecule of water ($-COO^-$ in one amino acid and $-NH_3^+$ in another join up to make CO and NH plus H_2O). An amino acid incorporated into the chain is sometimes referred to as a "residue." The same amino acid may be repeated within the chain, which might have 30 residues, of 15 different amino acids. The chain is called a polypeptide, and a short chain, with fewer than 20 residues, is called a peptide.

Archaebacteria: A class of *prokaryotic* organisms which generally live in extreme environments, and which in physicochemical and genetic terms are only distantly related to *eubacteria.* This distance, as well as certain features of their membrane structure, defines them as a class in their own right. One of the three main kingdoms of living things (the kingdom itself is called Archaea).

Base: The chemical element that distinguishes one *nucleotide* from another and therefore carries the genetic information. It is made of an "aromatic cycle," or ring of carbon and nitrogen atoms. There are five main bases: two purines, adenosine (A) and guanine (G), and three pyrimidines, cytosine (C), thymine (T) (found only in *DNA*), and uracil (U) (found only in *RNA*). In the double helix of DNA in the *chromosomes,* the bases face each other in complementary pairs joined by hydrogen bonds: A pairs with T, G with C, C with G, and T with A. The length of a *gene,* a *chromosome,* or a *genome* is given as a number of base pairs, kilobases (kb), or megabases (Mb).

Chloroplast: An *organelle* in the cells of the green parts of plants, containing chlorophyll. It is responsible for fixing CO_2 from the air and producing oxygen by photosynthesis. Chloroplasts derive from bacteria of the cyanobacteria type (also known as blue-green bacteria, and formerly as blue-green algae), which have become established within the cells. Their *chromosome* includes only a few *genes:* the rest have been transferred to the *nucleus.*

Chromosome: The *nucleus* of the cell (in *eukaryotes*) or the cell itself (in *prokaryotes*) contains the physical material of heredity, in the form of one or several chromosomes. As their Greek name indicates (χρῶμα, *chroma,* means "color"), these structures can be stained to make them visible with a microscope, and this is how they were discovered, well before their function was understood. A chromosome is made of a *DNA* molecule, whose sequence of *bases* constitutes the organism's *genome.* The very long molecule of DNA is folded up in a complicated fashion and is associated with all sorts of *proteins* responsible for its compaction, *replication,* and *transcription.*

Codon/anticodon: A set of three successive *nucleotides.* Codons in *messenger RNA* are read in phase with the start codon and are recognized by a complementary triplet of nucleotides (anticodon) of a *transfer RNA,* which carries the specific *amino acid* for that codon. With four nucleotides, permutations of three out of four give 64 codons, of which 61 code for specific amino acids, using the rule of the *genetic code;* one (AUG) codes for an amino acid (methionine) and is also a signal for the start of *translation;* and three are "stop" codons, with no corresponding amino acid.

Cytoplasm: The internal cell medium surrounding the *nucleus* and *organelles.* More like a gel than a watery solution, its organization is not very well understood, especially in bacteria.

DNA: DNA, or deoxyribonucleic acid, is a macromolecule made up of linked *nucleotides* and forming the *chromosomes.* It is usually a double strand but can be single-stranded in some organisms (viruses). The backbones of the two strands twist around each other to form the famous double helix, with the *bases* on the inside of the helix, facing each other in complementary pairs: A pairs with T, G with C, C with G, and T with A. In the backbones, a phosphate group forms a phosphodiester bond —C—O—(O)P(O)—O—C— between the deoxyribose (sugar) molecules of two successive nucleotides.

Endoplasmic reticulum: A membranous network in the *cytoplasm* of *eukaryotic* cells; generally associated with *ribosomes* during *translation.*

Enzyme: An enzyme is a catalyst that is specific to a particular chemical reaction. The molecules found in cells can react with each other in an infinite variety of ways. However, these reactions do not usually happen spontaneously, or happen only very slowly, because of a particular chemical constraint: the need to overcome an activation energy barrier. Imagine two ponds, one higher than the other, and separated by a dam. For the water to flow from one to the other, the dam has to be lowered. The role of an enzyme is both to lower the activation energy barrier and to align the molecules involved in the reaction (the substrates) in their correct positions to make the new products concerned. Enzymes are so specific that the same two substrates can result in different products, depending on the nature of the enzyme, because this aligns them differently in relation to each other. Most enzyme names end in –ase, (e.g., DNA polymerase) but some end in –in (e.g., subtilisin).

Eubacteria: A class of *prokaryotic* organisms, including the most familiar bacteria. One of the three kingdoms of living things (kingdom name Bacteria). Originally all organisms without a *nucleus* were thought to belong to the same group, but now eubacteria are considered to be as distinct from *archaebacteria* as from *eukaryotes.*

Eukaryote: A living organism whose cells have a *nucleus.* The processes of DNA *replication,* its *transcription* into *RNA,* and *translation* into *proteins* are therefore physically separated. One of the three main kingdoms of living things (kingdom name Eukarya), this group includes some single-celled organisms, such as the yeasts, and most multicellular organisms.

Exon/intron: The *genes* of *eukaryotes* contain sections called exons and introns: *ex*ons are the part of the *DNA* that is *ex*pressed; *int*rons are stretches of DNA that *int*ervene between the coding parts. Introns are *transcribed* along with the exons but are later removed during the formation of a mature *messenger RNA*. The role of introns is not well understood, but they are almost certainly not "junk." They certainly have a regulatory role, a role in the accuracy of the DNA replication process, and may act as timers or spacers.

Gene: The gene is the unit of heredity. It defines a product (either *RNA* or a *protein*) as well as control elements required for the organized synthesis of its product or products. The definition of a gene has varied considerably (the original definition based on Mendel's work was an operational one, and made no reference to a physical structure). It is often easy to define the coding part of a gene: this is the part of the *chromosome* that codes for a product, usually a protein. It is much more difficult to define its physical limits. A great deal of debate, often heated, has resulted from this. These are not just academic quarrels, but a crucial problem if efficient databases are to be set up to manage biological knowledge. Precise definitions are essential if information is to be handled by computer.

Genetic code: The rule by which the triplets of ribonucleotides *(codons)* in *RNA* are *translated* into the *amino acids* in *proteins*. Virtually universal for all organisms, the code uses 61 *codons* to specify 20 amino acids (the code is redundant, or "degenerate"). One codon represents both an amino acid (methionine) and the beginning of *translation*. Three codons (UAA, UAG, and UGA) have no amino acid counterpart and cause translation to stop. The genetic code must not be confused with the concept of "lines of code" in a computer program.

Genome: The genome is the organized collection of all an organism's *DNA*, regarded as a text written in a four-letter alphabet. It includes not only all the *genes* but also a large number of intergenic regions with multiple functions, particularly regulatory and architectural functions.

Genotype: The concept of genotype was invented well before the nature of *DNA*, the hereditary material, was understood, especially its nature as an alphabetic text. An abstract concept, genotype refers just to the set of *genes* of an organism, taking account of the fact that in a given species, the same gene can have several different variants *(alleles)*. It does not take into account the way they are organized into a coherent text, and for this reason it is likely to become obsolete quite quickly (except in the genotype/*phenotype* distinction), to be replaced by *genome* (which is, however, a more concrete concept and refers to the text of the genetic program).

Membrane: A structure formed from a double layer of asymmetrical molecules (lipids—hydrophobes—with a hydrophilic head) and *proteins,* and which separates the cell compartments. The cytoplasmic membrane separates the inside of the cell from the outside. It may be enclosed in a more complex envelope, with associated structures that give the cell a firm shape, for instance the rod shape of some bacteria (bacilli).

Metabolism: The sum of all the physicochemical changes that take place within a living organism. Most of the reactions involved are produced by the action of *enzymes.* Metabolism stops with death. There is an intermediate state, sometimes called dormancy, in which the organism's vital activity is suspended. It cannot be said to be alive until metabolism begins again.

Mitochondria: These are *organelles* found in most *eukaryotic* cells, and responsible for energy management, via the use of oxygen. Mitochondria are symbiotic bacteria that have degenerated (Paul Portier, *Les Symbiotes,* 1918), and their *genome* has been reduced to a very small number of *genes.* The rest have been transferred to the *nucleus.* The singular is mitochondrion.

Molecular chaperone: An auxiliary *protein,* of a family whose members are involved in the correct folding of the *amino acid* chain of most proteins.

Nucleotide: The basic component of nucleic acids. Each nucleotide is made up of a sugar molecule with five carbon atoms (ribose for *RNA* and deoxyribose for *DNA*), one of five *bases* composed of carbon and nitrogen, and one to three phosphate groups (each group has one phosphorous atom in the center of a tetrahedron of four oxygen atoms). The number of phosphate groups determines how energy-rich the nucleotide is. There are four deoxyribonucleotides, written dA, dC, dG, and dT, and four ribonucleotides, rA, rC, rG, and rT, but the *d* or *r* is omitted when there is no ambiguity (most of the time). A string of a few nucleotides is called an oligonucleotide; a long string is called a polynucleotide.

Nucleus: An *organelle* found in the cells of *eukaryotic* organisms, formed from a complex envelope, and containing the *chromosomes.*

Operon: In *prokaryotes, transcription* can lead to the synthesis of a *messenger RNA* coding for several *proteins,* not just one as is almost always the case with *eukaryotes.* A transcription unit like this, with its regulation system, is called an operon.

Organelle: *Eukaryotic* cells contain organelles, structures visible with an optical microscope and generally easy to isolate using appropriate physicochemical means (especially centrifugation). The most important ones are the *nucleus* (which contains the *chromosomes*), the *mitochondria* (which contain a chromosome whose *genome* codes for only a few *genes*), and, in plant cells, *chloroplasts* (which also contain a chromosome). There are many other, more varied types of organelles, whose functions are less universal, and which are not mentioned in this book. The *ribosomes,* very small organelles made of *RNA* and *proteins,* and which are the site of *translation,* have long been recognized in all cells, not just *eukaryotes.* Ribosomes are mainly visible under electron microscopy, as they have a diameter of about 20 nm (20 millionths of a millimeter).

Phenotype: The explicit manifestation of a *genotype* in a given individual. The phenotype is produced by all the individual's *genes* working together in combination with the effects of the environment. Skin color, for instance, is the result of the activity of at least eight genes (without counting those responsible for building the cells of the epidermis), producing all the variety and gradation of color seen in human skin. The difference between the concepts of genotype and phenotype is illustrated when we get a suntan: the same set of genes (the same genotype) can lead to different phenotypes—very pale or very dark skin—depending on the amount of exposure to ultraviolet radiation. It is therefore important not to identify a genotype through a specific phenotype, or to attempt to predict a phenotype on the basis of explicit knowledge of just one gene.

Prion: Diseases in the spongiform encephalopathy family ("mad cow disease," for instance) seem to be caused by an "unconventional" infectious agent, a *protein* called a prion. All mammals, and even much simpler cells such as brewer's yeast, have a protein called PrP (prion protein). There are two forms of this, the usual nonpathogenic form, and an abnormal form (the prion itself), which has a different shape and clumps together easily, forming plaques that destroy nerve cells. This abnormal form induces the normal form of PrP to convert to the pathogenic form. Although the mechanism of the final stages of the disease seems to be well understood, the contagion mechanism is not clearly established.

Prokaryote: An organism without a *nucleus,* usually single-celled. This group covers both *eubacteria* and *archaebacteria. Replication, transcription,* and *translation* take place in the same cell compartment.

Protein: A chain of *amino acids,* folded up in three dimensions. The amino acids are linked by the expulsion of a water molecule between the carboxylate residue ($-COO^-$) and the amino residue ($-NH_3^+$) of each amino acid. A short string of amino acids linked in this way is called a peptide. A polypeptide is a long string of amino acids. Several different levels of protein architecture can be distinguished. The amino acids forming the polypeptide chain make up its primary structure. This chain then folds up, producing a small number of different types of basic elements: helices, sheets, and turns. This is the protein's secondary structure. These elements combine with each other to form a three-dimensional conformation, the tertiary structure. For instance, the *prion* protein has two different tertiary structures, with different helices and sheets, one of which is normal and functional, the other toxic. Proteins often form functional complexes made up of several individual polypeptide chains, and the spatial form of these structures made of several chains is a quaternary structure. Proteins carry out numerous functions. As *enzymes,* they are first of all the catalysts in the *metabolism* of small molecules and molecules involved in *replication, transcription,* and *translation.* They are also responsible for the transport of metabolites. Because of their shape, they have an essential architectural role, as can be seen with silk proteins or those of the hair and nails. They also play a crucial role as control elements: the factors that determine whether a certain *gene* is to be transcribed, and when, are normally proteins. Almost all the functions of an organism are thus based on proteins.

Replication: Duplication of the chromosomal *DNA* molecule, by separating the two strands of the double helix and building a new complementary strand for each one, using the correspondence rule A -> T, T -> A, G -> C, and C -> G. Each new DNA molecule thus contains one old strand and one new strand.

Restriction enzyme: An *enzyme* that cuts *DNA* at specific sites (strings of *nucleotides*)—for example, the enzyme EcoRI cuts after the G in the sequence

GAATTC—and is used in making recombinant DNA. A restriction map of a *chromosome* is a physical map that shows the position of sites recognized by restriction enzymes. When several restriction enzymes are used in combination, separation of the resulting fragments after partial restriction may make it possible to reconstruct the map. However, this is a very difficult and sometimes impossible exercise, and now that we have direct access to the genome sequence it is no longer very useful.

Ribosome: An *organelle* in the *cytoplasm* made of several kinds of *RNA* and about 50 *proteins. Messenger RNA* is drawn through it and *translated* in protein synthesis.

RNA: RNA, or ribonucleic acid, is a macromolecule of a single strand of *nucleotides.* As in *DNA,* the nucleotides are made up of a phosphate group, a sugar, and a *base,* joined together by phosphodiester bonds; but in RNA the sugar is a ribose, and the bases are A, C, G, and U. There are different kinds of RNA, among them:

- Messenger RNA (mRNA): a template that carries the genetic information from the DNA, where it is stored, to the *ribosomes,* where it is *translated* into *proteins.*
- Transfer RNA (tRNA): the adaptors that establish the correspondence between the sequence of each messenger RNA and that of the *amino acids* of the *protein* it specifies. tRNA forms a cloverleaf shape, of which one loop has an *anticodon,* which decodes a corresponding *codon* of the *messenger RNA,* and the stem of the leaf is loaded with the specific amino acid for that codon.
- Ribosomal RNA (rRNA): different kinds of rRNA with specific sequences (related from one organism to another) make up the core of the ribosomes.
- RNA molecules sometimes have a catalytic activity similar to that of protein enzymes. In this case they are called ribozymes.
- Viruses can have a genome made entirely of RNA.

Transcription: Rewriting of a stretch of *DNA* into *RNA,* using the correspondence rule A -> U, T-> A, C -> G, and G-> C. This correspondence is not exactly the same as in *replication,* where A corresponds to T.

Translation: The rewriting of an *mRNA* in the form of a string of *amino acids* (a *protein*). Successive *codons* of mRNA are read in phase with a start codon (usually AUG), and the corresponding amino acid is added to the string in accordance with the *genetic code.*

Vectors, phages, and plasmids: In genetic engineering, a vector is an autonomous *replicating* unit into which scientists insert (clone) fragments of *DNA* in order to amplify them. The vector is then introduced into a host cell (usually *E. coli,* but also brewer's yeast and other organisms), where it replicates together with the cloned DNA it carries. There are two main types of vectors. They may be minichromosomes called plasmids, most of which are double-stranded circles of DNA. These are not essential to the life of the cell, but they give it particular properties, such as resistance to an antibiotic. Or they may be viruses, and when these infect bacteria they are called bacteriophages, or phages for short. There are two types of phages. Virulent phages infect their host and multiply until they burst the cell open, killing it (this is a lytic cycle). Temperate phages do not always kill their host, and many of them have an original method of multiplication. They can choose either to act like virulent phages for a time, setting off a lytic cycle (and killing their host), or they can remain hidden in their host in the form of a plasmid, or even integrate themselves into the host chromosome in the form of a prophage. In this case the lytic cycle will be set off only if particular events occur in which the survival of the phage is paramount. For instance, if the host's survival is threatened by its environment, the lytic cycle is set off, allowing the virus to escape instead of dying in the prophage state.

ACKNOWLEDGMENTS

This book grew out of a deep reflection, not about some reality far from my everyday life, but about one that takes up the best part of my time. So it also grew out of the questioning and the work of those with whom I set out on the singular adventure of sequencing a whole bacterial genome. It owes a debt to everyone who took part in this project: I cannot mention them all, but especially to Philippe Glaser, Alain Hénaut, Frank Kunst, Claudine Médigue, Ivan Moszer, and Alain Viari. André Goffeau also had a share in it, through his determination in developing genome programs, his generosity, and his kindness. So, too, did Piotr Slonimski, with his fertile intellect and, together with Alain Hénaut, through their support and friendship when the doubt was greatest. It owes a great deal, too, to those who have regularly taken part in discussions with me, on Thursday evenings: Nicolas Aumonier, Paul Brey, Geneviève Milon, Elena Presecan, Agnieszka Sekowska, and many others. Jérôme Ségal found me some fascinating documents on the beginnings of the theory of information, for which I owe him thanks. Raphaël and Pétronille Danchin took the trouble to read it all, and found a number of errors to correct, not all of them cosmetic. Daniel Fixari and Juliette Blamont, who are not biologists, read the whole text, corrected it, and suggested many improvements. They made a substantial contribution toward making the concepts I develop here easier to understand. Odile Jacob also suggested improvements, and took the risk of first publishing this essay in French, and I would like to express my gratitude to her. Finally, Alison Quayle made it accessible in English, not only by translating it faithfully, but also by making the transpositions required by the move from a Latin culture to an Anglo-American one. I am infinitely grateful to her for this. However, the book would never have seen the light of day at all, if one of life's improbable encounters had not enabled me to restore a raison d'être to life. This book reveals, conceptually, the strange vindication I have found even in the horror of existence. This book is pessimistic: it shows that life has no meaning. It is optimistic: it shows that over the dunghill, forever irreducible, the rose still blooms.

NOTES

1. Exploring the First Genomes

1 See, for example, H. Daudin, *De Linné à Lamarck: Méthodes de la classification et idée de série en botanique et en zoologie (1740–1790)* (1927; reprint, Paris: Editions des archives contemporaines, 1983).

2 Paul Feyerabend, *Against Method* (New York: Humanities Press, 1975).

3 Richard M. Burian, Jean Gayon, and Doris Zallen, "The Singular Fate of Genetics in the History of French Biology, 1900–1940," *Journal of the History of Biology* 21 (1988): 357–402.

4 Brenner gives a vivid and intelligent justification of his choices in Sydney Brenner, *My Life in Science* (BioMed Central, online at http://www.biomedcentral.com/info/whatis.asp, 2001).

5 "Formal" here has nothing to do with its everyday meaning of conventional or official. It means "relating to form" and has its roots in Plato's idea that every object is a temporary expression of an eternal, ideal "form," which he thought more real than the object itself. *Webster's* dictionary defines "formal" in this sense as relating to "the essential constitution of a thing as distinguished from the matter composing it (the formal nature of a square is a relation of lines and angles rather than a matter of space or solidity)." This is how scientists and mathematicians use the word, so "formal genetics" does not mean "official genetics," but the study of the essential nature and relationships of genes, quite separately from the study of the material they are made of.

6 Thomas D. Brock, *The Emergence of Bacterial Genetics* (Cold Spring Harbor, N.Y.: Cold Spring Harbor Laboratory Press, 1990).

7 Renato Dulbecco, "A Turning Point in Cancer Research: Sequencing the Human Genome," *Science* 231 (1986): 1055–56.

8 Ibid.

2. The Alphabetic Metaphor of Heredity

1 It was also an example of the social behavior of scientists, in contrast to the usual "solitary genius" stereotype, and is described by James Watson himself in *The Double Helix* (New York: Atheneum, 1968). The book links Watson and Crick's work with that of biochemists such as Oswald Avery and Erwin Chargaff and physicists like Max Delbrück, through the twists and turns of a tale that is not without its share of amusing and sometimes grating anecdotes.

2 P. Rabinow, *Making PCR: A Story of Biotechnology* (Chicago: University of Chicago Press, 1996).

3 Douglas R. Hofstadter, *Gödel, Escher, Bach: An Eternal Golden Braid* (New York: Basic Books, 1979).

4 J.-P. Benzécri, *L'Analyse des données* (Paris: Dunod, 1990).

3. What Genomes Can Teach Us

1 Many forces come together to make science what it is. Some of the socio-economic constraints have already been discussed, with the birth of genome initiatives. But authors such as Imre Lakatos provide a deeper understanding of the complex mixture of scientific and sociological concepts that make up the sociopolitical world. See Imre Lakatos, *The Methodology of Scientific Research Programmes* (Cambridge: Cambridge University Press, 1980).

2 Karl Popper, *The Logic of Scientific Discovery* (New York: Basic Books, 1959) (Popper's own translation of *Logik der Forschung*) and *Conjectures and Refutations* (New York: Basic Books, 1962).

3 Rudolf Carnap, *The Logical Structure of the World and Pseudo-problems in Philosophy* (Berkeley: University of California Press, 1967), translated by Rolf A. George from *Der logische Aufbau der Welt* (1928); Charles S. Peirce, *Collected Papers* (Cambridge, Mass.: Harvard University Press, 1934).

4 Thomas Kuhn, *The Structure of Scientific Revolutions* (Chicago: University of Chicago Press, 1962).

5 Jean-Paul Benzécri, *L'Analyse des données* (Paris: Dunod, 1990).

6 John Donne, Devotion XVII (1624), in *Devotions upon Emergent Occasions* (London: Simpkin, Marshall, Hamilton, Kent, n.d.).

7 The protons are represented by the notation H^+: H because a proton is the nucleus of the hydrogen atom, and $^+$ because the hydrogen atom has one electron, which carries a negative charge, and the loss of this electron when the hydrogen atom is reduced to a proton means that the proton is positively charged.

8 Eduardo Rocha, Agnieszka Sekowska, and Antoine Danchin, "Sulphur Islands in the *Escherichia coli* Genome: Markers of the Cell's Architecture?" *FEBS Letters* 476 (June 2000): 1–2, 8–11.

9 Editorial, "A Minimalist Approach to Life," *Science,* December 10, 1999.

4. Information and Creation

1 Once again, this is "formal" in its original sense, meaning "relating to form."

2 Niels Bohr, "Light and Life," *Nature* 131 (1933): 421–423 and 457–459.

3 Francis Crick, "Central Dogma of Molecular Biology," *Nature* 227 (1970): 561–563.

4 Lewis Wolpert describes vitalism as "an idea which assigns to human life, particularly consciousness, a special quality which must forever remain outside conventional science. Vitalism is usually associated with an antireductionist stance, the view being that life cannot be reduced to mere physics and chemistry, and that a more holistic approach is required." However, we should note that it is possible to be both reductionist (looking for a single basic physicochemical process to account for life) and vitalist (looking for a "life principle") at the same time; hence the confusion in many discussions. Lewis Wolpert, *The Unnatural Nature of Science* (Cambridge, Mass.: Harvard University Press, 1992).

5 For a discussion see Rolf Landauer's incisive comments in "Fashions in Science and Technology," *Physics Today* 50, no. 12 (1997): 61–62.

6 See, for instance, studies such as H. Jacobsen, H. G. Busse, and B. Havsteen, "Spontaneous Spatio-temporal Organization in Yeast Cell Suspension," *Journal of Cell Science* 43 (1980): 367–377, which have been used over and over again, and criticized, for example, in Arthur Winfree, *When Time Breaks Down* (Princeton: Princeton University Press, 1987).

7 *Random House College Dictionary* (New York: Random House, 1991); *Webster's Third New International Dictionary of the English Language* (Springfield, Mass.: G. & C. Merriam, 1961); *Shorter Oxford English Dictionary* (Oxford: Oxford University Press, 1959) and *New Shorter Oxford English Dictionary,* 1996 CD-ROM edition.

8 Alain Rey, *Le Robert électronique (sur la base du Grand Robert),* CD-ROM ed. (Paris: Dictionnaires Le Robert, 1997).

9 Claude E. Shannon and Warren Weaver, *The Mathematical Theory of Communication* (Urbana: University of Illinois, 1949).

10 J. S. Almeida et al., "Analysis of Genomic Sequences by Chaos Game Representation," *Bioinformatics* 17 (2001): 429–437.

11 Ivar Ekeland, "La Théorie des catastrophes," in *La Recherche* 301 (September 1997): 89.

12 Jack Cohen and Ian Stewart expose the shallowness of these ideas in their refreshing book *The Collapse of Chaos* (New York: Viking, 1994). Illuminating, except that I disagree with their simplistic view of emergent properties.

13 Charles Darwin, *The Descent of Man and Selection in Relation to Sex* (London: John Murray, 1871).

14 Erwin Schrödinger, *What Is Life?* (Cambridge: Cambridge University Press, 1945).

15 Antoine Danchin, preface to Erwin Schrödinger, *Qu'est-ce que la vie?* (Paris: Christian Bourgois, 1986), pp. 7–28.

16 Rolf Landauer, "Inadequacy of Entropy and Entropy Derivatives in Characterizing the Steady State," *Physical Review A* 12 (1975): 636–638.

17 *Οὐδέν χρῆμα μάτην γίνεται, ἀλλὰ πάντα ἐκ λόγου τέ καὶ ὑπ' ἀνάγκης* (*Ouden chrema maten ginetai, alla panta ek logou te kai hup' anangkes*); Antoine Danchin, "Order and Necessity," in *From Enzyme Adaptation to Natural Philosophy: Heritage from J. Monod*, ed. E. Quagliariello, G. Bernardi, and A. Ullmann (New York: Elsevier Sciences, 1987), pp. 187–196, proceedings of the symposium J. Monod and Molecular Biology, Yesterday and Today, Trani, Italy.

18 Norbert Wiener, *Cybernetics* (Cambridge, Mass.: MIT Press, 1948).

19 "A Conversation with Claude Shannon on Man's Approach to Problem Solving," *Cryptologia* 9 (1985): 167–175.

20 Shannon and Weaver, *The Mathematical Theory of Communication.*

21 Wiener, *Cybernetics.*

22 Alan Sokal and Jean Bricmont, *Fashionable Nonsense: Postmodern Intellectuals' Abuse of Science* (New York: Picador, 1998).

23 Myron Tribus and E. C. McIrvine, "Energy and Information," *Scientific American* 225, no. 3 (1971): 179–184.

24 Claude E. Shannon, "The Bandwagon," *IRE Transactions on Information Theory* 2 (1956): 3.

25 Jules Romains, *Les Hommes de bonne volonté,* vol. 4: *Eros de Paris* (Paris: Flammarion, 1932).

26 Quoted in Charles H. Bennett, "Notes on the History of Reversible Computation," *IBM Journal of Research and Development* 32 (1988): 16–23.

27 Schrödinger also comments on this: "incredibly small groups of atoms, much too small to display exact statistical laws, do play a dominant role in the very orderly and lawful events within a living organism"; *What Is Life?* p. 20.

28 Wojciech H. Zurek, "Thermodynamic Cost of Computation, Algorithmic Complexity and the Information Metric," *Nature* 341 (1989): 119–124.

29 Leo Szilard, "On the Decrease of Entropy in a Thermodynamic System by the Intervention of Intelligent Beings," *Behavioral Science* 9 (1964): 301–310; originally published in *Zeitschrift für Physik* 53 (1929): 840–852.

30 Léon Brillouin, *Science and Information Theory* (London: Academic Press, 1962).

31 See, for example, discussions and references in R. Landauer, "The Role of Fluctuations in Multistable Systems and in the Transition to Multistability," in *Bifurcation Theory and Applications in Scientific Disciplines: Annals of the New York Academy of Sciences* 316 (1979): 433–452; and in C. H. Bennett, "How to Define Complexity in Physics, and Why," in *Complexity, Entropy and the Physics of Information*, ed. W. H. Zurek, SFI Studies in the Sciences of Complexity, no. 8 (Reading, Mass.: Addison-Wesley, 1990), pp. 137–148.

32 Douglas R. Hofstadter, *Gödel, Escher, Bach: An Eternal Golden Braid* (New York: Basic Books, 1979).

33 John Myhill, "Some Philosophical Implications of Mathematical Logic. Three Classes of Ideas," *Review of Metaphysics* 6 (1957): 165–198.

34 Gregory J. Chaitin, *Algorithmic Information Theory* (Cambridge: Cambridge University Press, 1987); Jean-Paul Delahaye, *Information, complexité et hasard* (Paris: Hermès, 1994); Hubert P. Yockey, *Information Theory and Molecular Biology* (Cambridge: Cambridge University Press, 1992); Henry Quastler, ed., *Information Theory in Biology* (Urbana: University of Illinois Press, 1953).

35 Thomas Kuhn, *The Structure of Scientific Revolutions* (Chicago: University of Chicago Press, 1962).

36 Jean-Paul Benzécri, *L'Analyse des données* (Paris: Dunod, 1990).

37 Charles H. Bennett, "Logical Depth and Physical Complexity," in *The Universal Turing Machine: A Half-Century Survey*, ed. R. Herken (Oxford: Oxford University Press, 1988).

38 John von Neumann, *The Computer and the Brain* (New Haven: Yale University Press, 1958).

39 Empedocles, *Fragments*, from *Presocratic Fragments and Testimonials*, ed. James Fieser (Internet release, 1996), adapted from passages in John Burnet's *Early Greek Philosophy* (1892).

40 Cohen and Stewart, *The Collapse of Chaos.*

41 Charles Bonnet, *Oeuvres d'histoire naturelle et de philosophie* (Neuchâtel: Samuel Fauche, 1779).

5. What Is Life?

1 Erwin Schrödinger, *What Is Life?* (Cambridge: Cambridge University Press, 1945). See, e.g., H. F. Judson, *The Eighth Day of Creation* (New York: Simon and Schuster, 1979).

2 This gives the lie to the common belief that life "prefers" symmetry. This mistaken idea becomes dangerous when it is used to support eugenic or other discriminatory attitudes, suggesting that what is not "symmetrical" (variously interpreted) is not "normal," and should be eliminated.

3 Louis Pasteur, "Sur les corpuscules organisés qui existent dans l'atmosphère. Examen de la doctrine des générations spontanées. (Leçon professée à la société chimique de Paris, le 19 mai 1861)," in *Lecons de chimie et de physique professées en 1861 à la société chimique de Paris* (Paris: Hachette, 1862), pp. 219–254.

4 Dorothy L. Sayers and Robert Eustace, *The Documents in the Case* (London: Victor Gollancz, 1930).

5 Claude Bernard, *An Introduction to the Study of Experimental Medicine* (New York: Macmillan, 1927); translated by Henry Copley Greene from *Introduction à l'étude de la médecine expérimentale* (Paris : Ballière, 1865).

6 Claude Bernard, *Leçons de pathologie expérimentale* (Paris: Baillière, 1872).

7 Freeman J. Dyson, *Origins of Life* (Cambridge: Cambridge University Press, 1985).

8 This is explained in my book *Une Aurore de pierres. Aux origines de la vie* (Paris: Seuil, 1990) or its summary in English, "Homeotopic Transformation and the Origin of Translation," *Progress in Biophysics and Molecular Biology* 54 (1989): 81–86.

9 *Lucretius: On the Nature of Things* (London: Sphere, 1969), translated by Martin Ferguson Smith from *De Rerum Natura.*

10 Jean Baptiste Poquelin de Molière, *The Imaginary Invalid (Le malade imaginaire* (1673), in *Comedies (The School For Husbands, The Misanthrope, Tartuffe—or—The Impostor, The Miser, The Would-Be Gentleman, The Imaginary Invalid)* (Franklin Center, Pa.: Franklin Library, 1985).

11 See in particular the work of the mathematician René Thom, *Structural Stability and Morphogenesis* (New York: W. A. Benjamin, 1975); translated by David H. Fowler from *Stabilité structurelle et morphogenèse* (1972).

12 Jean Piaget, *Epistémologie des sciences de l'homme* (Paris: Gallimard, 1972).

13 François Jacob, *The Logic of Life* (New York: Pantheon, 1973); translated by Betty E. Spillman from *La Logique du vivant* (Paris: NRF Gallimard, 1972).

14 François Jacob, *Of Flies, Mice, and Men* (Cambridge, Mass.: Harvard Univer-

sity Press, 1998); translated by Giselle Weiss from *La Souris, la mouche et l'homme* (Paris: Odile Jacob, 1996).

15 Life combines water and many types of molecules at a temperature at which most quantum effects are smoothed out, except when dealing with crystals or other very strictly organized structures. "Uncertainty," or even "indeterminacy," another name by which Heisenberg's principle is known, does not imply indeterminism in the sense that one cause could have several different effects. Similarly "cannot be predicted" does not at all imply "indeterministic"; it means that even if we know the causes, we cannot know the consequences *in advance.*

16 Jacques Monod, *Chance and Necessity* (New York: Alfred A. Knopf, 1971); translated by Austryn Wainhouse from *Le Hasard et la nécessité* (Paris: Seuil, 1970).

17 St. Thomas Aquinas, *Summa Theologica,* trans. Father Laurence Shadpole, rev. Daniel J. Sullivan, Great Books of the Western World, no. 19 (Chicago: Encyclopaedia Britannica, 1952).

18 Jared Diamond, "Outcasts of the Islands," review of *The Island of the Colorblind* by Oliver Sacks, *New York Review of Books,* March 6, 1997.

19 Ibid.

20 Claudine Devauchelle et al., "Rate Matrices for Analyzing Large Families of Protein Sequences," *Journal of Computational Biology* 8 (2001): 381–399.

INDEX

7 9/03